D0208106

R.W. Barnwell M.Y. Hussaini
Editors

# Natural Laminar Flow and Laminar Flow Control

With 125 Illustrations

Springer-Verlag
New York Berlin Heidelberg London Paris
Tokyo Hong Kong Barcelona Budapest

R.W. Barnwell
Space Systems Division
NASA Langley Research Center
Hampton, VA 23665
USA

M.Y. Hussaini
ICASE
NASA Langley Research Center
Hampton, VA 23665
USA

Library of Congress Cataloging-in-Publication Division
Natural laminar flow and laminar flow control / [edited by] M.Y. Hussaini.
p. cm. -- (ICASE/NASA LaRC series)
Includes bibliographical references.
ISBN 0-387-97737-6 (New York). -- ISBN 3-540-97737-6 (Berlin)
1. Laminar flow. I. Hussaini, M. Yousuff. II. Series.
TL574.L3L36 1991
629.132'32--dc20 91-36758

Printed on acid-free paper.



Production managed by Karen Phillips; Manufacturing supervised by Jacqui Ashri.
Camera-ready copy provided by the editors.
Printed and bound by Braun-Brumfield, Ann Arbor, MI.
Printed in the United States of America.

9 8 7 6 5 4 3 2 1

ISBN 0-387-97737-6 Springer-Verlag New York Berlin Heidelberg
ISBN 3-540-97737-6 Springer-Verlag Berlin Heidelberg New York

# Preface

Research on laminar flow and its transition to turbulent flow has been an important part of fluid dynamics research during the last sixty years. Since transition impacts, in some way, every aspect of aircraft performance, this emphasis is not only understandable but should continue well into the future. The delay of transition through the use of a favorable pressure gradient by proper body shaping (natural laminar flow) or the use of a small amount of suction (laminar flow control) was recognized even in the early 1930s and rapidly became the foundation of much of the laminar flow research in the U.S. and abroad. As one would expect, there have been many approaches, both theoretical and experimental, employed to achieve the substantial progress made to date. Boundary layer stability theories have been formulated and calibrated by a good deal of wind tunnel and flight experiments. New laminar now airfoils and wings have been designed and many have been employed in aircraft designs. While the early research was, of necessity, concerned with the design of subsonic aircraft interest has steadily moved to higher speeds including those appropriate to planetary entry. Clearly, there have been substantial advances in our understanding of transition physics and in the development and application of transition prediction methodologies to the design of aircraft.

The involvement of the NASA Langley Research Center in lami-nar now and transition physics as applied to airfoil and wing design dates from the early 1930s. Most identify Eastman Jacobs as the initiator, if not the originator, of much of the research which led to the development of the 2- through 7-series laminar flow airfoils and the low turbulence tunnels in which they were tested. This work as well as a number of other studies concerning laminar flow control were carried out in the late thirties and early to mid forties; by 1950 laminar airfoil research was phased out to devote more resources to supersonic research.

Interest in laminar flow and transition research was rekindled at the Langley Research Center following the oil embargo in 1973. By the end of the 1970s a wide variety of laminar flow research activities were under wav in stability theory, airfoil/wing design, and flight testing aimed at both laminar- wing operational problems and transition physics. Revolutionary improvements in computer and materials technologies provided important incentives.

Research pertaining to laminar flow control continued throughout the eighties with many significant advances. In the light of progress made in the last decade, there is a need to document in one place some representative samples of the research accomplished and, where possible, to put it in context with earlier work and future opportunities. This is the purpose of the present volume.

Included in this volume are eight papers/chapters by well-known authorities in the field. The first article by J.N. Hefner contains a review of past and present laminar flow research programs at the NASA Langley Research Center and some suggestions as to the course of future work. Particular emphasis is placed on the potential of hybrid laminar flow control (HLFC), which is a combination of laminar flow produced through airfoil/wing shaping and that controlled by suction. Flight research, past and present, is the subject of the second article by R.D. Wagner et al. The third paper, authored by B.J. Holmes and C.J. Obara, concentrates on natural laminar flow (NLF) flight research. It contains valuable discussions of progress in test techniques and measurement technology and provides a listing of lessons learned.

D.M. Somers is the author of the fourth article, which is devoted to the design of subsonic natural laminar flow airfoils. A historical background of laminar airfoil development is given, and design philosophy and experimental validation are discussed. The next paper by W. Pfenninger treats transonic laminar flow control. A design philosophy for long range LFC transports is outlined.

The paper by P. Hall discusses the effect of wave interactions on the growth of distrubances in shear flows. Such interactions may have a significant impact on LFC predictions and have not been previously considered. The paper by D.M. Bushnell provides a succinct review of supersonic laminar flow control. He concludes that immediately available tools and experience should establish LFC as a viable and important component of SST/HSCT technology. The last paper, by P.J. Bobbitt et al., concerns the background of and results from the laminar flow control experiments carried out during the 1980s in the NASA Langley Research Center 8-foot Transonic Pressure Tunnel on a 23 degree swept airfoil model with a 7-foot chord. Details of the instrumentation, suction system, wind tunnel flow quality treatments and model fabrication are given. Results in the form of suction requirements, transition location, drag coefficients and boundary-layer stability calculations are presented.

The editors would like to take this opportunity to thank the staff of Springer-Verlag for their patience and cooperation in the preparation of this volume.

R.W. Barnwell
M.Y. Hussaini

# Contents

# LAMINAR FLOW CONTROL: INTRODUCTION AND OVERVIEW

*Jerry N. Hefner*

NASA Langley Research Center
Hampton, VA 23665

## 1. Introduction

Research in the area of laminar flow control (LFC) dates back to the 1930's when early applications of stability theory led to the observation that laminar boundary layers can be stabilized by either favorable pressure gradients or small amounts of wall suction. (An excellent summary of this work is presented in References 1 and 2.) Research was performed in many countries to explore approaches for achieving extensive laminar flow with these concepts. Stabilization of boundary-layer disturbances and instabilities by pressure gradient and shaping became known as natural laminar flow (NLF), and NACA research led to the development of the six-series NLF airfoil. International research on stabilization by suction, referred to as LFC with suction, was intensive at the same time and culminated in the United States in the 1960's with flight tests of a relatively unswept suction glove on an F-94 aircraft (Reference 3) and X-21 flight tests (References 4-7) of a totally new swept LFC wing on a reconfigured WB-66 aircraft.

Little laminar flow research was conducted from the mid-1960's to the mid-1970's. However, as a result of the increased aircraft fuel costs caused by the Arab Oil Embargo of the early 1970's, NASA again resumed LFC research in 1976 as part of the Aircraft Energy Efficiency Program (ACEE); this research, was later continued under the NASA Research and Technology Base Program (see Figure 1).

In the flight tests of the X-21 aircraft, laminar flow with full-chord slotted suction surfaces was achieved repeatedly over almost all of the intended laminar upper wing area to chord Reynolds numbers of approximately $20 \times 10^6$ (see Figure 2). Extensive laminar flow was also achieved on a nonroutine basis to wing chord Reynolds numbers as high as $47 \times 10^6$. Although this flight experiment showed that extensive laminar flow could be achieved in flight with slotted suction surfaces, unresolved concerns regarding maintenance and reliability of LFC systems prevented serious consideration of LFC as a design option for aircraft at that time. Principal concerns were the practicality of producing wing surfaces sufficiently smooth and wave-free to meet laminar-flow criteria and maintaining the wing surface quality in normal airline service operations.

NASA research since 1976 has addressed the maintenance and reliability concerns that were unresolved in the X-21 flights and is now focused on developing the technology for application of LFC to transport aircraft. This paper provides an overview of the laminar flow research in the United States — its status and its future direction.

## 2. Status of Ongoing U. S. Laminar Flow Research

NASA has provided the leadership for the current laminar flow research in the United States. Under the NASA Laminar Flow Program, computational, wind-tunnel, flight, and systems research is being focused to provide the required data base and design methodology to reduce the risks associated with both near- and far-term applications of laminar flow technology. Two primary approaches are being emphasized to delay the transition process and maintain laminar flow beyond the usual transition Reynolds numbers of $4 \times 10^6$ or less (Figure 3); i.e., NLF has the advantage of being a passive approach; however, it may be limited to sweep angles of approximately 20° or less and chord Reynolds numbers of less than $20 \times 10^6$. LFC with suction is more complex but will probably be required to some extent to achieve extensive laminar flow beyond chord Reynolds numbers of about $20 \times 10^6$ and for wing sweep angles in excess of about 20°-25°.

The multiplicity of factors affecting laminar flow (shown in Figure 4 taken from Reference 7) has made the Laminar Flow Program a high-risk research undertaking. The most fundamental of these factors are the Reynolds number at which laminar flow becomes turbulent, the degree of wing sweep used, and the airfoil geometry. Understanding the importance of these and the many other factors illustrated on the figure and how they relate to each other continues to be a critical part of the research program.

The NASA Laminar Flow Program consists of four major elements. These include: (1) aerodynamic design tools and methodology including the essential transition criteria; (2) wind-tunnel and flight calibration and validation of these transition criteria, design tools and methodology; (3) design and integration of LFC systems into advanced wing structures; and (4) performance and reliability of laminar flow concepts in the "real-world" environment. In this research program, NASA has worked closely with both industry and universities to ensure technology readiness for LFC in the 1990's. The papers in the present volume discuss many of the results obtained from this research program.

**NASA Airfoil Development.** Development and advanced computational design tools including stability theories and transition criteria (e.g., References 9-20 and Papers 3 and 4 in the present volume) have enabled a new class of low-drag laminar flow airfoils to be developed. NLF airfoils for a wide range of applications and suction LFC airfoils for larger transport aircraft have been designed and tested (see Figure 5). Some applications are discussed in References 21-26. The NLF (1)-0414F, HSNLF (1)-0213F and SCLFC (1)-0513F airfoils are discussed in References 21-23 and 26-27.

**Natural Laminar Flow.** The general aviation industry has enthusiastically accepted the new NASA NLF airfoil concepts and is already incorporating them into their advanced designs. Some of these airfoils have recently been tested on a Swearingen SX-300, a Mooney 301, and a Cessna 210. Results of the flight tests with the NLF (1)-0414 airfoil on the Cessna 210 airplane (Figure 6) are reported in Reference 28. In these flight tests, laminar flow was achieved on the upper and lower wing surfaces to approximately 70-percent chord. Loss of laminar flow did not significantly degrade the lift performance of the wing, and the flight experiments validated both the predicted performance and that obtained in wind-tunnel tests.

**Advanced Transition Measurement Techniques.** The calibration and validation of aerodynamic design tools and methodology for NLF and LFC applications require the ability to accurately determine the extent of laminar flow or the location where the laminar boundary layer undergoes transition. A number of advanced transition detection and measurement techniques have been developed to provide the required definitive data in both flight and wind-tunnel investigations. Four such techniques are shown on Figure 7 and are discussed in References 29 and 30.

**Advanced LFC Airfoil Development.** Subsonic/transonic laminar flow and transition research to validate stability theory, transition prediction criteria, and the aerodynamic LFC design tools continues to be conducted in wind-tunnel facilities at NASA Langley Research Center and in both university and industry laboratories. The most complex and difficult of these experiments (Figure 8) has been the supercritical LFC airfoil experiment that was conducted in the Langley Research Center 8-Foot Transonic Pressure Tunnel (TPT) (Reference 31). Extensive modifications to the 8' TPT were necessary for this experiment. These included modifications to reduce tunnel turbulence levels and installation of a honeycomb and five screens in the settling chamber and a sonic choke ahead of the diffuser. A contoured liner was installed in the test section to produce an infinite swept-wing flow over the model surface.

This experiment employed an advanced LFC airfoil incorporating the latest supercritical airfoil technology with features intended to simplify the achievement of laminar flow. The airfoil, shown on Figure 8, had supercritical flow on both the upper and lower surfaces and a drag divergence Mach number comparable to advanced turbulent supercritical airfoils, but with laminar flow, had nearly an order of magnitude higher lift-to-drag ratio. Full-chord suction with either slotted surfaces or perforated surfaces was investigated. Results of this research are discussed in References 26 and 27. The tests with both the slotted and perforated suction surface show that supercritical technology can be successfully combined with LFC technology to produce a supercritical LFC airfoil having at least 60-percent less drag than the comparable turbulent supercritical airfoil.

**Flight Experiments.** Flight research is a very important extension of the wind-tunnel research being conducted in the Laminar Flow Program. Since the maintenance of laminar flow control is a boundary-layer stability program, it is crucial that definitive laminar flow research be conducted at the appropriate unit and chord Reynolds numbers, Mach number, and in the correct disturbance environment. In wind tunnels, the disturbance environment is generally not representative of that in flight; wind tunnels typically have high turbulence levels that can adversely affect the transition phenomena. Also, to achieve large chord Reynolds numbers in wind tunnels, the unit Reynolds number must be large to compensate for relatively small models; this exacerbates the allowable roughness and waviness requirements associated with fabricating the smooth model.

Five important flight experiments are illustrated on Figure 9. The Lear 28/29 flight tests have provided access to the transonic flight environment for NLF research, for fundamental transition research, and for evaluation of advanced transition measurement techniques. The 757 NLF flight tests provided near-field acoustic data on a transport at cruise for the first time and showed that laminar flow can be maintained on wings near wing-mounted engines. The F-111 flight tests and the F-14 Variable Sweep Transition Flight Experiment (VSTFE) have provided the data base essential to the evaluation of sweep, Mach number, and Reynolds number on transition and NLF at transonic speeds. The OV-1 NLF nacelle experiments have provided data to validate acoustic theory and to assess the feasibility of NLF on nacelles. The research with the 757, F-14, and OV-1 is discussed in References 32 to 34; research on the F-111 and Lear 28/29 is discussed in References 35 and 36, respectively.

**LFC Wing Panel Development.** Wing structural design is the central problem in the definition of a practical large commercial LFC transport. Under the ACEE Program, contracts were awarded to the Lockheed-Georgia Company and the Douglas Aircraft Company to develop LFC wing panel concepts and evaluate their feasibility.

The Lockheed-Georgia Company design (Figure 10) employed a ducting network integrated into the primary wing structure and extensively used graphite epoxy composite materials. The details of this concept are discussed in Reference 37). The main feature of the Lockheed concept was that it employed slotted suction through a titanium skin with fluid-dispensing slots in the leading edge for de-icing and protection from insect contamination; it was found in earlier work (Reference 8) that insects do not adhere to wet leading edges. Suction was applied on both the upper and lower surfaces of the wing.

The Douglas Aircraft Company concept (shown on Figure 10 and discussed in Reference 38) used perforated suction strips in the titanium skin with less ducting in the primary structure. A retractable Krueger device in the leading edge served as a line-of-sight shield for protection from insects; as a supplement, spray nozzles behind the shield were used to wet the leading edge. Suction was applied only on the upper wing surface.

Since these concepts appeared so promising, they were used in the subsequent Leading-Edge Flight Test (LEFT) Program on the NASA Jetstar aircraft flown from the Ames-Dryden Flight Research Facility. Currently, new concepts for fabricating LFC panels are being explored; these include superplastic forming and diffusion bonding. Also, methods of fabricating panel joints which can meet the smoothness criteria for laminar flow over the airplane life-cycle are being developed.

**LFC Leading-Edge Flight Test.** The most formidable problem facing practical LFC applications is the integration of the deicing, insect protection, and suction systems into the wing leading edge. Since the Lockheed-Georgia Company and the Douglas Aircraft Company had developed concepts for integrating these systems into the wing structure of transport aircraft, contracts were awarded to these companies to design, fabricate, and install practical leading-edge test articles on the NASA Jetstar. Flight research on the NASA Jetstar (Figure 11) was conducted to evaluate the performance and reliability of these systems in the "real-world" environment. This flight research program has successfully demonstrated that practical and reliable leading-edge systems can be designed and fabricated, and that these systems perform extremely well in an airline environment without any unusual maintenance requirements. Results of this research are discussed in References 39 to 43.

**Hybrid LFC Flight Experiment.** The next step for subsonic/transonic laminar flow is to evaluate hybrid laminar flow control (HLFC). HLFC combines suction LFC and NLF to achieve extensive laminar flow and is probably most applicable for wing chord Reynolds numbers to $40 \times 10^6$ and wing sweep angles between 20°-30°. The concept of particular interest for near-term transport application uses suction in the leading edge ahead of the front spar with a slightly favorable or roof-top pressure gradient over the wing box (see Figure 12); the goal is to maintain laminar flow to approximately 60-percent chord. HLFC has the advantages of being less complex than full-chord LFC, requiring less suction, and allowing the use of a more conventional wing box structure.

Figure 13 taken from Reference 49 shows the potential fuel savings for future transport aircraft that might utilize NLF, LFC, and HLFC. The potential benefits of HLFC, although not as large as the more complicated LFC, are seen to be sizable for the vehicle class flying typically 2,000-5,000 nautical miles. Therefore, the performance of HLFC at practical Reynolds numbers and Mach numbers in the "real-world" environment must be determined. A cooperative NASA, USAF, and industry HLFC Flight Research Experiment is underway. This experiment is to be conducted on a partial-span HLFC test article mounted on a Boeing 757 transport aircraft wing at chord Reynolds numbers approaching $40 \times 10^6$. The goals of this research include: HLFC and perforated suction performance at high Reynolds numbers, environmental effects, off-design performance, and design tool and methodology validation. Since the Jetstar LEFT Program only evaluated the leading-edge problem, the HLFC flight experiment will be designed to achieve extensive laminar flow at realistic flight Reynolds numbers.

**Supersonic LFC.** LFC may be more important to supersonic cruise that it is to subsonic/transonic cruise because of its potential impact on the critical areas shown on Figure 14. Unfortunately, whether extensive laminar flow can be practically maintained at supersonic speeds has not yet been established. Therefore, using the experience and knowledge gained from the subsonic/transonic laminar flow program, the next major thrust in the NASA Laminar Flow Program will be supersonic LFC.

Currently, the supersonic LFC program is being developed with problem areas and research directions being identified (see Paper 9 of this volume). The physics of supersonic transition and LFC are already being investigated including the effects of roughness and waviness, unit Reynolds number, acoustic environment, disturbance amplification through shocks, and suction through perforated surfaces. Stability theories and boundary-layer transition criteria, developed for compressible subsonic and transonic flow, are also being evaluated to determine their applicability to supersonic LFC (Reference 45). Navier-Stokes codes are being developed to simulate transition at supersonic speeds and to evaluate boundary-layer receptivity.

**Supersonic Low-Disturbance Tunnel.** Definitive data for evaluating LFC and transition physics and for developing and validating stability theories and transition criteria must be obtained in ground facilities with low background disturbance levels or in flight. Unfortunately, the turbulence levels in essentially all existing supersonic and hypersonic wind tunnels are sufficiently high to alter the transition phenomena in these facilities. For the past 10 years, research at the NASA Langley Research Center has been conducted to develop a low-disturbance Mach 3.5 wind tunnel. A pilot model of this facility has been built, and transition Reynolds numbers equivalent to those obtained in flight have been measured on cones in this facility (Reference 46); research to better understand supersonic transition physics is now being conducted in this pilot facility. (See References 47 and 48). Construction of the full-size Supersonic Low-Disturbance Tunnel is planned in the near future. An alternate Mach 6 nozzle has also been developed for hypersonic transition research and is being evaluated.

**Supersonic Flight Transition Measurements.** Supersonic swept-wing data of the quality necessary for exploring the transition phenomena, evaluating compressible flow transition criteria at supersonic speeds, and assessing the feasibility of obtaining significant laminar flow at supersonic speeds is almost non-existent. A window of opportunity for obtaining some of this much needed data became available in late 1985 and early 1986. Clean-up gloves to achieve the needed smooth surface finish were installed on the leading edge of an F-15 at the Ames-Dryden Flight Research Facility and on the leading edges of the wing and vertical tail of an F-106 at the Langley Research Center. Surface pressure and hot-film data were obtained in flight tests with both aircraft. Results of these exploratory investigations, which are helping to better define future supersonic LFC flight research, are discussed in Reference 49.

**Supersonic Transition and Laminar Flow Flight Experiments.** As a result of the exploratory transition experiments on the F-15 and F-106, supersonic swept-wing transition and LFC flight experiments on an F-16XL aircraft are now being developed. Suction through a slotted or perforated suction surface would be ultimately applied in the leading-edge region of the wing with a glove installed aft of the suction surface to provide the desired pressure distribution. Surface pressures, hot-film data, and liquid crystal flow visualization data would be employed to examine the extent of laminar flow achieved and the effect of suction in stabilizing supersonic laminar boundary layers. Measurements of the freestream disturbance environment (including ice particles, water vapor, and turbulence) and the acoustic environment on the wing test surface will be made.

These data are required to enhance the understanding of supersonic transition physics, as well as to calibrate and validate the transition prediction and LFC design tools under development. Flight experiments will also validate the performance and reliability of supersonic LFC systems in the "real-world" environment just as being done for HLFC at transonic speeds.

### 3. Concluding Remarks

Over the past decade, research conducted in the United States in the areas of both natural laminar flow and laminar flow control with suction has produced some very impressive results. The general aviation industry has already begun to incorporate natural laminar flow into their new aircraft designs, and it is anticipated that designs of future advanced subsonic commercial transports will soon include laminar flow control. Hybrid laminar flow control, which incorporates suction ahead of the wing front span and shaping/favorable pressure gradient over the wing box, appears to be a relatively simple, low-risk near-term approach. The next major step for laminar flow research is to develop the technology for application to supersonic transport aircraft. Early analyses indicate that supersonic laminar flow control could provide significantly more important benefits for a future high-speed civil transport than subsonic laminar flow control will provide for future subsonic transport aircraft.

## References


[1] Bushnell, D. M. and Tuttle, M. H., 1979, *Survey and bibliography on attainment of laminar flow control in air using pressure gradient and suction*, NASA RP-1035, V. 1.

[2] Wagner, R. D., Maddalon, D. V., Bartlett, D. W., and Collier, F. S., Jr. 1988, *Fifty years of laminar flow flight testing*, SAE Paper 881393, Aerospace Technology Conference and Exposition, Anaheim, CA.

[3] Groth, E. E., Carmichael, B. H., White, R. C., and Pfenninger, W., 1957, *Low drag boundary layer suction experiments in flight on the wing glove of a F-94A airplane – Phase II: Suction*

*through 69 slots*, NA1-57-318, BLC-94 (Contract AF-33(616)-3168), Northrop Aircraft.

[4] Antonatus, P. P., 1966, *Laminar flow control concepts and applications*, **Astronautics and Aeronautics**, V. 4, No. 7, pp. 32-36.

[5] Nenni, J. P. and Gluyas, G. L., 1966, *Aerodynamic design and analysis on an LFC surface*, **Astronautics and Aeronautics**, V. 4, No. 7.

[6] White, R. C., Suddreth, R. W., and Weldon, W. G., 1966, *Laminar flow control on the X-21*, **Astronautics and Aeronautics**, V. 4, No. 7, pp. 38-43.

[7] Pfenninger, W. and Reed, V. D., 1966, *Laminar-flow research and experiments*, V. 4, No. 7, pp. 44-47, 49-50.

[8] Maddalon, D. V. and Wagner, R. D., 1986, *Operational considerations for laminar flow aircraft*, Laminar Flow Aircraft Certification Workshop, NASA SP-2413, pp. 247-266.

[9] Malik, M., 1987, *Stability theory applications to laminar flow control*, NASA Symposium on NLF and LFC Research, NASA CP 2487, pp. 219-244.

[10] Nayfeh, A., 1987, *Nonparallel stability of boundary layers*, NASA Symposium on NLF and LFC Research, NASA CP 2487, pp. 245-259.

[11] Hall, P., 1987, *Interaction of Tollmien-Schlichting waves and Görtler vortices*, NASA Symposium on NLF and LFC Research, NASA CP 2487, pp. 261-271.

[12] Kerschen, E., 1987, *Boundary-layer receptivity and laminar flow airfoil design*, NASA Symposium on NLF and LFC Research, NASA CP 2487, pp. 273-287.

[13] Kalburgi, V., Mangalam, S., Dagenhart, J., Tiwari, S., 1987, *Görtler instability on an airfoil*, NASA Symposium on NLF and LFC Research, NASA CP 2487, pp. 289-300.

[14] Nayfeh, A., 1987, *Effect of roughness on the stability of boundary layers*, NASA Symposium on NLF and LFC Research, NASA CP 2487, pp. 301-315.

[15] Bushnell, D. M., Hussaini, V., and Zang, T., 1987, *Sensitivity of LFC techniques in the non-linear regime*, NASA Symposium on NLF and LFC Research, NASA CP 2487, pp. 491-516.

[16] Harris, J., Iyer, V., and Radwan, S., 1987, *Numerical solutions of the compressible 3-D boundary layer equations for aerospace configurations with emphasis on LFC*, NASA Symposium on NLF and LFC Research, NASA CP 2487, pp. 517-545.

[17] Goradia, S. and Morgan, H., 1987, *Theoretical methods and design studies for NLF and HLFC swept wings at subsonic and supersonic speeds*, NASA Symposium on NLF and LFC Research, NASA CP 2487, pp. 547-575.

[18] Biringen, S. and Caruso, M., 1987, *Numerical experiments in transition control in wall-bounded shear flows*, NASA Symposium on NLF and LFC Research, NASA CP 2487, pp. 577-592.

[19] Maestrello, L., Parikh, P., and Bayliss, A., 1987, *Application of sound and temperature to a control boundary-layer transition*, NASA Symposium on NLF and LFC Research, NASA CP 2487, pp. 593-616.

[20] Rawls, J. W., 1987, *Near-field noise prediction of an aircraft in cruise*, NASA Symposium on NLF and LFC Research, NASA CP 2487, pp. 617-636.

[21] Viken, J., Viken, S., Pfenninger, W., Morgan, H., and Campbell, R., 1987, *Design of low-speed NLF(1)-0414F and high-speed HSNLF(1)-0213 airfoils with high-lift systems*, NASA Symposium on NLF and LFC Research, NASA CP 2487, pp. 637-671.

[22] Murri, D., McGhee, R., Jordan, F., Davis, P., and Viken, J., 1987, *Wind-tunnel results of the low-speed NLF(1)-0414F airfoil*, NASA Symposium on NLF and LFC Research, NASA CP 2487, pp. 673-726.

[23] Kolesar, C., 1987, *Design and test of a natural laminar flow large Reynolds number airfoil with high design cruise lift coefficient*, NASA Symposium on NLF and LFC Research, NASA CP 2487, pp. 727-751.

[24] Waggoner, E., Campbell, R., Phillips, P., and Hallissy, J., 1987, *Design and test of an NLF wing glove for the variable-sweep transition flight experiment*, NASA Symposium on NLF and LFC Research, NASA CP 2487, pp. 753-776.

[25] Maughmer, M. and Somers, D., 1987, *Design of an airfoil for a high altitude long-endurance, remotely piloted vehicle*, NASA Symposium on NLF and LFC Research, NASA CP 2487, pp. 777-794.

[26] Brooks, C. and Harris, C., 1987, *Results of LFC experiment on slotted, swept supercritical airfoil in Langley's 8-Foot Transonic Pressure Tunnel*, NASA Symposium on NLF and LFC Research, NASA CP 2487, pp. 453-469.

[27] Berry, S., Dagenhart, J., Brooks, C., and Harris, C., 1987, *Boundary layer stability analysis of LaRC 8-Foot LFC experimental data*, NASA Symposium on NLF and LFC Research, NASA CP 2487, pp. 471-489.

[28] Befus, J., Latas, J., Nelson, E., Carr, J., and Ellis, D., 1987, *In-flight measurements of lift, drag, and pitching moments on an advanced NLF wing on a Cessna Model T210*, NASA Symposium on NLF and LFC Research.

[29] Holmes, B., Carraway, D., Manuel, G., and Croom, C., 1987, *Advanced measurement techniques, Part I*, NASA Symposium on NLF and LFC Research, NASA CP 2487, pp. 317-340.

[30] Johnson, C., Stainback, P., Carraway, D., Stack, P., Yeaton, R., Dagenhart, J., Hall, R., and Lawing, P., 1987, *Advanced measurement techniques, Part II*, NASA Symposium on NLF and LFC Research, NASA CP 2487, pp. 341-419.

[31] Harvey, W. D. and Pride, J. D., 1981, *NASA Langley laminar flow control airfoil experiment*, NASA CP-2218, pp. 1-42.

[32] Runyan, L., Blelak, G., Rehbahani, R., Chen, A., and Rozendaal, R., 1987, *757 NLF glove flight test results*, NASA Symposium on NLF and LFC Research, NASA CP 2487, pp. 795-818.

[33] Meyer, R., Bartlett, D., and Trujillo, B., 1987, *F-14 VSTFE and results for the clean-up flight test program*, NASA Symposium on NLF and LFC Research, NASA CP 2487, pp. 819-844.

[34] Hastings, E., Faust, G., Mungur, P., Obara, C., Dodbele, S., Schoenster, J., and Jones, M., 1987, *Status report on a natural laminar flow flight experiment*, NASA Symposium on NLF and LFC Research, NASA CP 2487, pp. 887-921.

[35] Montoya, L. C., Steers, L. L., Christopher, D., and Trujillo, B., 1981, *F-111 TACT natural laminar flow glove flight results*, NASA CP-2208.

[36] Holmes, B. J., Croom, C. C., Gall, P. D., Manuel, G. S., and Carraway, D. L., 1986, *Advanced transition measurement methods for flight applications*, AIAA Paper 86-9786, AIAA/AHS/CASI/DGLR/IES/ISA/ITEA/SETP/SFTE 3rd Flight Testing Conference, Las Vegas, NV.

[37] Lange, R. H., 1987, *Design integration of laminar flow control for transport aircraft*, **Journal of Aircraft**, V. 21, pp. 612-617.

[38] Pearce, W. E., 1982, *Progress at Douglas on laminar flow control applied to commercial transport aircraft*, Proceedings of the 13th ICAS Congress and AIAA Aircraft Systems and Technology Conference, Seattle, WA, V. 2, pp. 811-817.

[39] Fisher, D. and Fischer, M., 1987, *The development flight tests of the Jetstar LFC leading-edge flight experiment*, NASA Symposium on NLF and LFC Research, NASA CP 2487, pp. 117-140.

[40] Powell, A., 1987, *The right wing of the L.E.F.T. airplane*, NASA Symposium on NLF and LFC Research, NASA CP 2487, pp. 141-161.

[41] Davis, R., Maddalon, D., and Wagner, R., 1987, *Performance of laminar flow leading-edge test articles in cloud encounters*, NASA Symposium on NLF and LFC Research, NASA CP 2487, pp. 163-193.

[42] Maddalon, D., Fisher, D., Jennett, L., and Fischer, M., 1987, *Simulated airline service experience with laminar flow control leading-edge system*, NASA Symposium on NLF and LFC Research, NASA CP 2487, pp. 195-218.

[43] Montoya, L. and Maddalon, D., 1987, *Suction discontinuities in LFC leading-edge surfaces*, NASA Symposium on NLF and LFC Research.

[44] Kirchner, M. E., 1987, *Laminar flow: Challenge and potential*, NASA Symposium on NLF and LFC Research, NASA CP 2487, pp. 25-44.

[45] Bushnell, D. M. and Malik, M., 1987, *Supersonic laminar flow control*, NASA Symposium on NLF and LFC Research, NASA CP 2487, pp. 923-946.

[46] Beckwith, I., Chen, F., and Malik, M., 1987, *Design and fabrication requirements for low noise supersonic/hypersonic wind tunnels*, NASA Symposium on NLF and LFC Research, NASA CP 2487, pp. 947-964.

[47] Morrisette, L. and Creel, T., *Effects of wall surface defects on boundary layer transition in quiet and noisy supersonic flow*, NASA Symposium on NLF and LFC Research, NASA CP 2487, pp. 965-980.

[48] Creel, T., Malik, M., and Beckwith, I., 1987, *Experimental and theoretical investigation of boundary layer instability mechanisms on a swept leading edge at Mach 3.5*, NASA Symposium on NLF and LFC Research, NASA CP 2487, pp. 981-995.

[49] Collier, F., Rose, O., Miller, D., and Johnson, C., 1987, *Supersonic boundary-layer transition on the LaRC F-106 and DFRF F-15 aircraft*, NASA Symposium on NLF and LFC Research, NASA CP 2487, pp. 997-1024.

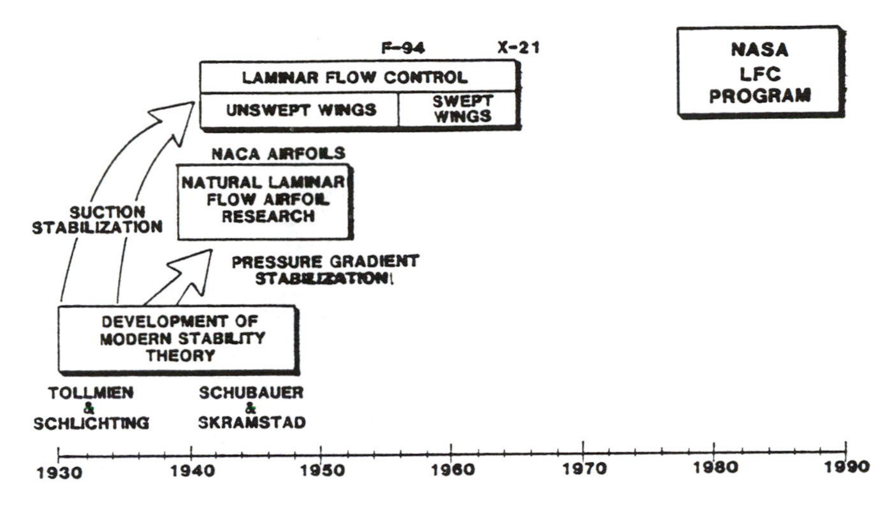


Figure 1. Chronology of Laminar Flow Research.

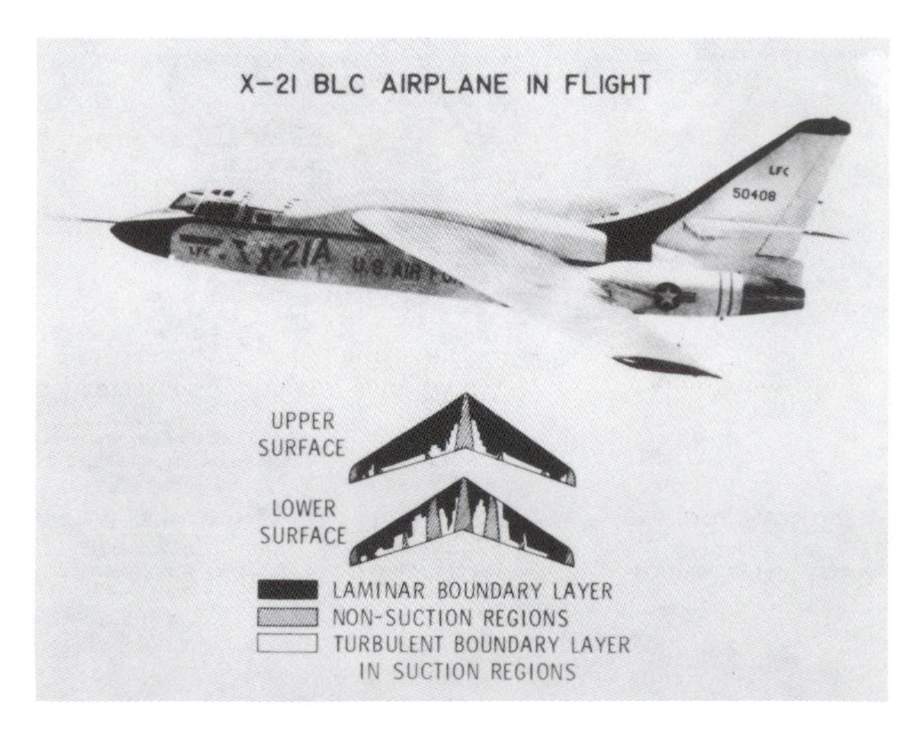


Figure 2. X-21 BLC Airplane in Flight.

MAINTENANCE OF LAMINAR FLOW
FOR VISCOUS DRAG REDUCTION

- Pressure gradient/shaping
- Suction through slotted or perforated surfaces

Laminar flow

Laminar flow

Laminar flow

Turbulent flow

Laminar flow

Turbulent flow

Natural Laminar Flow

Laminar Flow Control

Figure 3. Maintenance of Laminar Flow for Viscous Drag Reduction.

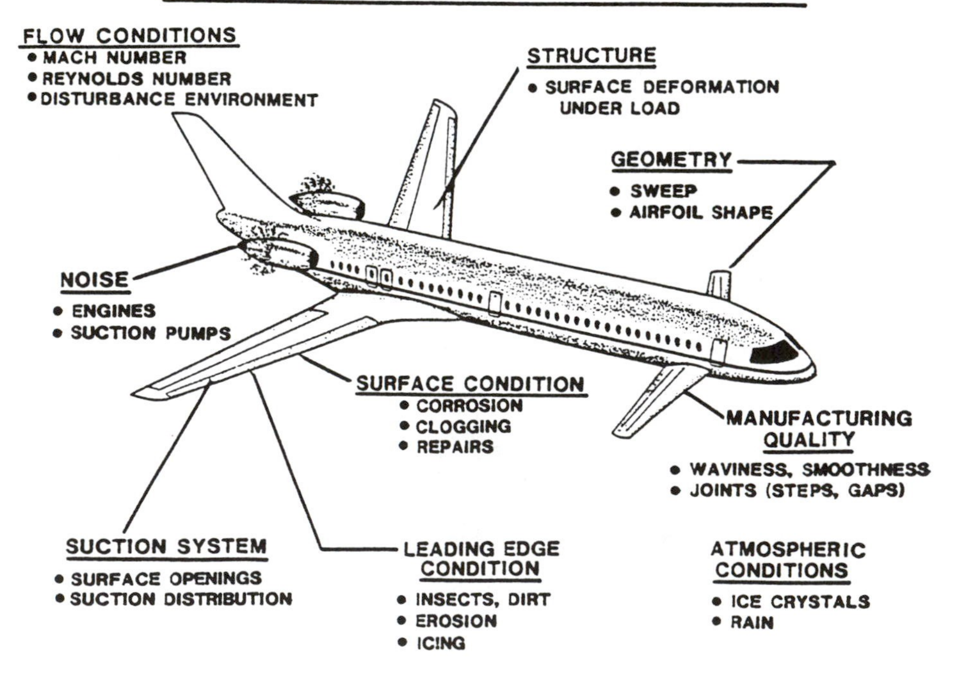


Figure 4. Factors Affecting Laminar Flow.

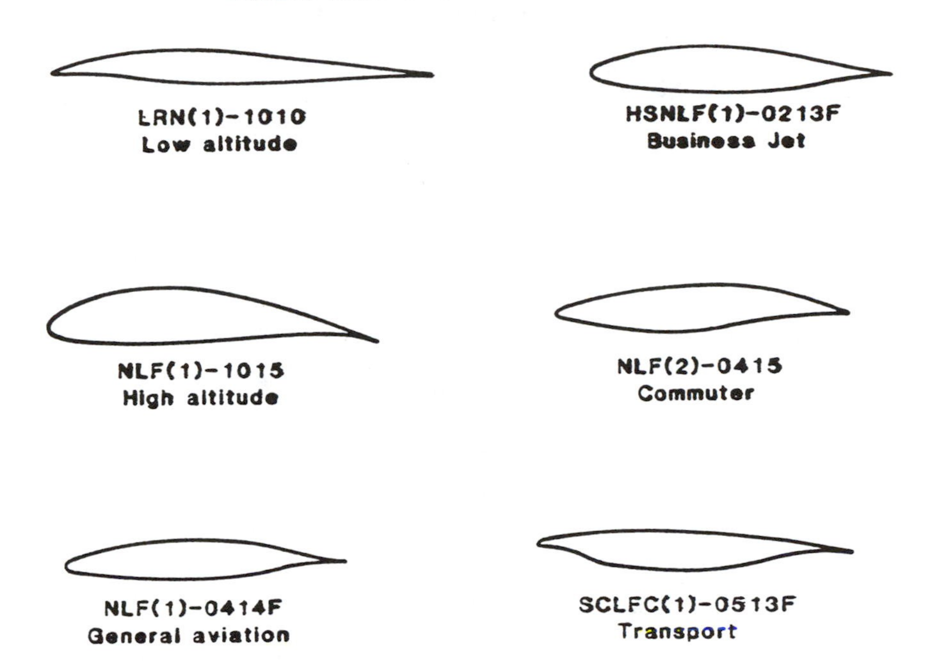


Figure 5. NASA Airfoil Development.

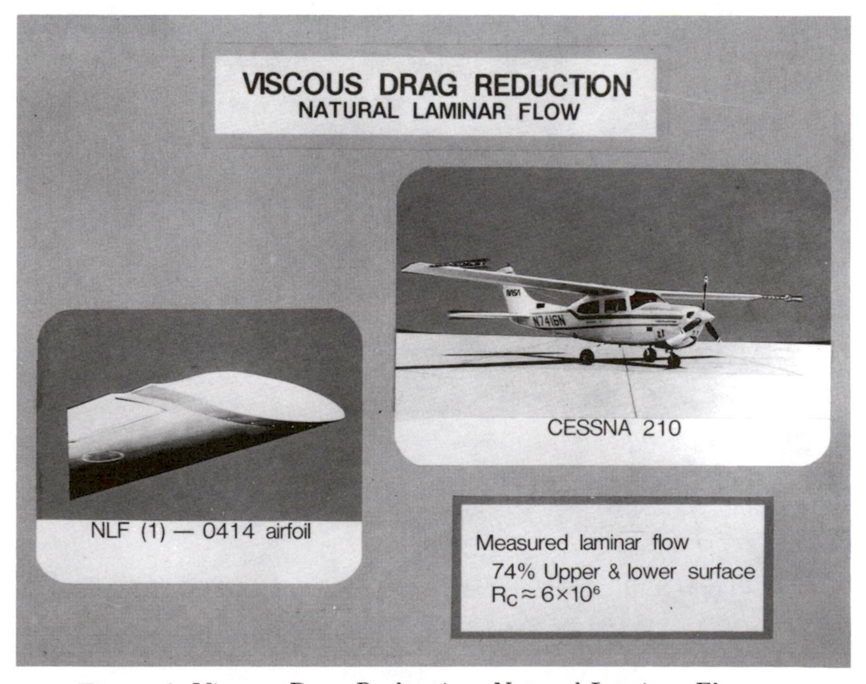


Figure 6. Viscous Drag Reduction. Natural Laminar Flow.

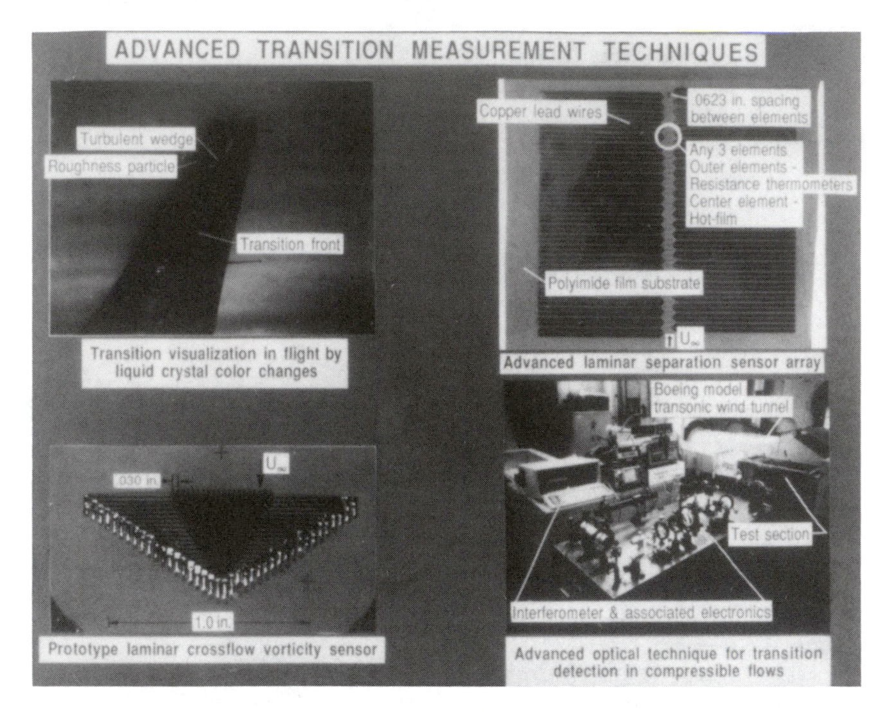


Figure 7. Advanced Transition Measurement Techniques.

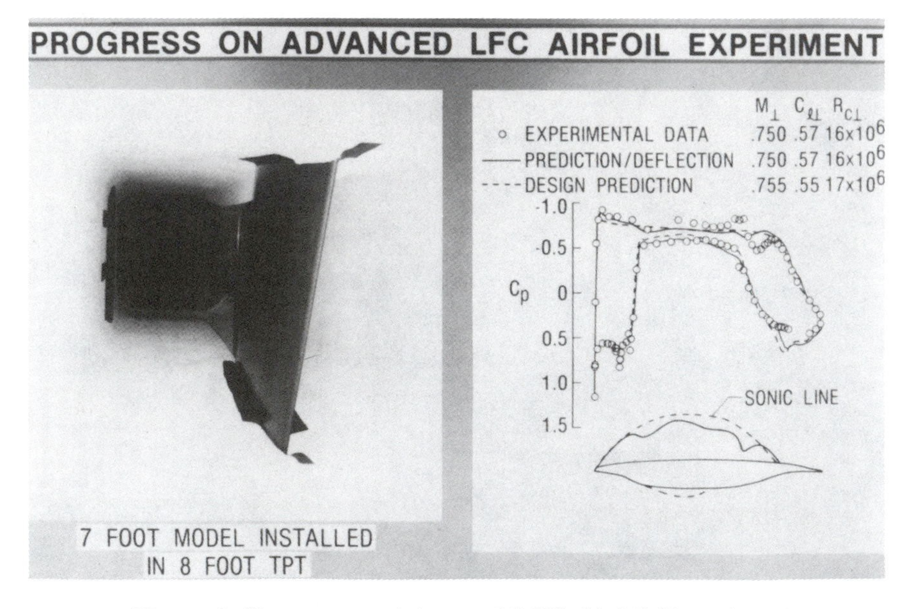


Figure 8. Progress on Advanced LFC Airfoil Experiment.

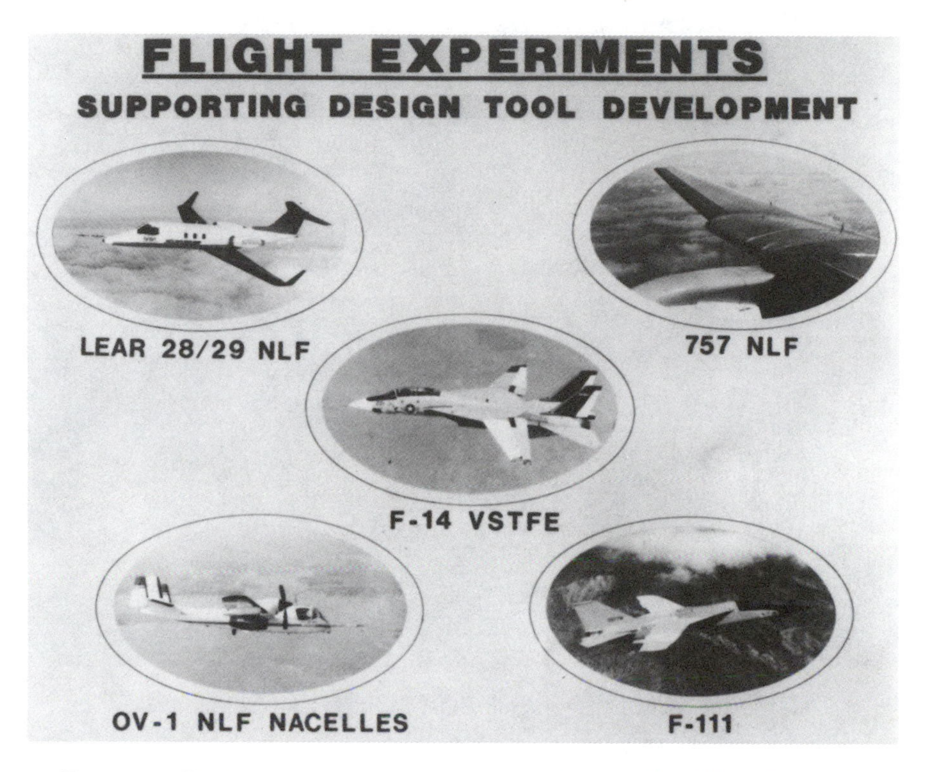


Figure 9. Flight Experiments. Supporting Design Tool Development.

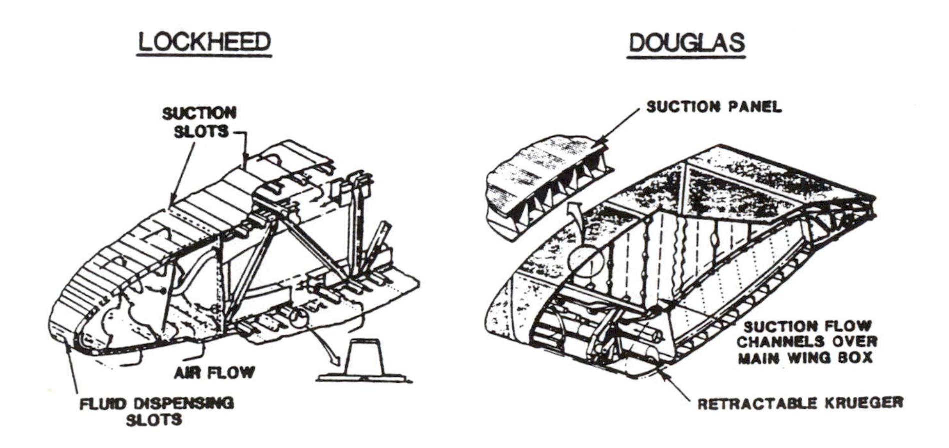


Figure 10. LFC Wing Panel Development.

Figure 11. NASA Jetstar. LFC Leading Edge Flight Test.

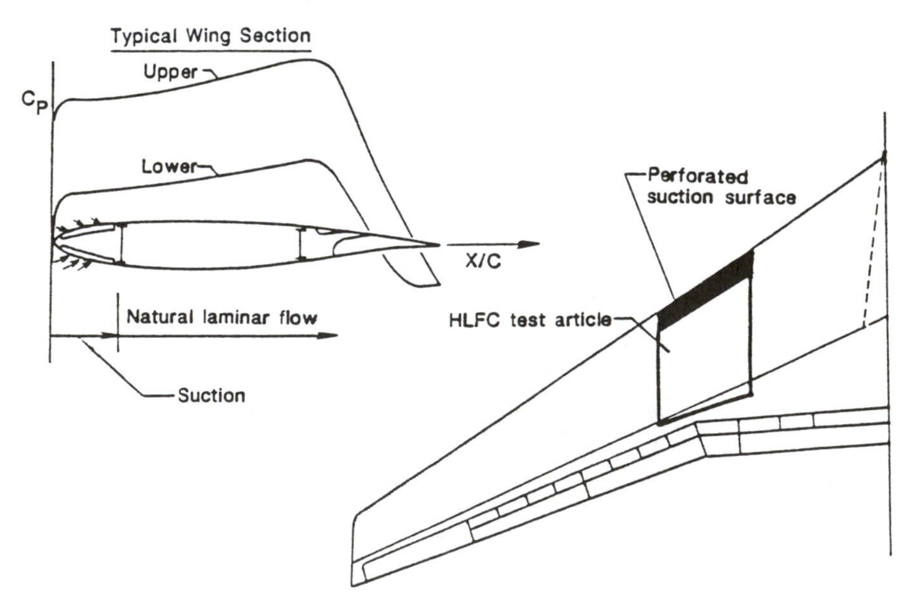


Figure 12. Hybrid LFC Flight Experiment.

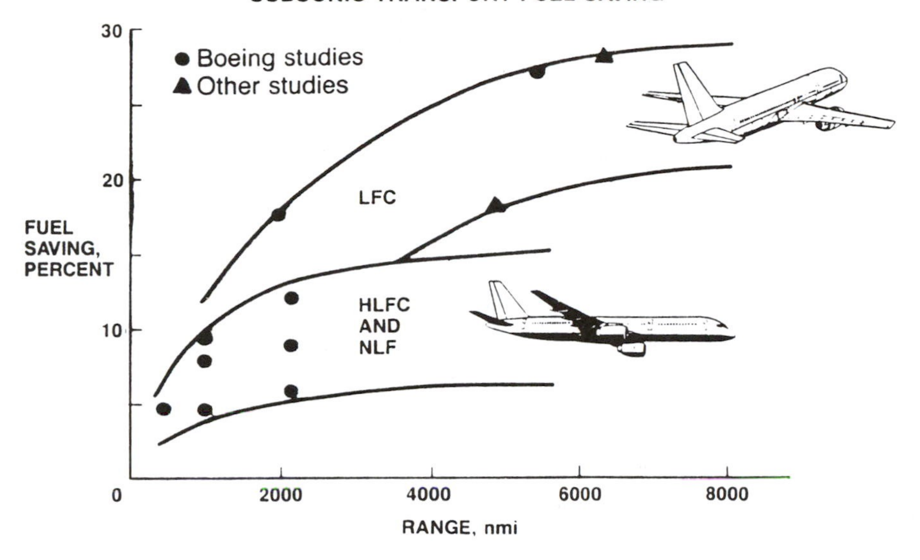


Figure 13. Laminar Flow Potential– Subsonic. Subsonic Transport Fuel Saving.

## POTENTIAL BENEFITS OF SUPERSONIC LFC

- INCREASED L/D
- REDUCED SURFACE TEMPERATURE
- REDUCED GROSS WEIGHT
- REDUCED SONIC BOOM
- INCREASED SEAT-MILES PER GALLON OF FUEL

Figure 14. Potential Benefits of Supersonic LFC.

# LAMINAR FLOW FLIGHT EXPERIMENTS - A REVIEW

*R. D. Wagner, D. V. Maddalon, D. W. Bartlett, and F. S. Collier, Jr.*
National Aeronautics and Space Administration
Langley Research Center, Hampton, VA. 23665

*A. L. Braslow*
Analytical Services and Materials, Inc.
Hampton, VA 23666

## 1. Introduction

On December 17, 1937, B. Melvill Jones presented the first Wright Brothers' lecture at Columbia University in New York (Ref. 1). His lecture, which was entitled "Flight Experiments on the Boundary Layer," dealt specifically with the first British flight observations of transition of the boundary layer from laminar to turbulent flow. These data, Jones concluded, showed that it is possible to retain a laminar layer over at least one-third of the whole wing surface even when the chord Reynolds number is as high as 8 millions. In the 50 years since this presentation, much flight research has been performed to explore the potential of laminar flow control for drag reduction. Both passive control and active control by suction (designated as natural laminar flow and laminar flow control, respectively) have been researched and impressive results achieved. The successes of the early natural laminar flow (NLF) flight testing were remarkable, with the achievement of an extent of laminar flow and transition Reynolds numbers which were not to be exceeded in flight for over 40 years. Nevertheless, mid-century manufacturing capabilities were such that insufficiently smooth or wave-free wing surfaces led to failure of attempts to transfer this technology to practice. The experience with laminar flow control (LFC) nearly paralleled that of NLF. LFC was recognized as a potentially more powerful means for achieving extensive laminar boundary layer flow, although admittedly more complex from the systems standpoint. LFC flight research began in the 1940's and peaked in the 1960's with the USAF/Northrop X-21 program, the most ambitious LFC flight test to date, which attempted to achieve full chord and full span laminar flow on a swept wing (Ref. 2). Two WB-66 aircraft were fitted with new, full chord suction controlled laminar flow wings and flight tested over 3 years. The main result of the program was that laminar flow control was observed to be aerodynamically achievable, but surface quality and structural complexity still appeared formidable barriers to LFC applications.

For a period of almost 10 years, research in NLF and LFC was dormant. The energy crisis of the early 1970's revived interest in the technology and flight testing resumed. Today, the prospects for a practical technology are brighter than ever. We have a greater understanding of the phenomena involved and new, less-complex systems concepts are evolving. Critically important to this new outlook is the fact that our manufacturing capabilities have dramatically advanced since the 1960's and the needed wing surface quality appears within our reach.

In this paper, the flight testing conducted over the past 50 years will be reviewed. In the more recent flight testing, emphasis will be placed on those conducted under the NASA LFC Program (Ref. 3) which has been directed towards the most challenging technology application, the high subsonic speed transport. To place these recent experiences in perspective, earlier important flight tests will first be reviewed to recall the lessons learned at that time.

## 2. Acronyms and Abbreviations

| | |
|---|---|
| TACT | Transonic aircraft technology |
| NACA | National Advisory Committee on Aeronautics |
| NPL | National Physics Laboratory |
| LFC | Laminar flow control |
| mac | Mean aerodynamic chord |
| LTPT | Low-Turbulence Pressure Tunnel |
| Max | Maximum |
| PW | Pratt and Whitney |
| NLF | Natural laminar flow |
| LE, TE | Leading edge, trailing edge |
| OASPL | Overall sound pressure level |
| rpm | Revolutions per minute |
| dB | Decibel |
| N-OASPL | Normalized overall sound pressure level |
| Hz. | Hertz |

| | |
|---|---|
| HLFC | Hybrid laminar flow control |
| Alt. | Altitude |

## 3. Symbols

| | |
|---|---|
| $C_L$ | Lift coefficient |
| $C_p$ | Pressure coefficient |
| $\bar{c}$ | Mean chord |
| $c$ | Chord |
| $F$ | Frequency |
| $\ell$ | Liter |
| $M_\infty$ | Freestream Mach number |
| $M_{LOCAL}$ | Local Mach number |
| $M_{FAN}$ | Fan Mach number |
| $N$ | Engine rpm |
| $N_{TS}$ | N-factor for Tollmien-Schlichting (TS) calculation |
| $N_{CF}$ | N-factor for crossflow (CF) calculation |
| $R_c, R_x, R_T$ | Reynolds number based on $c$, $x$, and $x_T$; respectively |
| $R_{CF}$ | Reynolds number based on $W_{max}$ and $\delta_{0.01}$ |
| $T$ | At transition |
| $t$ | Time |
| $W$ | Crossflow velocity |
| $x$ | Distance along chord in streamwise direction |
| Greek | |
| $\delta_{0.01}$ | Distance from wall above $W_{max}$ where $W = 0.01W_{max}$ |
| $\rho$ | Density |
| $\mu$ | Laminar coefficient of viscosity, or microns |
| $\Lambda$ | Sweep angle |
| $\eta$ | Span station |
| $\beta$ | Side slip |

## 4. Early Laminar Flow Flight Research

A brief review of some of the most significant past efforts on laminar flow is beneficial in understanding the needs for further research. Examples of some of the most pertinent flight tests will be presented to highlight the knowledge gained. In such a review, wind tunnel tests cannot be ignored, because of the often elucidative impact they had.

The earliest known attempts to attain extensive regions of laminar flow in flight were made in the late 1930's and early 1940's. Both NLF with favorable chordwise pressure gradients and active LFC with boundary layer suction were investigated.

The B-18 flight test by the NACA in 1939 (Ref. 4) was a major milestone in the development of NLF. Therein, an attempt was made to prolong the run of laminar flow to higher Reynolds number than had previously been achieved by flight testing a 17-foot chord, 10-foot span wooden NLF glove on the wing of the test aircraft (Figure 1). An exceptional effort was made to evaluate surface quality effects by working the wing to previously unattained smoothness and fairness. The flight test clearly displayed the importance of surface discontinuities and finish. In fact, the adverse effect of surface disturbances (surface waves, two-dimensional type steps, and three-dimensional type roughness) was the most pervasive factor observed in the early tests and continued to be the principal cause of limited laminar flow in most future flight and wind tunnel tests. Although the severity of the surface disturbances was always shown to be aggravated by increased unit Reynolds number, it was not until considerable research (made possible by the development of the Langley Low-Turbulence Pressure Tunnel , LTPT) had been completed in the late 1940's and 1950's that an understanding of this phenomenon was developed, resulting in a quantitative ability to predict the magnitude of permissible three-dimensional disturbances (Ref. 5). It was not until the 1960's that a quantitative ability to predict permissible waviness and two-dimensional type disturbances was developed.

The maximum transition Reynolds number, based on freestream conditions and distance to the position of transition, attained in the B-18 flight tests was about 11.3 million with laminar flow to 42.5-percent chord on an NACA 35-215 section with only a 3-percent chord loss of laminar flow due to engine and propeller noise. This NLF transition Reynolds number was not to be exceeded in flight for over 40 years, until the NASA F-111/TACT NLF glove flight tests to be discussed later in this paper. The B-18 flight test was very encouraging in its time, because it indicated that the flight environment was possibly more benign for laminar flow than wind tunnels which until then had achieved laminar flow only at lower Reynolds number. The wind tunnel tests were highly compromised because the higher unit Reynolds numbers of the tunnels exacerbated the roughness problem. Later wind tunnel tests, in the quiet LTPT

(Ref. 6), suggested that the B-18 maximum length of NLF was constrained by the glove dimensions or achievable aircraft unit Reynolds number, and even higher transition Reynolds numbers might be obtained in flight. Indeed, the wind-tunnel experiments performed by Braslow (Refs. 6 and 7) showed natural transition Reynolds numbers of 14 to 16 million for 6-series airfoils.

During World War II, several military aircraft were built with NACA 6-series airfoils, which were designed to achieve extensive natural laminar flow. Perhaps the most notable of these airplanes was the P-51 Mustang. But it is doubtful that much laminar flow was achieved on these aircraft, because attention was not given to the surface quality that was required to maintain laminar flow. These aircraft flew in a harsh environment for obtaining laminar flow (i.e., at high speeds and low altitudes such that the unit Reynolds number was high) which placed stringent demands on surface smoothness and fairness. But after the war, attempts were made to see if NLF technology could be reduced to practice. The flight tests of King Cobra and Hurricane aircraft reported in References 8 through 10 are examples of such efforts (see Figures 2 and 3). The King Cobra used production wing surfaces that were highly polished and filled to reduce waviness. The Hurricane employed an NLF section in a special, "low-drag construction" wing thought to be suitable for the maintenance of laminar flow. With highly polished surfaces good NLF performance was achieved on these aircraft, but underlying concerns with the practicality of the wing surface tolerances and maintenance defeated these efforts. Now, some 40 years later, the general aviation industry is just beginning to explore the use of NLF on aircraft for which the Reynolds number capability was more than demonstrated by the early NLF flight testing. Many general aviation aircraft now fly at higher altitudes, where unit Reynolds numbers are lower, and recent advancements in wing fabrication techniques now offer the possibility of routinely producing small aircraft with sufficient surface smoothness and fairness.

Active laminar flow control with boundary layer suction has been used in attempts to extend the laminar flow into the region of adverse chordwise pressure gradient, which is not possible to any appreciable degree with NLF. Suction through porous materials, multiple slots, and perforations were tried with various degrees of success (e.g. Refs. 11-15). Three Vampire aircraft (Figure 4 and Refs. 12 through 14) with a number of suction surface configurations (continuous porous, perforated, and porous strips) and an F-94 aircraft (Figure 5 and

Ref. 15) with suction through multiple slots were flight tested in the mid-1950's. The F-94 tests were very encouraging. With 69 slots between 41-to 95-percent chord, full-chord laminar flow to length Reynolds numbers of 36.4 million was obtained on the F-94. The addition of slots and suction in the favorable gradient (x/c less than 41 percent) was found to significantly broaden the lift-coefficient range for low drag with laminar flow achievement. Laminar flow was lost behind shock waves on the F-94 when the aircraft speed was increased to the point where the local Mach number on the airfoil surface exceeded about 1.09. An important observation of the F-94 flight program was that the remains from insect impacts at low altitudes became subcritical at high speeds and altitudes above 20,000 feet for the boundary- layer flow of the unswept, F-94 wing. Unfortunately, this experience did not prevail in later flight test of swept wings, for which smaller critical roughness height has been observed in the regions of boundary-layer crossflow (Ref. 16). The Vampire aircraft tests experienced unusual surface roughness difficulties. Continuous suction (from 6 to 98-percent chord) through a porous panel, monel metal cloth (covered with nylon) or through 0.007 inch diameter perforations proved nearly as successful as the slotted F-94 surface, but each of these surfaces was thought to be impractical to manufacture and maintain. Two, "more practical" surfaces were tested, but with poor results. One incorporated porous strips of suction with sintered metal inserts; the other had perforations, 0.020 inch in diameter (the smallest holes then thought to be practical for manufacturing). The metal inserts caused surface discontinuities under flight loads, and the larger perforations caused transition by introducing unstable secondary flow in the boundary layer.

In the early 1960's, the most ambitious LFC flight program to date was undertaken by the Northrop Company. Under U.S. Air Force sponsorship, two WB-66 aircraft were modified with slotted suction wings and designated X-21 experimental aircraft (Figure 6 and Ref. 2). At the end of the program, full-chord laminar flow with suction was routinely obtained at Reynolds number of about 20 to 25 million. This was only after a long and difficult effort to improve performance through the systematic isolation and solution of problems, many due to wing sweep.

The most troublesome phenomenon encountered with the X-21 involved leading-edge turbulence contamination, a problem unique to swept wings. On the X-21, and at about the same time on a swept slotted-suction wing mounted vertically on the fuselage of a

Lancaster bomber (Figure 7 and Ref. 14 and 17), the significance of this problem became apparent. Although previous small-scale wind tunnel and flight experimentation by the British (Refs. 18 and 19) had indicated the existence of the spanwise turbulence contamination problem, its significance had gone unrecognized until the large scale flight tests. Subsequent flight and wind tunnel tests indicated that leading-edge scale was a predominant factor and that proper treatment of the inboard wing leading edge could prevent turbulence contamination of the swept wing from disturbances that propagate down the wing leading edge along the attachment line (e.g. Ref. 16). Although this phenomenon is now understood, it requires careful attention in the design of large LFC aircraft.

Another adverse effect of wing sweep on the ability to attain laminar flow had been found earlier during flights by the British with an Armstrong Whitworth A.W.52 airplane in 1951 (Ref. 18) with a natural laminar flow airfoil. A series of tests were performed where transition was shown to occur very close to the leading edge as a result of the formation of streamwise vortices in the laminar boundary layer. Later, an inability to obtain laminar flow in the last 20-percent chord of a Vampire trainer aircraft (Ref. 14) was attributed to the forward sweep of the trailing edge. This sweep-induced boundary-layer instability was caused by the large crossflows resulting from strong, favorable or adverse chordwise pressure gradients on swept wings. Research prior to the X-21 program (Ref. 20) showed that the proper application of suction is effective in controlling this crossflow instability; a result borne out in the X-21 flight tests.

Although structural flaws in the X-21 wing design produced surface waves and discontinuities that required liberal use of filler for smoothness, extensive laminar flow was routinely obtained at cruise altitudes of 40,000 ft. at Mach 0.75. A composite of the best wing surface laminar flow performance is shown in Figure 8 with remarkably good results obtained on the upper and, to a lesser degree, the lower wing surfaces. Flexing of the wing in flight continually deteriorated the surface quality due to the filler loss, but a series of 12 flights showed good repeatability even with major surface discrepancies on the last flight (Figure 9). Nonetheless, the X-21 wing structure was just not good enough to provide the surface quality needed for a convincing demonstration that LFC was ready for application. The poor laminar flow performance at lower altitudes and higher chord Reynolds number was undoubtedly due to the aggravated effects of poor surface quality at higher unit Reynolds numbers (Figure 10). Still, maximum-length laminar-flow Reynolds numbers up to 45.7 million were observed in some areas.

Another phenomenon adverse to achieving laminar flow was also realized and investigated during the X-21 program. It was noted that flight through visible cirrus clouds, and sometimes very light haze, caused loss of laminar flow. At cruise altitudes, cirrus clouds are composed mainly of ice crystals; entrainment of the crystals in the boundary layer produced local turbulence leading to the loss of laminar flow (Figure 11). Turbulent vortices shed by ice particles in the boundary layer were thought to trigger transition for certain combinations of particle size, concentration, and residence time in the boundary layer. At the termination of the X-21 program, concerns about this phenomenon and other unanswered issues on the operation of LFC aircraft were high.

In summary, when interest in laminar flow technology was rekindled by the energy crisis in the early 1970's, the fundamental aerodynamic concepts of both passive and active laminar flow control had been well established, verified in wind tunnel tests and demonstrated in various flight tests. The aerodynamics of the technology appeared to be well in hand. Laminar flow to Reynolds numbers up to 16 million had been observed on two-dimensional NLF sections, and it was not clear that an upper bound on the transition Reynolds number had been reached. Suction control had been demonstrated for boundary layers in adverse pressure gradients and on swept wings at Reynolds numbers well above 16 million; specifically, full chord laminar flow to about 36 million chord Reynolds for the former and 46 million for the latter. Yet, no practical application had been made with any suction method. The ability to manufacture and maintain aircraft surfaces with admissible tolerances, considerably smaller than required for turbulent aircraft, and at acceptable cost was still viewed as a formidable challenge. Neither suction slots nor perforations could be manufactured economically within required tolerances and the latter were believed to generate disturbances that adversely affected the ability to attain laminar flow at large length Reynolds numbers. Criteria for the proper design of slots were greatly improved during the X-21 flight program. With respect to perforated surfaces, early research indicated the need for hole diameters smaller than could be practically fabricated at that time. Porous surfaces with the required structural characteristics and aerodynamic smoothness were not available.

Over the past decade, NASA and the aircraft industry launched programs to continue the development of this technology and to provide the information needed for objective decisions on its application to new aircraft. The flight tests to be reviewed subsequently have been an integral part of those efforts.

## 5. F-111/TACT NLF Glove Flight Experiment

The NACA 6-series airfoils were originally developed for low-drag, NLF applications. In actuality, these airfoils were used on many of the early jet aircraft because they had very good performance as turbulent airfoils. However, modern supercritical airfoil technology has since led to improved airfoils with greatly enhanced turbulent performance (i.e., drag divergence Mach number, thickness ratio and lift coefficient capability). For this reason, in the late 1970's, the Boeing Company designed a supercritical, NLF airfoil in a NASA contract study to evaluate NLF for transport aircraft applications (Ref. 21). With the Boeing airfoil as a starting point, a supercritical NLF airfoil was also designed at the Langley Research Center and flight tested at the Ames-Dryden Flight Research Facility on the F-111/TACT aircraft (Refs. 22-24). The objective of the flight test program was to investigate natural laminar flow at transonic speeds.

A supercritical, natural laminar-flow airfoil glove was installed on the right wing panel of the F-111/TACT aircraft (Figure 12). The glove was made of fiberglass skins with an inner core of polyurethane foam and bonded to the metal wing skin. For symmetry, an uninstrumented glove was also installed on the left wing panel. The glove had a 6-foot span, a 10-foot chord, and was finished to "sail-plane" quality. The glove airfoil design pressure distribution (Figure 13) had a favorable gradient that extended to about 70-percent chord on the upper surface ( $dC_p/d(x/c) = -0.4$ ) and to about 50-percent chord on the lower surface ( $dC_p/d(x/c) = -0.8$ ). The airfoil design lift coefficient was 0.5 at a Mach number of 0.77 and a Reynolds number of 25 million. On the upper surface at this condition, supersonic flow extended from about 20-percent chord to 70-percent chord where the favorable gradient terminated in a weak shock. The glove was installed on the airplane to achieve the design pressure distribution at 10 degrees of leading-edge sweep; however, wind-tunnel tests had indicated that the pressure distributions at the higher sweep angles (up to 26 degrees) were acceptable for obtaining transition data at these conditions (i.e., no leading edge peaks or premature adverse gradients). In hindsight, the low design sweep angle of the glove was very conservative, but at that time, some studies had been very pessimistic regarding the amount of laminar flow that could be obtained at even moderate sweep angles and Reynolds numbers approaching 30 million (Ref. 21).

Results from the flight-test program (Ref. 24) indicate that the maximum extent of laminar flow was about 55-percent chord on the upper surface at 10 degrees of sweep for a chord Reynolds number of 28 million. However, as the wing sweep was increased to 26 degrees, the transition location moved forward to the 10 to 20-percent chord range (Figure 12). On the lower surface at 28 million chord Reynolds number, the maximum extent of laminar flow was about 50-percent chord (the start of the adverse gradient) and this was achieved to sweep angles as high as 15 degrees.

The wind-tunnel pressure distributions on the glove upper surface were much smoother than those obtained in flight (Figure 13), particularly at the higher sweep angles. Although the majority of the wind-tunnel results have not been published, stability analyses are presented for five cases in Reference 25. Based upon the wind-tunnel pressure distributions, these analyses predicted transition locations significantly further aft than those measured in flight on the upper surface. The irregularities in the flight upper-surface pressure distributions, which led to premature transition, were apparently caused by shocks propagating onto the glove from the inboard wing, and not by surface waves in the glove skin. In retrospect, the 6-ft. span of the glove was too small to isolate the glove from the flow on the remainder of the basic F-111/TACT wing over a broad range of conditions. Even for the design point at 10 degrees of sweep, there was a weak shock wave on the glove near 55-percent chord that limited the extent of laminar flow to this point instead of further aft near the pressure minimum (Figure 13). Since the lower-surface flow was subcritical, the lower-surface flight pressures were much smoother than those obtained on the upper surface, and in several cases laminar flow was obtained to the pressure minimum (approximately 50-percent chord). However, the steeper favorable gradient on the lower surface (Figure 13) was not suitable for achieving large runs of laminar flow at the higher sweep angles because of increased crossflow instability.

The F-111/TACT NLF experiment was brief, and consequently the transition data were very limited. However, the results were very encouraging. The maximum transition Reynolds numbers of about 15 million on the upper surface for 10 degrees of sweep, and 14 million on the lower surface at 15 degrees of sweep were significantly higher than values obtained in previous NLF flight tests. The closest comparable flight test had been conducted over 40 years earlier on the B-18 bomber previously discussed. During that test, a maximum transition Reynolds number of about 11.3 million was obtained.

## 6. F-14 Variable Sweep Transition Flight Experiment

Since the F-111/TACT NLF glove pressure distributions had not been designed to minimize cross-flow at the higher sweep angles, and since the maximum extent of laminar flow on both the upper and lower surface was determined by adverse pressure gradients, even larger transition Reynolds numbers at moderate sweep angles seemed possible. In addition, the techniques for fabricating and bonding large and very smooth foam and fiberglass test surfaces or gloves to metal wings had been developed and proven acceptable for flight testing. Consequently, the F-111/TACT NLF experiment paved the way for a follow-on program that could provide a much broader transition data base.

The F-14 Variable Sweep Transition Flight Experiment was initiated in 1984 (Refs. 26 through 30) with flight tests being completed in 1987. These tests were conducted with an F-14 (Figure 14) on loan to NASA from the Navy. Obtaining transition data was the primary objective of the program - not airfoil design verification. Therefore, only the upper surface of the wing was gloved in order to provide a laminar flow test surface. The gloves extended from about 10-percent chord on the lower surface to about 60-percent chord on the upper surface (spoiler hinge line) and covered the majority of the variable-sweep outer panel (Figure 15). Four rows of flush static pressure orifices and three arrays of hot-films were distributed along the span for determination of the local wing pressure distributions and transition locations. These data and the other associated flight parameters were monitored in real-time on the ground during all the testing.

Two gloves were flight-tested during the program: one was a "clean-up" or smoothing of the basic F-14 wing (modified NACA 6-series airfoils), while the second involved significant contour modifications to the basic F-14 wing. The second glove, designed at NASA Langley (Ref. 27), provided more moderate favorable pressure gradients than the "clean-up" glove, and achieved more of a two-dimensional type flow (straighter isobars) over a larger part of the span. Both gloves were constructed of fiberglass skins with an inner core of polyurethane foam (Ref. 29). Measurements taken on the gloves with a mechanical deflection gauge having support feet two inches apart indicated wave amplitudes no larger than 0.002 in. Representative pressure distributions at several Mach numbers are presented in Figures 16 and 17 for both gloves. The Langley glove design provided a wide variety of pressure distributions with different favorable gradients to about 50-percent chord over a broad Mach number range (0.6 to 0.8).

Transition locations for the gloves are presented in References 28 and 30. To compare various transition or laminar flow experiments, transition Reynolds number is an appropriate parameter, therefore, the maximum transition Reynolds number observed in several of the more significant flight and wind tunnel experiments are presented in Figure 18. For this figure, transition Reynolds number is based on freestream conditions, rather than local conditions. In addition to the F-111 and F-14 experiments, included in Figure 18 are results from several natural laminar flow tests: the B-18 flight test (Ref. 4); the King Cobra flight test (Refs. 8 and 9); the 757 NLF glove flight test conducted by the Boeing Company (Refs. 31 through 33); and low-speed wind-tunnel tests conducted in the 12-Foot Tunnel at NASA-Ames Research Center (Ref. 34) and the LTPT at the Langley Research Center (Refs. 6 and 7). Prior to the F-111 and F-14 flight tests, the highest NLF transition Reynolds numbers for airfoils or wings had been obtained in the LTPT at Langley and the 12-Foot Tunnel at Ames. These are very quiet tunnels and only until recently have airplanes (i.e., jet-powered aircraft) been able to match the Reynolds number capability of these facilities. More importantly, very few large aircraft have had the capability of providing large runs of laminar flow.

As previously discussed, results from the F-111/TACT NLF Glove Experiment had exceeded the prior maximum values for natural laminar flow transition Reynolds numbers that had been obtained in flight on the B-18 and King Cobra. Results obtained during the F-14 VSTFE indicate maximum transition Reynolds number values exceeding F-111/TACT and wind tunnel values up to 30 degrees of sweep. For the F-14 VSTFE, a maximum transition Reynolds number of about 17.6 million was obtained at 15 degrees of sweep, 13.5 million at 20 degrees, and 12 million at 25 degrees. Beyond 25 degrees of sweep, maximum transition Reynolds number decreased rapidly to about 5 million at 35 degrees of sweep. It should be pointed out that for all the maximum transition Reynolds number cases on both the F-111 and F-14, the amount of laminar flow was limited by either adverse pressure gradient or shock wave location. This suggests that even higher transition Reynolds numbers are possible in flight.

In comparison to the NLF tests, as would be expected, maximum transition Reynolds numbers for most suction or laminar flow control experiments are much higher. As previously discussed, transition Reynolds numbers of about 30 to 36 million were obtained in flight on the Vampire and F-94, and a value of about 46 million was obtained in a small area of the X-21 wing. However, with suction only in the leading-edge region of swept wings, the transition Reynolds number for natural laminar flow designs can be significantly increased - possibly doubled. This concept, called hybrid laminar flow control (HLFC), is discussed later in the paper.

In stability analyses underway at Langley, the F-14 transition data are being examined with the maximum amplification option of Reference 35, but these analyses are also including surface and in-plane streamline curvature effects (Refs. 36, 37 and 38). The initial efforts have concentrated on transition data for conditions where Tollmien-Schlichting wave growth is small and crossflow-like disturbances dominate the transition process. Stability analyses are performed for both stationary and nonstationary crossflow disturbances.

Illustrated in Figures 19 through 22 are results for two typical flight conditions on the "clean-up" glove of the F-14 aircraft. These flight conditions produce strong, favorable pressure gradients (Figures 19 and 21) that lead to little or no Tollmien-Schlichting wave growth and dominance of crossflow-like disturbances. The crossflow Reynolds number development (Figures 19 and 21) is indicative of strong crossflow vortices, and transition occurs for both conditions when this parameter exceeds 400, a value somewhat higher than the 175 to 300 range observed at low speeds. Previous analyses (Ref. 39) indicate that the effects of compressibility on crossflow-like disturbances are small; comparison of the N factors for stationary and nonstationary crossflow vortices, with and without compressibility effects, confirms this, as shown in Figures 20 and 22. Also, stationary crossflow vortices are not the most highly amplified disturbances for these conditions, but indeed, nonstationary vortices with frequencies of about 2000-to 3000-hertz are more highly amplified. In the absence of significant compressibility effects, the incompressible code developed by Malik and Poll (Ref. 36) has been used to examine surface and inplane curvature effects on the disturbance development. These effects, shown in Figures 20 and 22, are quite significant; the N factors of the most highly amplified waves are reduced from about 15 to around 10 at the measured transition location. The magnitude of the curvature effects in all the flight data analyzed to date gives rise to concerns over any previous attempts to correlate transition data with stability codes ignoring curvature effects and questions the generality of those correlations. Immediate plans are to begin examination of data for flight conditions with stronger Tollmien-Schlichting wave growth to explore the possibility of interactions of Tollmien-Schlichting waves and crossflow vortices.

## 7. 757 Wing Noise Survey and NLF Glove Flight Test

In 1985, under a NASA contract, the Boeing Company performed a flight test to measure the acoustic environment in cruise on the wing of a 757 aircraft and to determine of the potential effects of the acoustic environment on boundary layer transition (Refs. 31, 32, and 33). Prior to this flight test, there were no extensive measurements of the noise environment on the wing of a commercial transport with wing-mounted, high-bypass-ratio turbofan engines. Engine noise concerns had led to conservatism in LFC aircraft design studies, with designs restricted to aft engine placement with a potentially severe adverse impact on performance and a degradation of LFC fuel savings potential. A major part of the 757 flight test was an attempt to achieve a limited amount of laminar flow over the wing and measure the impact of the engine noise intensity on the extent of laminar flow. Although the primary goals differed, an interesting parallel exists between the 757 and the B-18 tested some 45 years earlier. As with the B-18, the 757 experiments yielded encouraging results with regard to engine noise effects on laminar flow.

Boeing removed a leading edge slat on the 757 wing just outboard of the starboard engine and installed a 10-foot span, NLF glove constructed of dense foam with a fiberglass epoxy overlay to produce a smooth, nearly wave-free surface (Figure 23). The glove was designed to achieve laminar flow on both the upper and lower surface, with 20-to 30-percent chord laminar flow expected without adverse engine noise effects. This anticipated result was made possible in part by unsweeping the wing to 21 degrees at the glove location and by the favorable pressure distribution over the wing. A single microphone was installed on the glove leading edge and eight others were installed on each of the wing surfaces (upper and lower) - three on the glove and five distributed over the remaining wing surface (Figure 23). Hot films were used to detect the transition front on the upper and lower surfaces. Measurements were made over cruise altitudes of 25 to 41 thousand feet at Mach numbers of 0.63 to 0.83. The starboard engine was throttled from maximum continuous thrust to idle at cruise speeds and altitudes.

Some order in the acoustic data is achieved by normalization of the OASPL's with the ambient pressure over the altitude range of the test conditions. These data are shown for two microphone locations in Figure 24. Normalized OASPL's are presented for a microphone on the glove and one aft of the glove for various flight Mach numbers and engine power settings; the latter reflected by fan exhaust Mach number variations. Flight Mach number and engine power setting effects are measureably different on the upper and lower wing surface. The lower surface normalized OASPL's show a strong dependence on engine power setting with about a 20 dB. increase occurring from engine idle (fan Mach number equal to 0.7) up to maximum continuous thrust (fan Mach number equal to 1.28) when the lower surface acoustic characteristics seem dominated by engine noise. Engine power setting has little influence on the OASPL's on the upper surface, but significant variations occur with flight Mach number. The wing appears to effectively shield the upper surface from radiated engine noise and the dominant noise sources are presumed to be of aerodynamic origin. The data present strong evidence that the wing upper-surface flow field has a major influence on the radiated acoustic field, particularly at higher cruise speeds when shock waves occur on the wing. The supercritical flow over the upper surface inhibits forward radiation of sound from downstream sources, aerodynamic or engine related.

Attempts have been made to analyze the 757 acoustic data and make comparisons with theoretical predictions (Ref. 33). A procedure developed by the Lockheed Georgia Company under NASA contract has been used (Ref. 40). To the authors' knowledge this is the only code available to make near field noise predictions that include all the potentially relevant noise sources at flight cruise conditions. However, the theory lacks inclusion of the important effects of scattering, refraction, and reflection of sound fields due to the airframe or flow fields about it. For this reason, predictions for only the lower surface OASPL's and spectra have been made. These results are not discussed in any depth herein. Generally, the results indicate that our ability to predict the acoustic environment at high cruise speeds and altitudes is poor. Theory suggests that the lower wing surface noise should be dominated by the fan exhaust broad band shock noise at cruise thrust conditions, which is consistent with the observed correlation with the fan Mach number; but the predicted levels of OASPL are 10 to 40 dB. too high. Trailing edge noise is predicted to be an important aerodynamic noise source, particularly at aft wing locations; the data doesn't confirm this. Convective and dynamic amplification effects have large impacts upon the predictions. These effects or the methodology for their implementation are made suspect by the data. Clearly, more analyses of these acoustic data are needed to unravel the confused picture presented by the data and theory. With further analyses, the broad range of conditions for the 757 data could possibly permit useful calibration of the Lockheed code.

The amount of laminar flow obtained on the NLF glove was very encouraging. This result indicates that the acoustic environment may be benign enough to achieve extensive laminar flow on wings with wing-mounted engines. The results are summarized in Figure 25 wherein the design condition and conditions of maximum extent of laminar flow are shown. A maximum of nearly 30-percent chord laminar flow was obtained on both surfaces. At the design condition, best results were obtained on the upper surface, athough laminar flow was not uniform across the gloved span and was most extensive inboard. The upper surface pressures on the glove showed peaks in the outboard region which presumably led to earlier boundary layer transition. Transition was more uniform across the lower surface with 26-percent chord laminar flow achieved when the aircraft was sideslipped to reduce the leading edge sweep by 6.8 degrees.

On the upper wing surface, the extent of laminar flow was essentially unaffected by engine power setting. Since the power setting had no effect on the upper-surface noise levels, the unchanging extent of laminar flow is not surprising. On the lower surface, however, the noise levels varied over 20 dB., but almost imperceptibly small (2 to 3-percent chord decreases at most) changes in the extent of laminar flow were observed. Over the range of flight conditions, boundary layer stability analysis (Ref. 33) identified stationary crossflow vortices in the boundary layer to be highly unstable and possibly the dominant disturbances leading to transition. The small effect of variations in engine noise level on the transition location on the lower surface may indicate that engine noise does not have a significant effect on crossflow disturbances. If crossflow disturbance growth in the leading edge is controlled by suction, laminar flow much more extensive than achieved in this flight test could be possible even in the presence of engine noise. However, in an HLFC application, the Tollmien-Schlichting wave growth may be comparable or greater than the crossflow disturbance growth; engine noise might then be expected to limit the extent of laminar flow.

## 8. The NASA Leading-Edge Flight Test Program

Earlier in this paper, some of the key laminar flow flight programs that laid the foundation for today's knowledge were briefly reviewed. These flight tests removed any doubt that extensive laminar flow could be achieved in flight. They did not, however, resolve concerns relative to the practicality of producing surfaces sufficiently smooth and wavefree, and of maintaining the required surface quality during normal service operations. In the late 1970's, with the recent significant progress made in the development of new materials, fabrication techniques, analysis methods, and design concepts, a reexamination of these issues appeared warranted.

Previous experience had shown that the leading-edge region of the swept wing presented the most difficult aerodynamic problems associated with attainment of laminar flow. In addition, the leading edge is subject to foreign object damage, insect impingement, rain erosion, icing, and other contaminants. Also, an anti-icing system, an insect protection system, and a suction and perhaps purge system must all be packaged into a relatively small leading-edge box volume. Most of these problems are common to all the concepts under consideration for the achievement of extensive laminar flow and solutions are needed to establish the practicality of laminar flow for various types of aircraft.

In 1980, the NASA Leading-Edge Flight Test (LEFT) Program was initiated as a flight validation of two leading-edge LFC concepts then under development in NASA contract efforts with industry. The flight program objectives were to: (1) demonstrate that required leading-edge systems can be packaged into a wing leading-edge section of a size representative of a commercial transport aircraft, and (2) demonstrate systems performance under operational conditions representative of subsonic commercial transport aircraft. Complete LFC leading-edge systems were installed in the leading-edge box of a

JetStar airplane (Figure 26). Descriptions of the systems illustrated in Figures 27 and 28 are provided in References 41 and 42. Two leading-edge test articles were built and flown using a perforated and a slotted suction concept. Each spanned about 6-foot of the wing and had the same external contour, dimensionally about equivalent to the leading-edge box of a DC-9-30 at the mean aerodynamic chord. The wing leading edge sweep was 30 degrees. Different systems were used in each test article. One used suction through approximately 1 million, 0.0025-inch diameter, 0.035-inch spaced, electron-beam perforated holes in a 0.025-inch thick titanium skin to maintain laminar flow on the test article upper surface. A Krueger-type flap served as a protective shield against insect impact on this leading edge. In future applications, the Krueger shield could also serve as a high-lift leading-edge device. A freezing-point depressant liquid, Propylene Glycol Methyl Ether (PGME) was sprayed on the perforated, wing upper surface from nozzles mounted underneath the shield to augment the insect shield protection and to provide an anti-icing capability. To prevent clogging of the perforations by the wetting fluid, a purging system was included to clear the LFC passages by pressurizing the subsurface and thus remove PGME fluid from the LFC ducts and surface. The second test article used suction through 27 narrow spanwise slots (about 0.004-inch wide) on both upper and lower titanium surfaces. This test article contained insect protection and anti-icing systems consisting of PGME fluid dispensed through dual purpose slots in the leading edge. Purge was also provided for this leading edge.

After an initial flight test program to optimize the system's performance, the LEFT systems were flight tested in a simulated airline service in different geographical areas, seasons, and weather conditions in the United States (Figure 29). During the simulated service, one-to four-flights per day were made from three "home base" airports (Atlanta, Pittsburgh, and Cleveland). A total of 62 flights to 33 airports were made. Flights were made from Atlanta in July 1985, Pittsburgh in September 1985, and Cleveland in February 1986. The weather experienced thus varied from severe summer to severe winter conditions. To realistically simulate typical transport operations, an on-off operation of all systems was imposed; no adjustments were made prior to or during flights. Transport cruise flight conditions were emphasized, but investigations were also made of the ability to attain laminar flow at other than cruise conditions. References 43 through 46 provide a summary description of the program results.

The emergence of electron-beam perforated titanium as a practical manufacturing surface which meets laminar flow waviness specifications with practical aircraft fabrication methods is considered a major development of the LEFT program. The perforated titanium leading-edge presented no difficult fabrication problems. This test article yielded clearly superior performance (relative to the slotted configuration) and was in virtually the same condition when flights ended in October 1987, as when flights began in November 1983. Four years of flying resulted in no degradation of laminar flow performance as a result of service, and no evidence of any deterioration in surface quality was observed. Essentially complete laminar flow on the test article was consistently obtained from 10,000 to 38,000 feet altitude with no need for any special maintenance.

The results obtained with the slotted-surface test article, however, were not as favorable. Fabrication of this configuration involved some extremely difficult problems that led to a suction surface that was only marginally acceptable with respect to surface smoothness and waviness. This was reflected in consistently poorer laminar flow flight performance than for the test article with the perforated surface. Still, as much as 80 percent of the slotted upper surface suction area was observed to be laminar in routine flight service.

Since no attempt was made to obtain laminar flow beyond the front spar, the LEFT tests should not be interpreted as showing that perforations are aerodynamically better than slots. Indeed, the perforated approach should be pursued with caution because additional flight testing is required to larger values of length Reynolds number. At higher Reynolds number, the experience of the early flight tests with larger holes (i.e., progressive performance deterioration with increased Reynolds number) could be repeated. Slots may, therefore, be preferred at higher Reynolds number. Accordingly, it is clear that more development of fabrication techniques for slotted suction surface configurations is required; some initial work in this direction has been undertaken (Ref. 47).

The LEFT program relaxed concerns about the operational loss of laminar flow when entering clouds or haze. It provided some confirmation of an extensive analysis of world-wide cloud-cover (based on 6250 flight hours of specially instrumented commercial aircraft) which resulted in an estimate of 6 percent for the amount of flight time spent in clouds and haze (Ref. 48). During the simulated service flights, measurements were also taken of the time spent in clouds and haze. These LEFT results, based on 6 hours and 52 minutes of data taken during 13 flights within the United States, showed that clouds and haze were encountered about 7 percent of the time (Ref. 46). No effort was made to avoid cloud encounters, and a sample of one flight including a cloud penetration is shown in Figure 30. As expected, laminar flow was lost during cloud penetrations, but was regained afterwards. The small percentage of time that clouds are encountered indicates that laminar flow loss during cloud penetrations in cruise will not appreciably decrease the large economic and fuel gains predicted for laminar flow transport aircraft. However, potential cloud-encounters en route and flight management to avoid clouds could be operational considerations for future aircraft.

To summarize the LFC systems performance during the simulated service, all operational experience was positive. No dispatch delays were encountered due to the LFC systems. There was no need to adjust suction system controls throughout the test range of cruise altitude, Mach number, and lift coefficient. Laminar flow was obtained after exposure to heat, cold, humidity, insects, rain, freezing rain, snow, ice, and moderate turbulence. The insect protection systems were required during descent as well as ascent and were effective when used. Perforated test article results indicated that the supplemental spray system is not necessary for LFC transport airplanes equipped with a properly designed insect shield/high-lift device, although the spray system may be necessary for anti-icing purposes. Ground deicing of the LFC test articles was no more difficult than normal deicing of commercial transports, and snow and ice accumulation was easily eliminated using hand-held deicing equipment. The NASA LEFT simulated airline service flights demonstrated that effective practical solutions for the problems of suction laminar-flow aircraft leading edges are available for commercial transport aircraft.

The LEFT Program has been the only laminar-flow flight test with suction since the X-21 ended in 1965. The original intent of the LEFT program was to examine systems suitable for future laminar-flow control aircraft, but these systems would be equally applicable to hybrid laminar-flow control aircraft that use suction only in the leading-edge box.

## 9. HLFC Flight Experiment

One of the most significant developments of the NASA research on laminar flow in the last few years, has been the recognition of a hybrid laminar flow control concept that integrates LFC and NLF and avoids the objectionable characteristics of each. The leading-edge sweep limitation of NLF is overcome by applying suction in the leading-edge box to control crossflow instabilities. Wing shaping for favorable pressure gradients to allow NLF over the wing box removes the need for inspar LFC suction and greatly reduces the system complexity. The possibility of achieving extensive laminar flow on commercial or military transport aircraft is offered with a system no more complex than that already proven in the Leading Edge Flight Test Program on the NASA JetStar. To explore this possibility, the NASA Langley Research Center, the Air Force Wright Aeronautical Laboratory, and the Boeing Commercial Airplane Company have initiated a cooperative flight program. A high Reynolds number HLFC Flight Experiment will be performed on a 757 aircraft equipped with a partial-span HLFC system for the upper surface of the left wing.

The test aircraft and test region are illustrated in Figure 31. A 20-foot span of the wing just outboard of the left engine pylon will be modified. A new leading-edge box will be installed with suction achieved through a perforated titanium surface. The structural concept will be similar to that used on the JetStar for the Leading Edge Flight Test and will include a leading-edge Krueger integrated into the full wing high-lift system and designed to also be an insect shield for the wing (Figure 32). The leading edge box will be contoured to achieve the desired pressure distribution over the test surface (Figure 33). Analyses indicate that this can be accomplished without changing the inspar contour of the 757. Indeed, measurements of the 757 production wing surface have shown that only minor shaving or filling of some rivets will be necessary to meet laminar flow smoothness and fairness criteria. The inspar production wing surface will thus serve as the test surface downstream of the new leading-edge box.

The practicality of the HLFC concept for realistic flight applications is untried to date in either flight or the wind tunnel, but will be evaluated to chord Reynolds numbers over 30 million at the cruise conditions of modern transport aircraft. An extended flight test program is planned for calendar year 1990 to achieve operational experience with HLFC and to fully evaluate the potential for future applications.[1] Success could lead to the long-awaited transfer of this technology to the drawing board and ultimately to practice.

---

[1]As of August 1990, the 757 HLFC flight test was successfully completed; documentation is forthcoming.

## 10. Concluding Remarks

The potential benefits of laminar flow technology have been so enticing that possibly no other technology has received such persistent attention in flight research over so long a time. The misgivings of the critics are fading with the accomplishments of this research. The aerodynamic issues seem nearly resolved, and the manufacturing capabilities of the airframe industry appear to have advanced to the point that the aerodynamic criteria for smooth, wave-free wing surfaces is a practical production goal. The current NASA, AFWAL, and Boeing HLFC Flight Experiment could provide the verification needed to place this technology in practice. Initial applications may provide only modest improvements, but with the confidence of success, bolder steps could revolutionize aircraft design.

## References

1. Jones, B. M.: Flight Experiments on the Boundary Layer. JAS, Vol. 5, No. 3, pp. 81-94. January 1938.

2. Whites, R. C.; Sudderth, R. W.; and Wheldon, W. G.: Laminar Flow Control on the X-21. Astronautics and Aeronautics, Vol. 4, No. 7, July 1966.

3. Wagner, R. D.; and Fischer, M. C.: Developments in the NASA Transport Aircraft Laminar Flow Program. AIAA-83-0090, January 1983.

4. Wetmore, J. W.; Zalovcik, J. A.; and Platt, R. C.: A Flight Investigation of the Boundary-Layer Characteristics and Profile Drag of the NACA 35-215 Laminar-Flow Airfoil at High Reynolds Numbers. NACA L-532, 1941.

5. von Doenhoff, A. E.; and Braslow, A. L.: Effect of Distributed Surface Roughness on Laminar Flow. Chapter of Boundary Layer and Flow Control-Its Principles and Applications, Volume 2, edited by G. V. Lachmann, Pergamon Press, New York, 1961.

6. Braslow, A. L.: Investigation of Effects of Various Camouflage Paints and Painting Procedures on the Drag Characteristics of

an NACA 65(421)-420, a = 1.0 Airfoil Section. NACA CB No. L4G17, July 1944.

7. Braslow, A. L.; and Visconti, F.: Investigation of Boundary-Layer Reynolds Number for Transition on an NACA 65(215)-114 Airfoil in the Langley Two-Dimensional Low-Turbulence Pressure Tunnel. NACA TN 1704, October 1948.

8. Gray, W. E.; and Fullam, P. W. J.: Comparison of Flight and Tunnel Measurements of Transition on a Highly Finished Wing (King Cobra). RAE Report Aero 2383, 1945.

9. Smith, F.; and Higton, D.: Flight Tests on King Cobra FZ.440 to Investigate the Practical Requirements for the Achievement of Low Profile Drag Coefficients on a "Low Drag" Aerofoil. R and M 2375, British A.R.C., 1950.

10. Plascoff, R. H.: Profile Drag Measurements on Hurricane II z. 3687 Fitted with Low-Drag Section Wings. RAE Report No. Aero 2153, 1946.

11. Zalovcik, J. A.; Wetmore, J. W.; and von Doenhoff, A. E.: Flight Investigation of Boundary-Layer Control by Suction Slots on an NACA 35-215 Low-Drag Airfoil at High Reynolds Numbers. NACA WR L-521, 1944.

12. Head, M. R.: The Boundary Layer With Distributed Suction. R and M No. 2783, British A. R. C., 1955.

13. Head, M. R.; Johnson, D.; and Coxon, M.: Flight Experiments on Boundary Layer Control for Low Drag. R and M No. 3025, British A. R. C., March 1955.

14. Edwards, B.: Laminar Flow Control - Concepts, Experiences, Speculations. AGARD Report No. 654. Special Course on Concepts for Drag Reduction, 1977.

15. Pfenninger, W.; Groth, E. E.; Whites, R. C.; Carmichael, B. H.; and Atkinson, J. M.: Note About Low Drag Boundary Layer Suction Experiments in Flight on a Wing Glove of a F94-A Airplane. Reference No. NAI-54-849 (BLC-69), Northrop Aircraft, Inc., December 1954.

16. Pfenninger, W.: Laminar Flow Control Laminarization. In *Special Course on Concepts for Drag Reduction*, AGARD-R-654, June 1977.

17. Landeryou, R. R.; and Porter, P. G.: Further Tests of a Laminar Flow Swept Wing with Boundary Layer Control by Suction. Report Aero. No. 192, College of Aeronautics, Cranfield (England), May 1966.

18. Gray, W. E.: The Effect of Wing Sweep on Laminar Flow, RAE TM Aero. 255, 1952.

19. Anscombe, A.; and Illingworth, L. N.: Wind Tunnel Observations of Boundary Layer Transition on a Wing at Various Angles of Sweepback. R and M No. 2968, British A. R. C., 1956.

20. Pfenninger, W.; Gross, L.; and Bacon, J. W., Jr.: Experiments on a 30 Degree Swept, 12 Percent Thick, Symmetrical, Laminar Suction Wing in the 5-Foot by 7-Foot Michigan Tunnel. Northrop Corp., Norair Division Report. NAI-57-317 (BLC-93), February 1957.

21. Boeing Commercial Airplane Company: Natural Laminar Flow Airfoil Analysis and Trade Studies, NASA CR-159029. May 1979.

22. Steers, L. L.: Natural Laminar Flow Flight Experiment. NASA CP-2172, pp. 135-144, October 1980.

23. Montoya, L. C.; Steers, L. L.; Christopher, D.; Trujillo, B.: F-111 TACT Natural Laminar Flow Glove Flight Results. NASA CP-2208, pp. 11-20, September 1981.

24. Boeing Commercial Airplane Company: F-111 Natural Laminar Flow Glove Flight Test Data Analysis and Boundary-Layer Stability Analysis. NASA CR-166051. January 1984.

25. Runyan, L. J.; and Steers, L. L.: Boundary-Layer Stability Analysis of a Natural Laminar Flow Glove on the F-111 TACT Airplane, Viscous Flow Drag Reduction, Vol. 72, Progress in Astronautics and Aeronautics, pp. 17-32, 1980.

26. Rozendaal, R. A.: Variable Sweep Transition Flight Experiment (VSTFE) - Parametric Pressure Distribution, Boundary-Layer Stability Study and Wing Glove Design Task. NASA CR-3992, 1986.

27. Waggoner, E. G.; Campbell, R. L.; Phillips, P. S.; and Hallissy, J. B.: Design and Test of an NLF Glove Wing for the Variable Sweep Transition Flight Experiment. NASA CP-2487, pp. 753-776, 1987.

28. Wagner, R. D.; Maddalon, D. V.; Bartlett, D. W.; Collier, F. S. Jr.; and Braslow, A. L.: Laminar Flow Flight Experiments. From Proceedings of the Transonic Aerodynamics Symposium held at NASA Langley Research Center. NASA CP 3020, 1988.

29. Bohn-Meyer, M.: Constructing Gloved Wings for Aerodynamic Studies, AIAA 88-2109, May 1988.

30. Meyer, R. R.; Trujillo, B. M.; and Bartlett, D. W.: F-14 VSTFE and Results of the Cleanup Flight Test Program. NASA CP-2487, pp. 819-844, 1987.

31. Boeing Commercial Airplane Company: Flight Survey of the 757 Wing Noise Field and Its Effect on Laminar Boundary Layer Transition. NASA CR-178216, Vol. I - Program Description and Data Analysis. March 1987.

32. Boeing Commercial Airplane Company: Flight Survey of the 757 Wing Noise Field and Its Effect on Laminar Boundary Layer Transition. NASA CR-178217, Vol. 2 - Data Compilation. March 1987.

33. Boeing Commercial Airplane Company: Flight Survey of the 757 Wing Noise Field and Its Effect on Laminar Boundary Layer Transition. NASA CR-178419, Vol. 3 - Extended Data Analysis. May 1988.

34. Boltz, F. W.; Kenyon, G. C.; and Allen, C. Q.: Effects of Sweep Angle on the Boundary-Layer Stability Characteristics of an Untapered Wing at Low Speeds. NASA TN D-338, 1960.

35. Malik, M. R.: Cosal-A Black Box Compressible Stability Anal-

ysis Code for Transition Prediction in Three-Dimensional Boundary Layers. NASA CR-165925, 1982.

36. Malik, M. R.; and Poll, D. I. A.: Effect of Curvature on Three-Dimensional Boundary Layer Stability. AIAA Journal, Vol. 23, No. 9, pp. 1362-1369. September 1985.

37. Collier, F. S., Jr.; and Malik, M. R.: Stationary Disturbances in Three-Dimensional Boundary Layers. AIAA Paper 87-1412, June 1987.

38. Collier, F. S., Jr.: Curvature Effects on the Stability of Three-Dimensional Laminar Boundary Layers. Ph.D. Dissertation, Virginia Polytechnic Institute and State University, May 1988.

39. Mack, L. M.: Boundary Layer Stability Theory. AGARD Report No. 709, Special Course on Stability and Transition of Laminar Flows, VKI, Brussels, 1984.

40. Swift, G; and Mangur P.: A Study of the Prediction of Cruise Noise and Laminar Flow Control Noise Criteria for Subsonic Air Transports. NASA CR-159104, August 1979.

41. Douglas Aircraft Company Staff: Laminar Flow Control Leading Edge Glove Flight Test Article Development. NASA CR-172137, 1984.

42. Etchberger, F. R.: Laminar Flow Control Leading Edge Glove Flight Aircraft Modification Design, Test Article Development, and Systems Integration. NASA CR-172136, 1983.

43. Fisher, D. F.; and Fischer, M. C.: The Development Flight Tests of the JetStar LFC Leading-Edge Flight Experiment. NASA CP-2487, pp. 117-140, 1987.

44. Maddalon, D. V.; Fisher, D. F.; Jennett, L. A.; and Fischer, M. C.: Simulated Airline Service Experience with Laminar Flow Control Leading-Edge Systems. NASA CP-2487, pp. 195-218, 1987.

45. Powell, A. G.; and Varner, L. D.: The Right Wing of the LEFT Airplane. NASA CP-2487, pp. 141-161, 1987.

46. Davis, R. E.; Maddalon, D. V.; and Wagner, R. D.: Performance of Laminar Flow Leading Edge Test Articles in Cloud Encounters. NASA CP-2487, pp. 163-193, 1987.

47. Goodyear, M. D.: Application of Superplastically Formed and Diffusion Bonded Aluminum to a Laminar Flow Control Leading Edge. NASA CR-178316, 1987.

48. Jasperson, W. H.; Nastrom, G. D.; Davis, R. E.; and Holdeman, J. D: GASP Cloud - and Particle Encounter Statistics, and Their Application to LFC Aircraft Studies (2 Vols.). NASA TM 85835, October 1984.

**B-18 FLIGHT TEST**

(CIRCA 1940)

44.8′

17′

Test Area

10′

Centerline of fuselage

- NACA 35-215 Airfoil
- Wooden Glove
  ( ± 1/1000 inch Waves)
- Transition Reynolds No.
  $\sim 11.3 \times 10^6$
  at 42.5% Chord
- Engine/Propeller Noise Reduced Laminar Run by 3% Chord

Figure 1. The B-18 NLF Glove Flight Test.

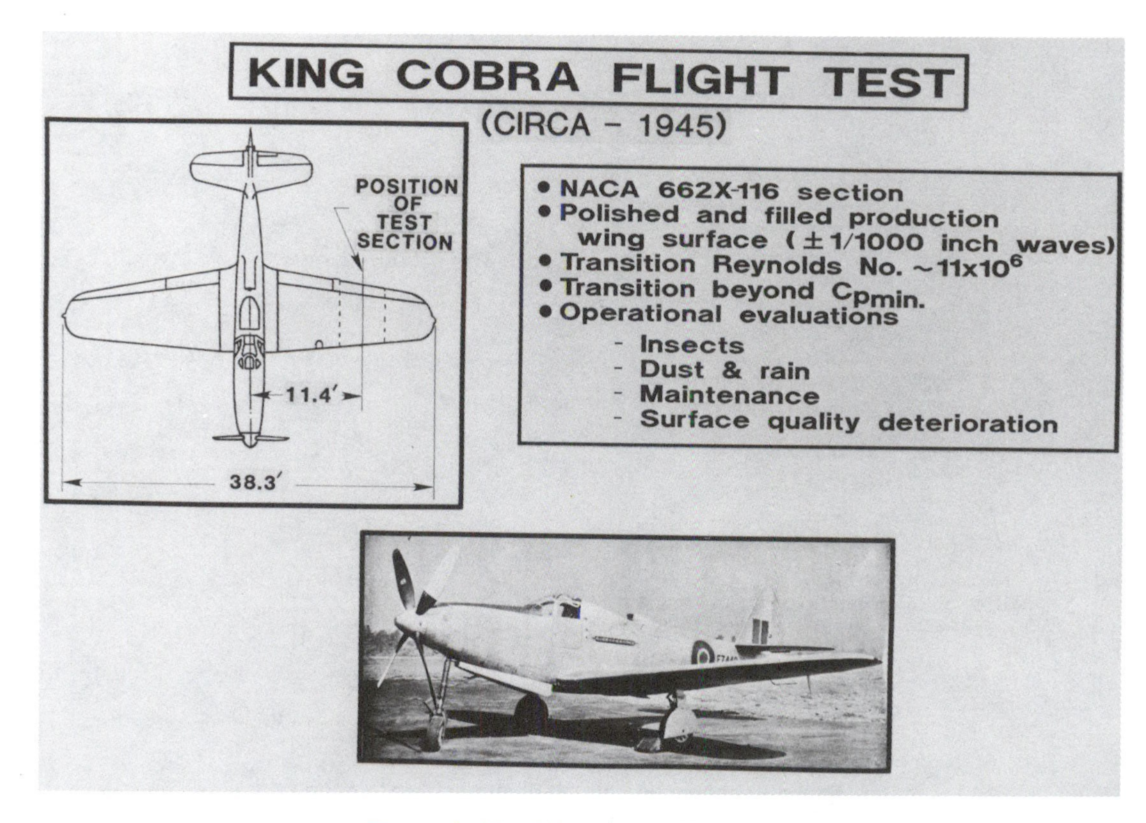


Figure 2. The King Cobra NLF Flight Test.

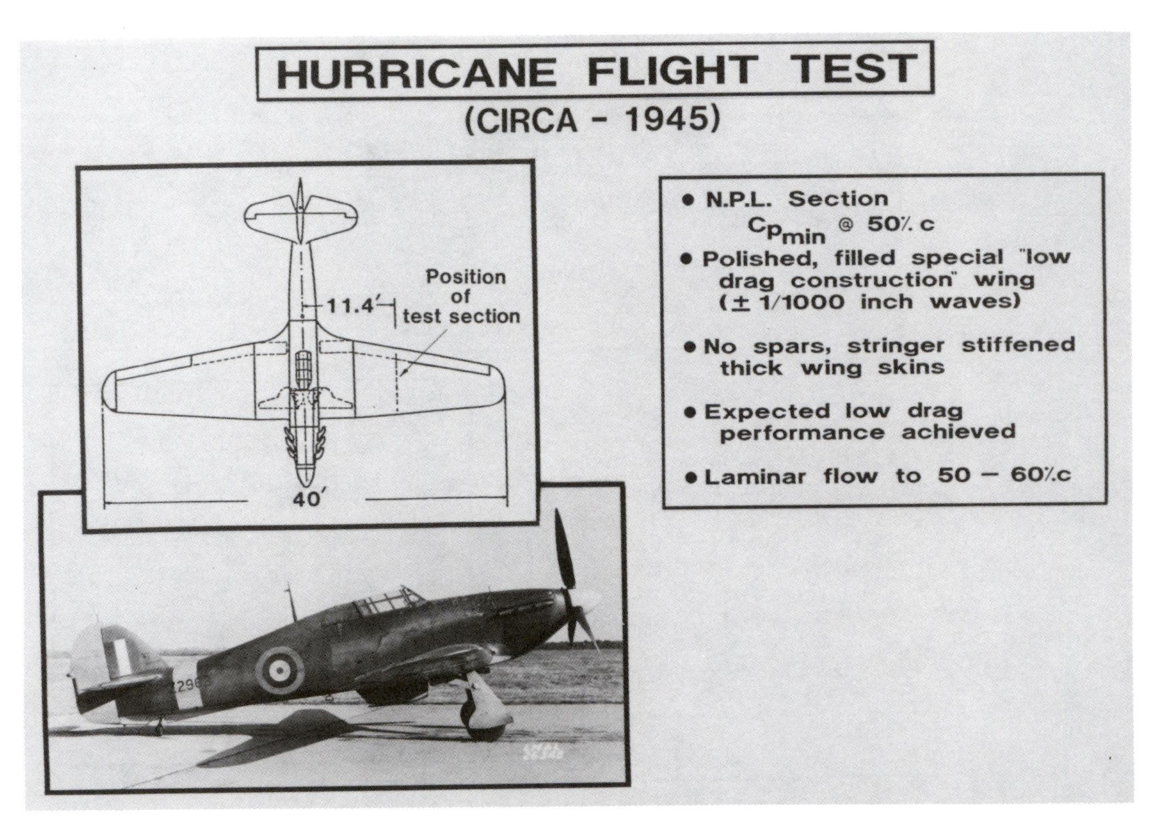


Figure 3. The Hurricane NLF Flight Test.

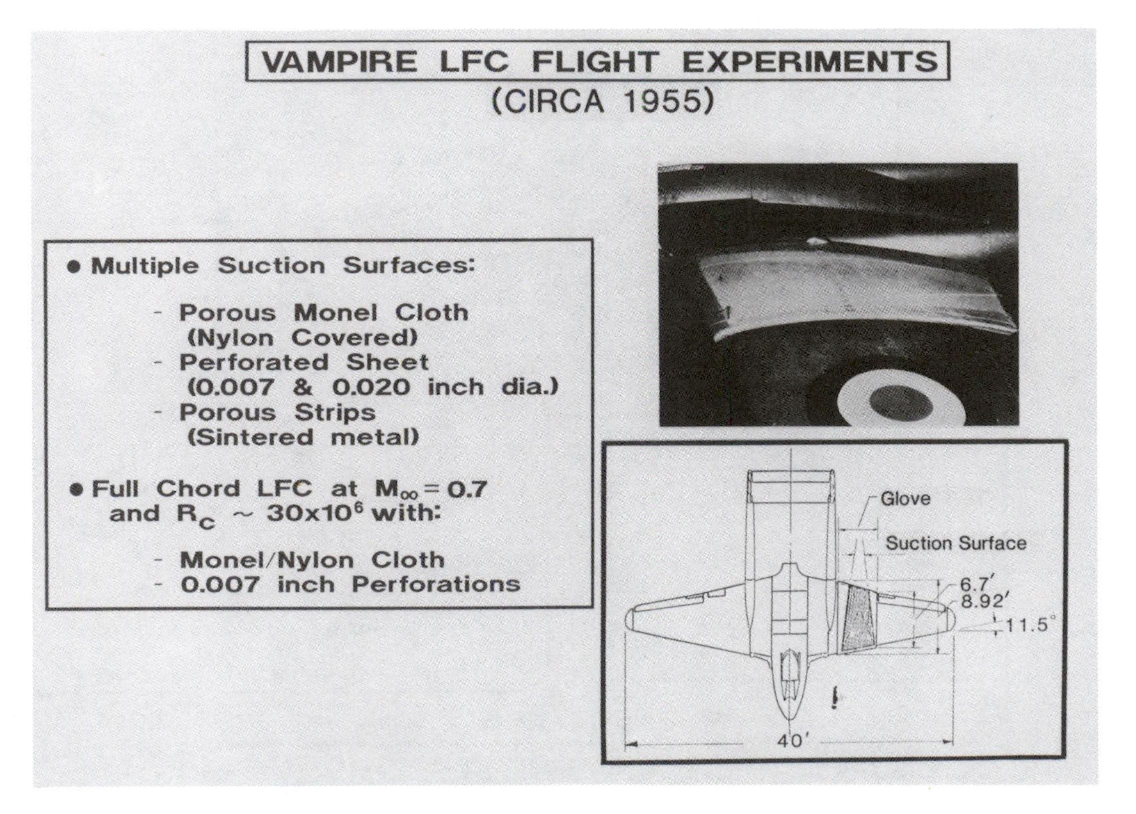


Figure 4. The Vampire LFC Flight Experiments.

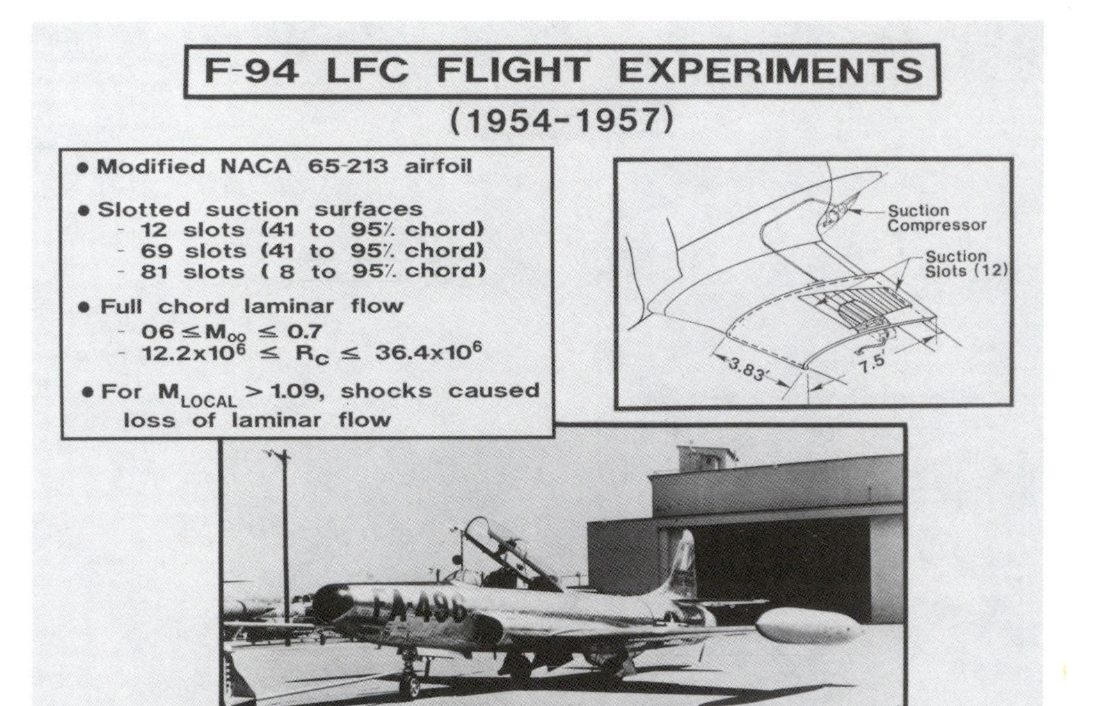


Figure 5. The F-94 LFC Flight Experiments.

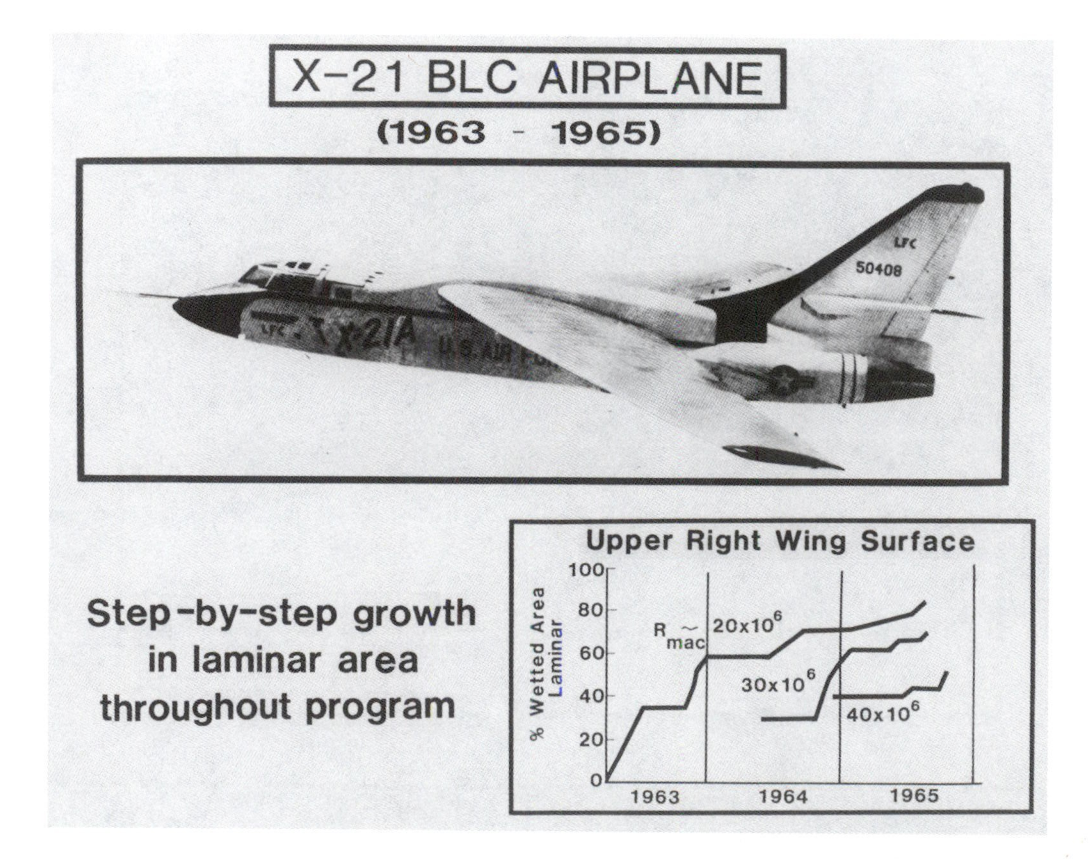


Figure 6. The X-21 LFC Flight Program.

LANCASTER SWEPT LFC WING EXPERIMENT
(CIRCA 1965)

Figure 7. The Lancaster Swept LFC Wing Flight Experiment.

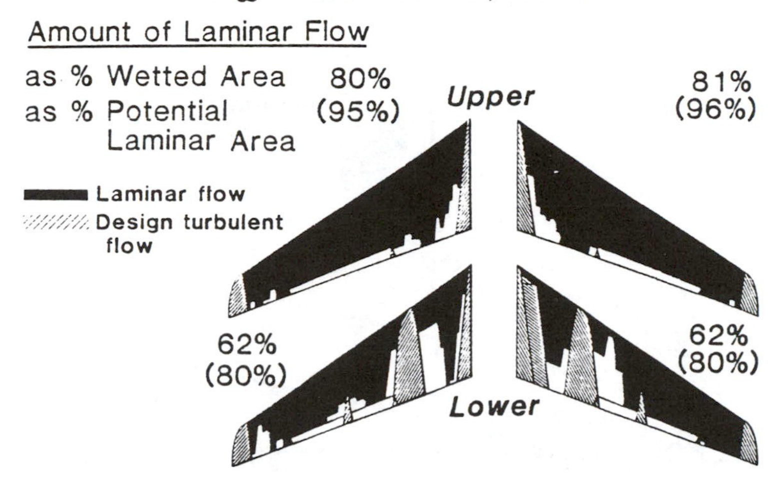


Figure 8. The X-21 Maximum Laminar Flow Areas.

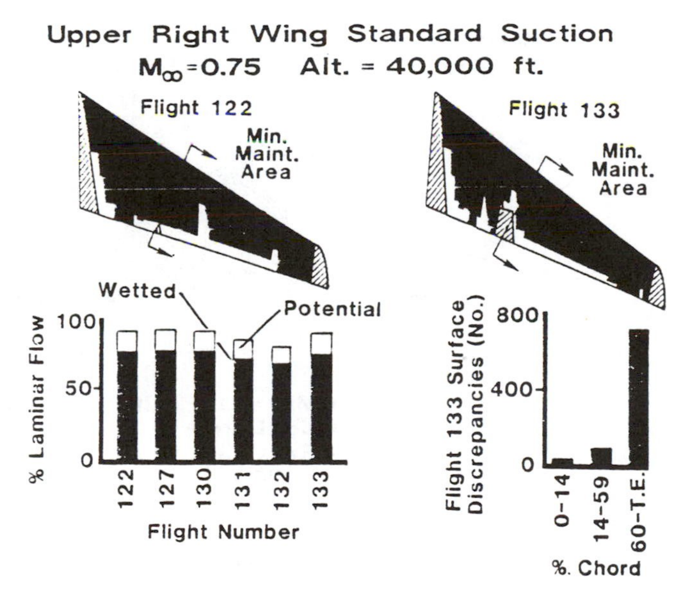


Figure 9. Maintenance Effects on the Repeatability of Laminar Flow on the X-21.

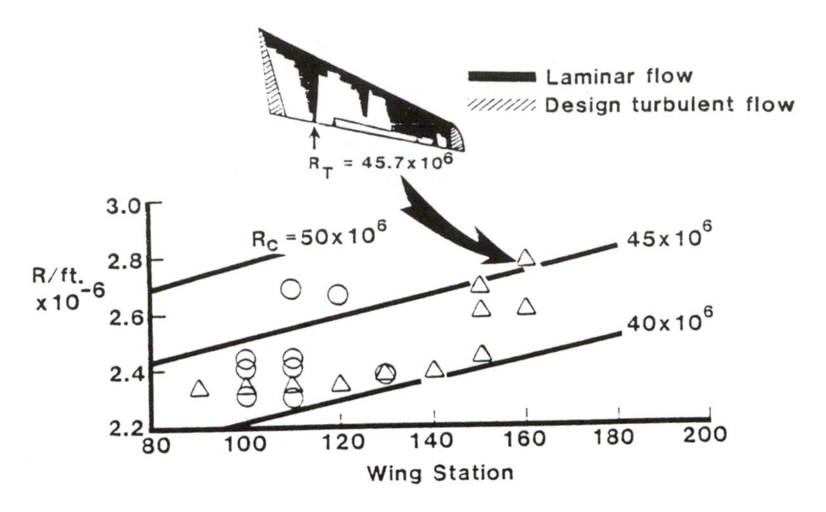


Figure 10. X-21 High Chord Reynolds Number Results.

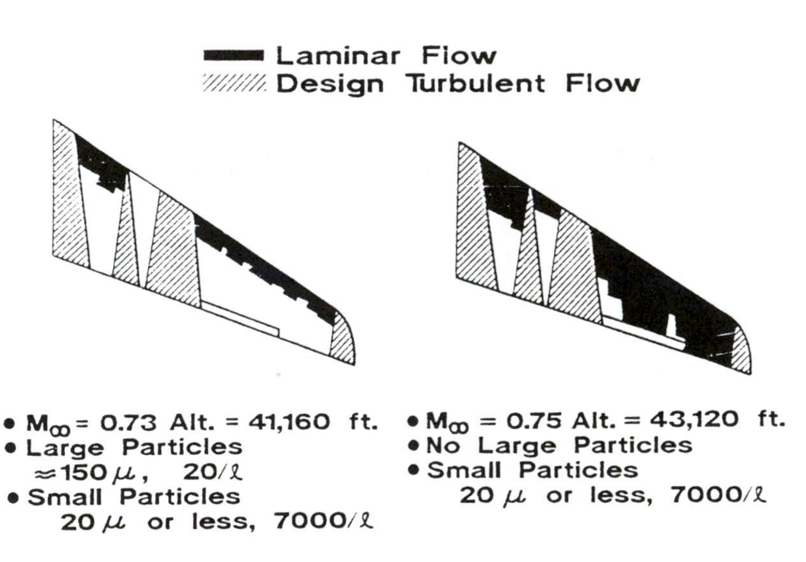


Figure 11. Effects of Ice Perticle Encounters on X-21 Results.

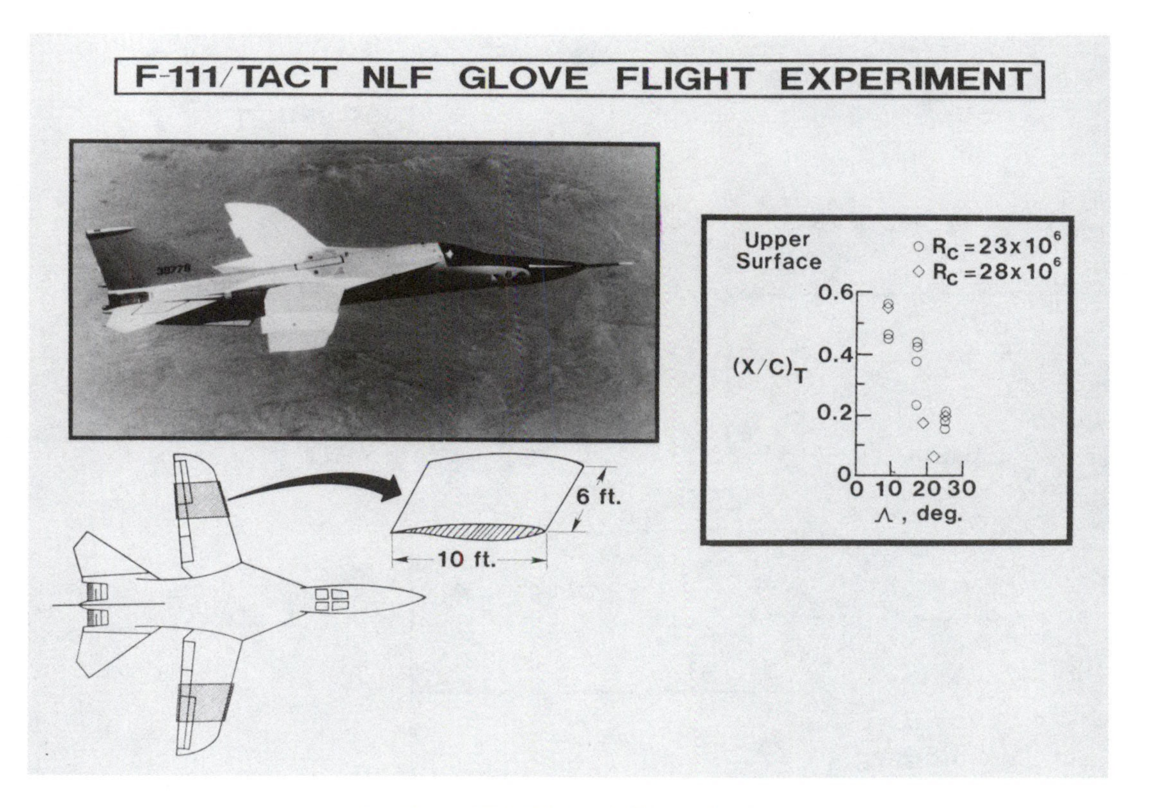


Figure 12. The F-111/TACT NLF Glove Flight Experiment.

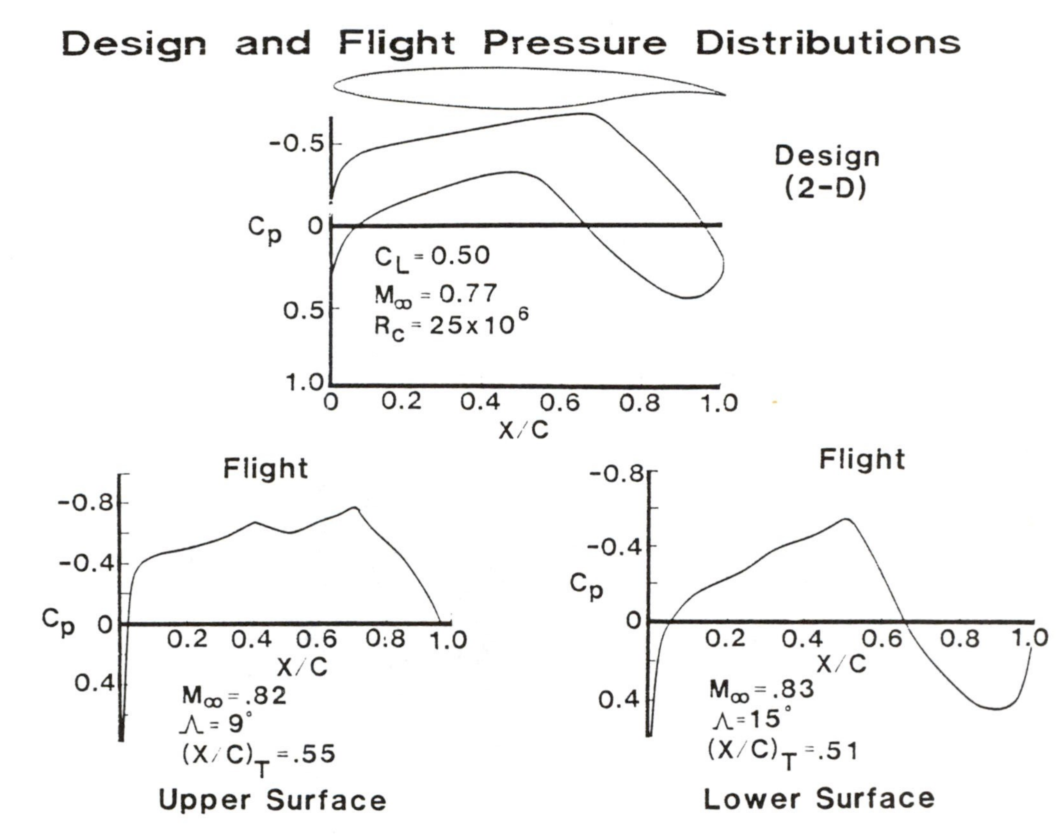


Figure 13. Design and Flight Pressure Distributions for the F-111/TACT NLF Glove.

Figure 14. F-14 Test-Bed Aircraft for the Variable Sweep Transition Flight Experiment

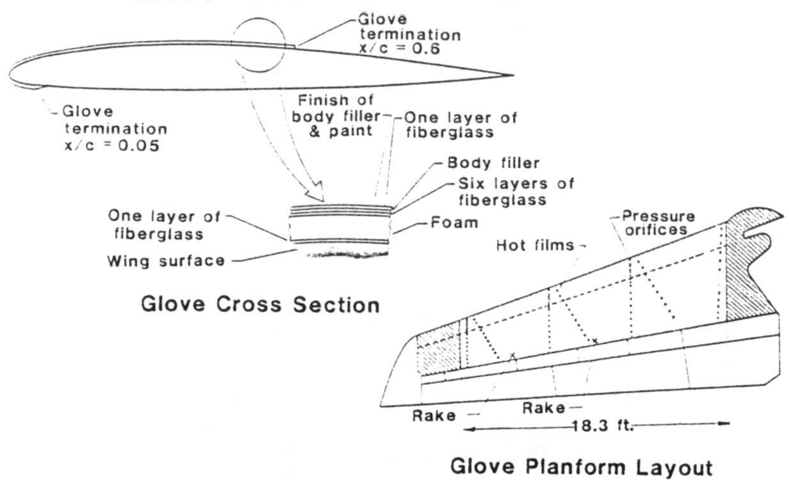


Figure 15. F-14 VSTFE Glove Details.

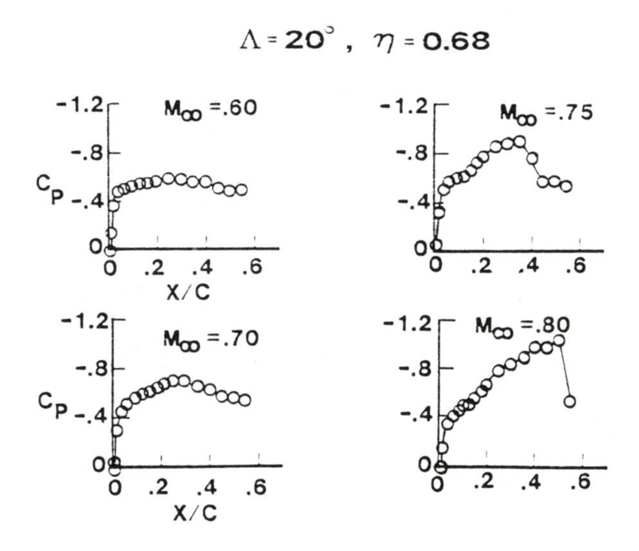


Figure 16. F-14 VSTFE "Clean-Up" Glove Pressure Distributions.

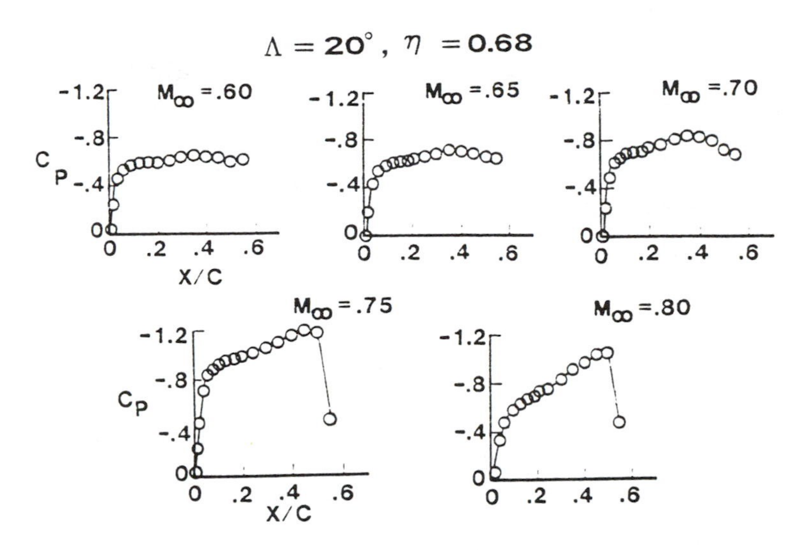


Figure 17. F-14 VSTFE "Langley-Design" Glove Pressure Distributions.

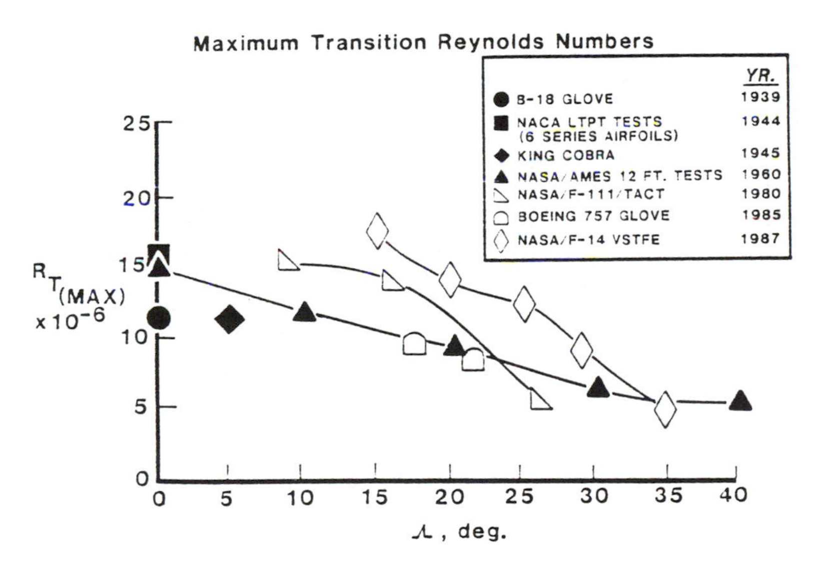


Figure 18. Maximum Transition Reynolds Number for Several Natural Laminar Flow Experiments.

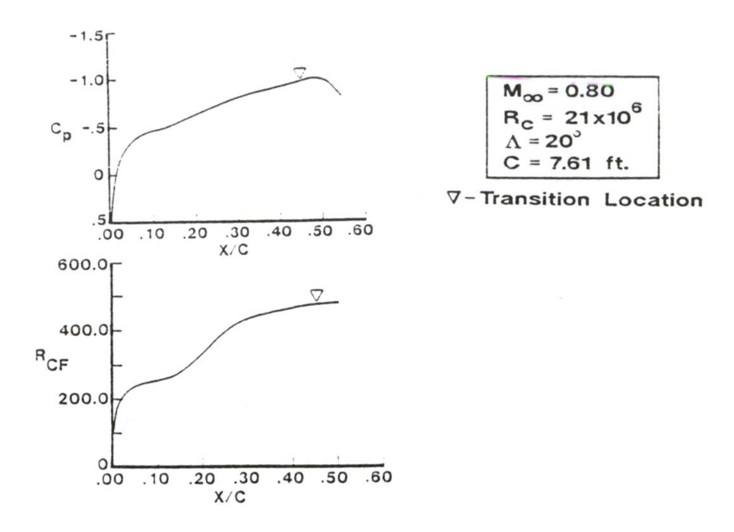


Figure 19. F-14 "Clean-Up" Glove Pressure and Crossflow Reynolds Number Distributions at a Mach Number of 0.8 and a Chord Reynolds Number of 21 Million.

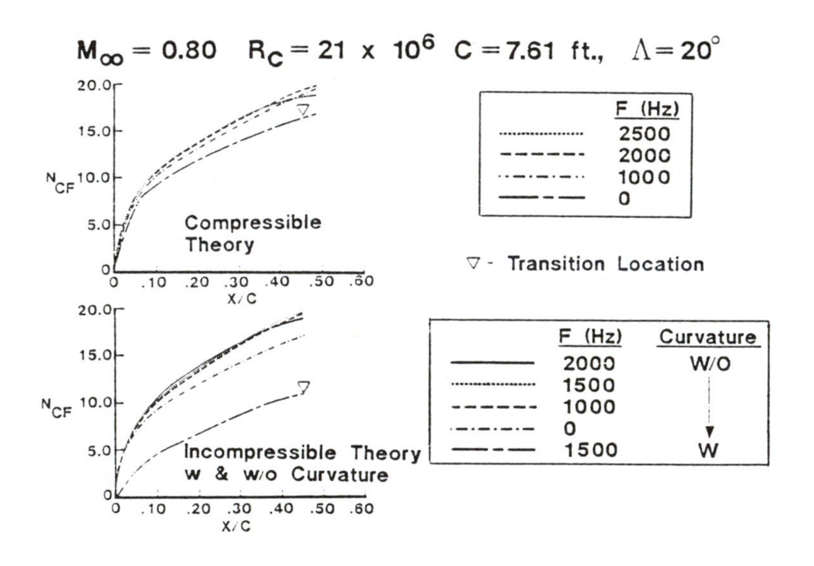


Figure 20. Crossflow N-Factors (With and Without Curvature Effects) for the F-14 "Clean-Up" Glove at a Mach Number of 0.8 and a Chord Reynolds Number of 21 Million.

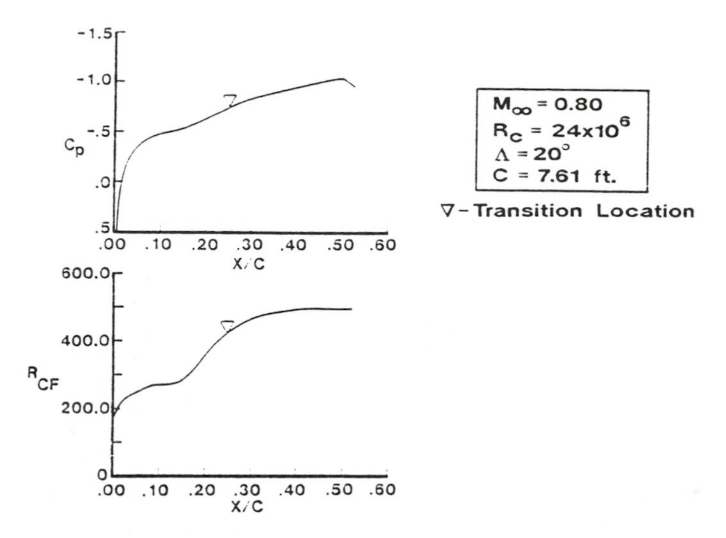


Figure 21. F-14 "Clean-Up" Glove Pressure and Crossflow Reynolds Number Distributions at a Mach Number of 0.8 and a Chord Reynolds Number of 24 Million.

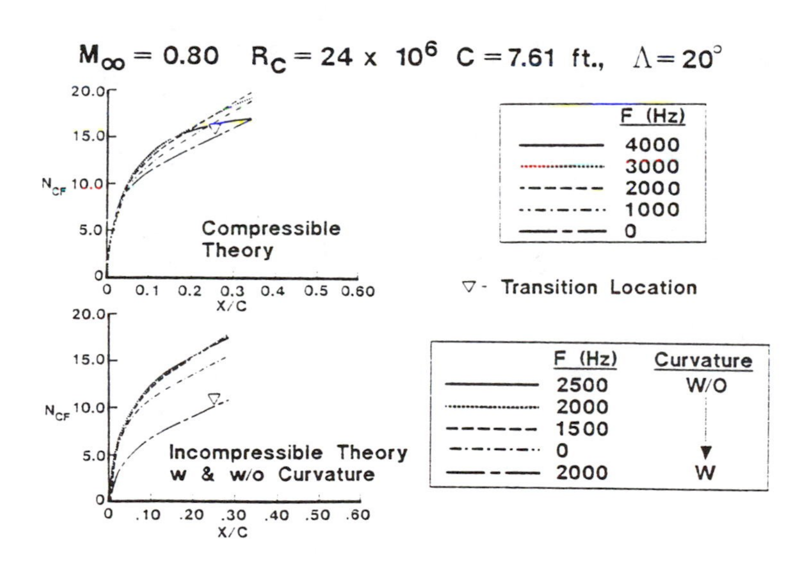


Figure 22. Crossflow N-Factors (With and Without Curvature Effects) for the F-14 "Clean-Up" Glove at a Mach Number of 0.8 and a Chord Reynolds Number of 24 Million.

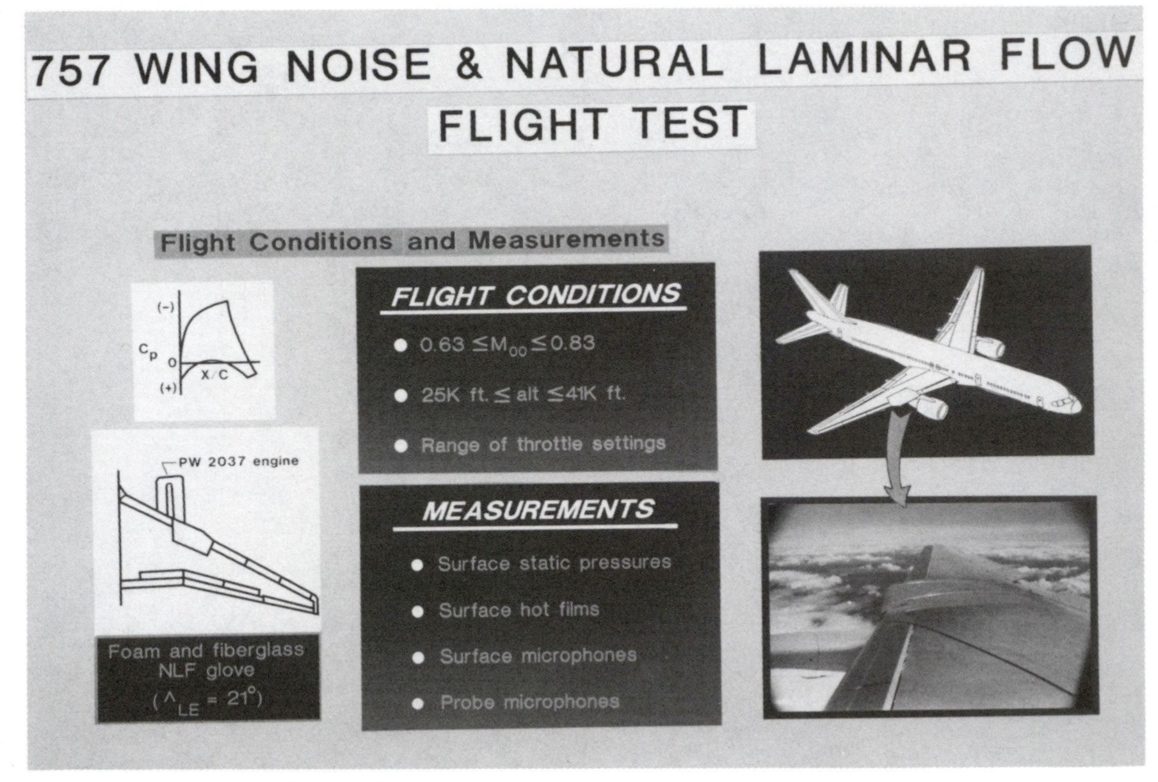


Figure 23. The 757 Wing Noise Survey and NLF Glove Flight-Test Program.

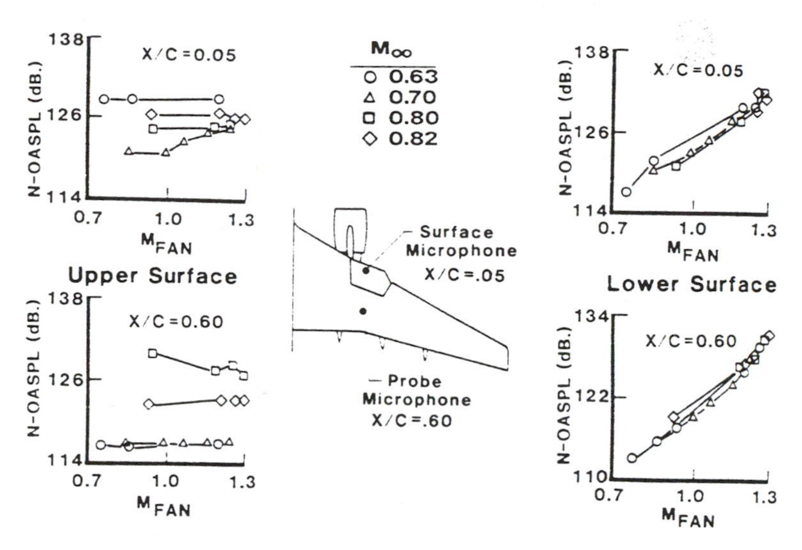


Figure 24. Normalized Overall Sound Pressure Level Versus Fan Mach Number for the 757 Wing.

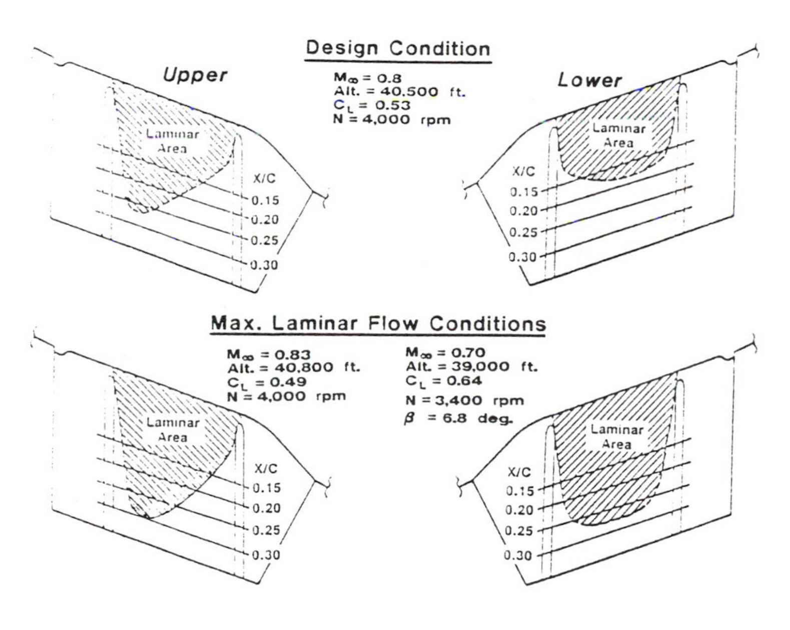


Figure 25. Measured Extent of Laminar Flow on the 757 NLF Glove.

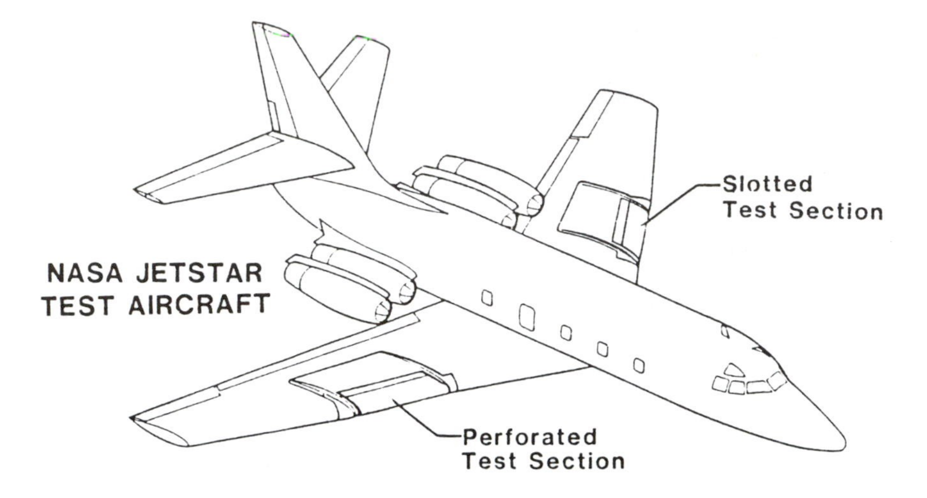


Figure 26. JetStar Test-Bed Aircraft for the NASA Leading-Edge Flight Test Program.

- Suction on upper surface only
- Suction through electron-beam-perforated skin
- Leading-edge shield extended for insect protection
- De-icer insert on shield for ice protection
- Supplementary spray nozzles for protection from insects and ice

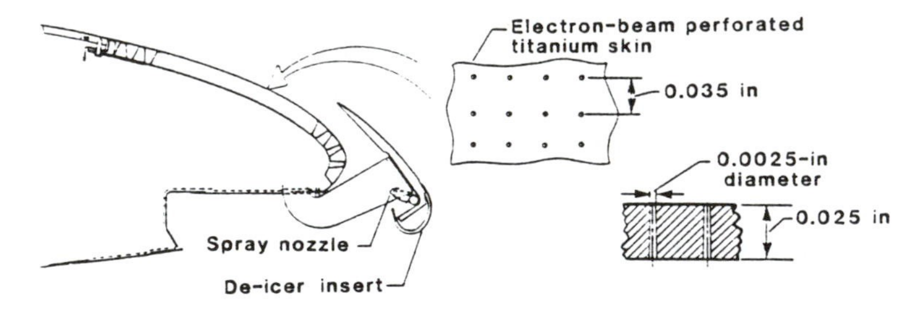


Figure 27. The Leading-Edge Flight Test Program Perforated Test Article.

- **Suction on upper and lower surface**
- **Suction through spanwise slots**
- **Liquid expelled through slots for protection from insects and icing**

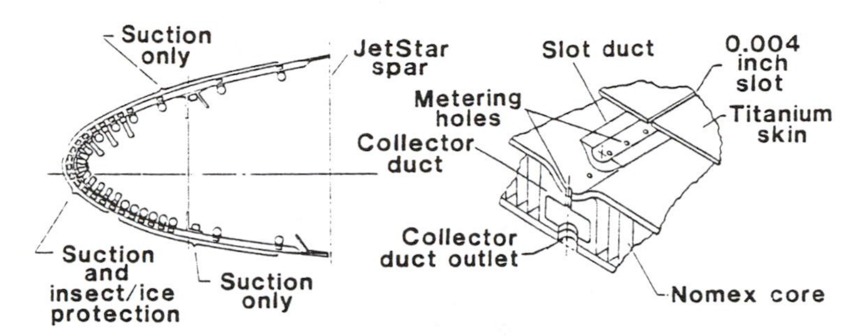


Figure 28. The Leading-Edge Flight Test Program Slotted Test Article.

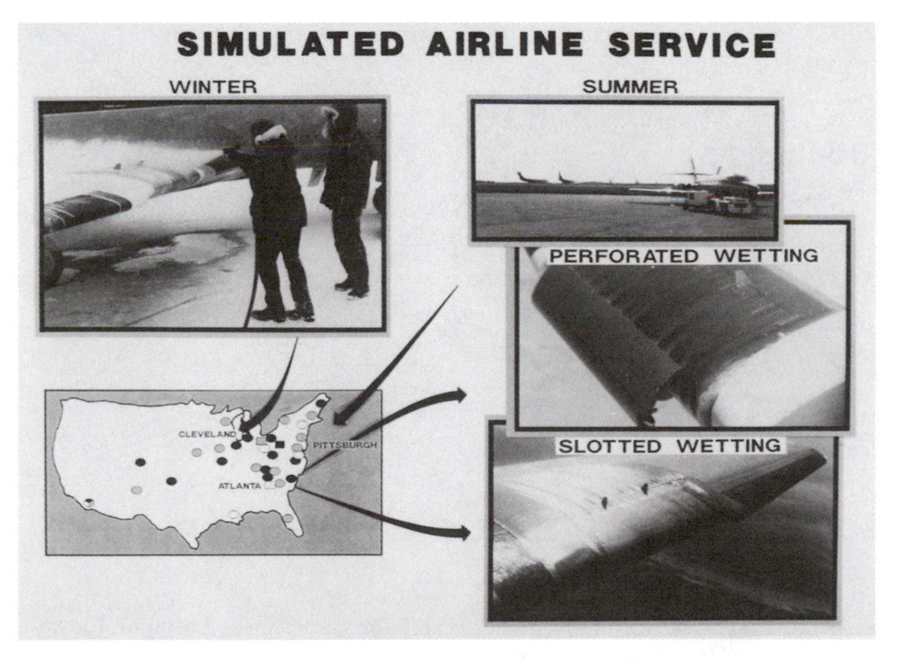


Figure 29. The Leading-Edge Flight Test Program Simulated Airline Service.

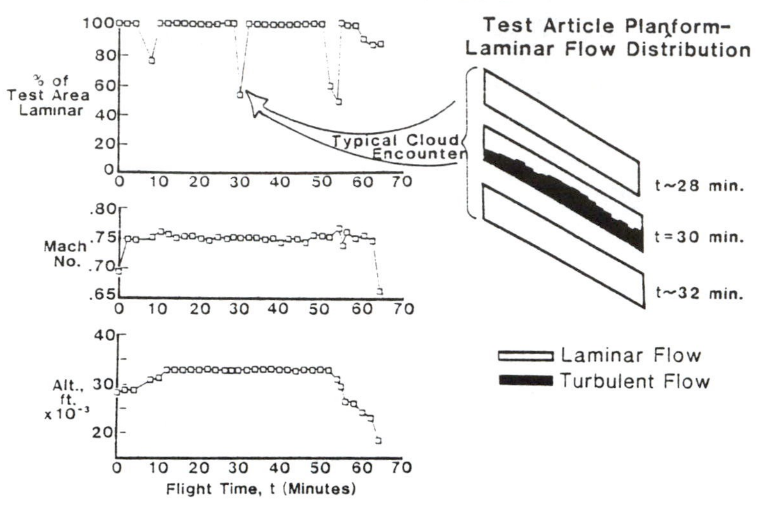


Figure 30. Typical Flight Profile From the LEFT Program Simulated Airline Service.

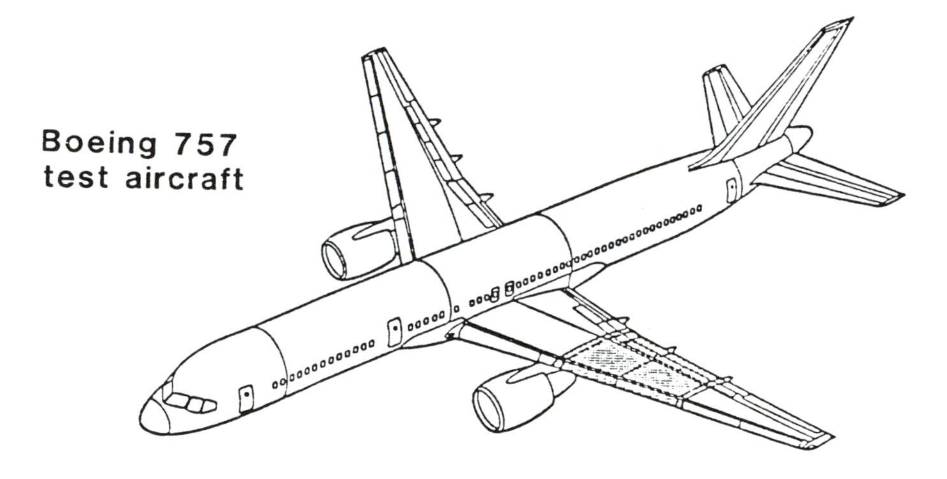


Figure 31. The 757 Test-Bed Aircraft for the Hybrid Laminar Flow Control (HLFC) Flight Experiment.

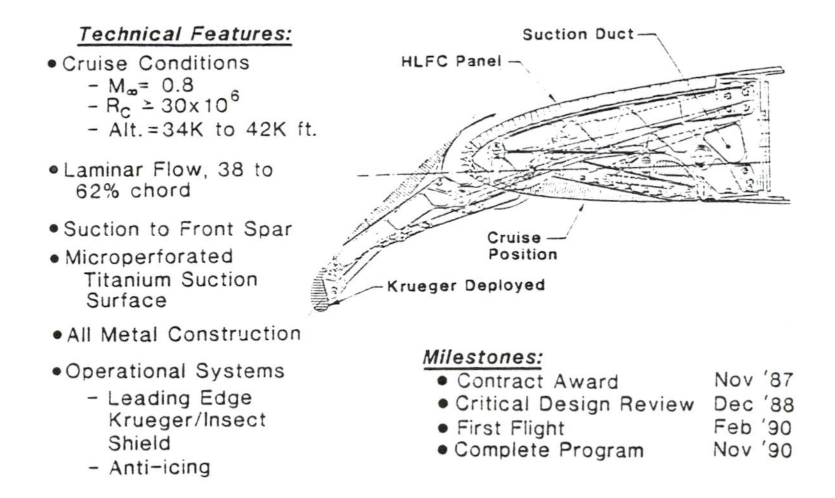


Figure 32. The HLFC Flight Experiment Technical Features and Milestones.

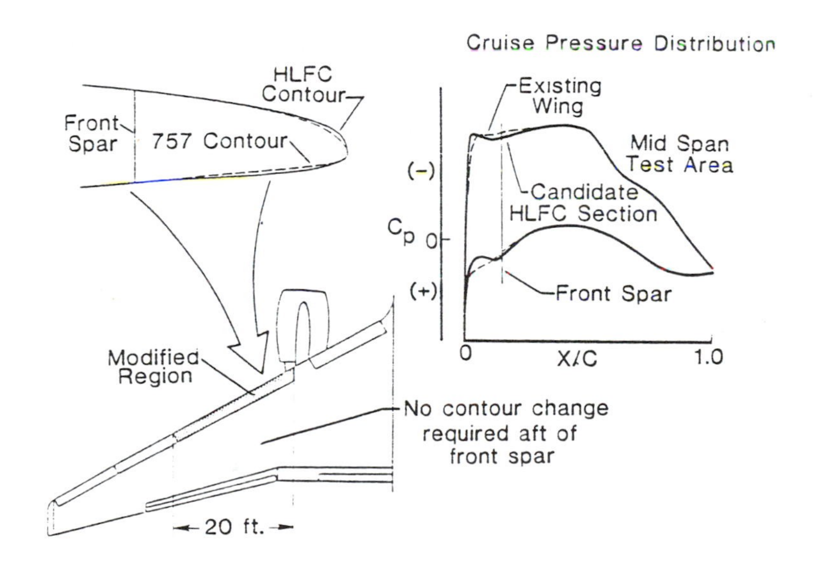


Figure 33. The 757 Wing Leading-Edge Modification for the HLFC Flight Experiment.

# FLIGHT RESEARCH ON NATURAL LAMINAR FLOW APPLICATIONS

*Bruce J. Holmes*
NASA Langley Research Center
Hampton, Virginia

*Clifford J. Obara*
Lockheed Engineering and Sciences Company
Hampton, Virginia

## Abstract

*Natural laminar flow (NLF) is clearly one of the most potentially attractive drag reduction technologies by virtue of its relative simplicity. NLF is achieved passively, that is, by design of surface shapes to produce favorable pressure gradients. However, it is not without its challenges and limitations. This chapter describes the significant challenges to achieving and maintaining NLF and documents certain of the limitations for practical applications. A brief review of the history and of more recent NLF flight experiments is given, followed by a summary of lessons learned which are pertinent to future applications. The chapter also summarizes important progress in test techniques, particularly in flow visualization and hot-film techniques for boundary-layer measurements in flight.*

## Nomenclature

| | |
|---|---|
| a | amplitude gain |
| c | local chord, ft |
| $C_d$ | section drag coefficient |
| $C_{D,c}$ | canard drag coefficient |
| $C_\ell$ | airplane lift coefficient |

| | |
|---|---|
| $C_{L,c}$ | canard lift coefficient |
| $C_p$ | pressure coefficient |
| h | altitude, ft |
| $h$ | wave height, in. |
| M | Mach number |
| n | logarithmic amplification ratio, $n=\ln(A/A_0)$ |
| R | Reynolds number |
| R' | unit Reynolds number, $ft^{-1}$ |
| $R'_1,R'_2,R'_3$ | reference unit Reynolds numbers, $ft^{-1}$, $R'_3>R'_2>R'_1$ |
| t/c | thickness to chord ratio |
| u | local velocity, ft/sec |
| $U_{oo}$ | freestream velocity, knots or ft/sec |
| $V_c$ | calibrated airspeed, knots or ft/sec |
| $V_i$ | indicated velocity, knots or ft/sec |
| x | longitudinal position, ft |
| z/c | vertical dimension in percent local chord |
| $\alpha$ | angle of attack, deg |
| $\Lambda$ | leading-edge sweep angle, deg |
| $\lambda$ | wave length, in. |

## Subscripts

| | |
|---|---|
| e | edge of boundary layer |
| max | maximum |
| min | minimum |
| t | transition |

## Notations

| | |
|---|---|
| CTA | constant temperature anemometer |
| FRM | flow reversal meter |
| NLF | natural laminar flow |
| PSD | power spectral density |
| RMS | root mean square |
| T-S | Tollmien-Schlicting |

## 1. Introduction

Five decades of flight experiences and many more years of theoretical and experimental research on the ground with natural laminar flow (NLF) have provided a basis of understanding how this technology can be used for reduction of viscous drag on modern practical airplanes. The classical concerns about the practicality of NLF have related to maintainability as well as achievability. The maintenance of NLF on modern, smooth wings requires that the surfaces be kept free from critical amounts of surface contamination (e.g., insect debris or ice), freestream disturbances (e.g., noise and turbulence), and surface damage. Compared to phenomena affecting the achievability of NLF, less is understood about maintainability of NLF under the wide ranges of Reynolds numbers, Mach numbers, meteorological conditions, flight profiles, and aircraft configurations that characterize the potential applications for NLF. It is generally true, however, that ease in maintenance of the NLF surface improves as the Reynolds number decreases.

The earliest efforts to achieve NLF in flight were uniformly successful on specially prepared and gloved airframe surfaces and unsuccessful on the production metal surfaces of the 1940's and 1950's era [1]. Transition near 65 percent chord was observed on the specially prepared wing section in the classic British Royal Aeronautical Establishment King Cobra experiments (see Figure 1) [2]. Although laminar flow was achieved over an extensive portion of the wing, the use of laminar flow on production aircraft was not practical. Close examination of those early wing fabrication methods reveals the shortcomings to have been excessive waviness between the ribs and stringers, excessive step heights or gap widths at the skin joints, and excessive heights of protuberances from certain riveting techniques (press-countersunk or dimpled rivets, for example).

Previous NLF flight experiments in which the transition location and/or section drag were determined are presented in References 3-26 and are summarized in Table 1 from Reference 1. These experiments included both unprepared (production) surfaces and specially prepared (filled and sanded) surfaces and airfoil

***Figure 1. King Cobra FZ 440 aircraft used for laminar flow research in the 1940's [2].***

gloves. The experiments on the production quality surfaces of that period resulted in little or no laminar flow due to the fabrication shortcomings noted above. However, on the specially prepared surfaces and gloves the transition locations and airfoil performance typically closely matched the theoretical predictions and the low-turbulence, wind-tunnel-model test results. The successes of the prepared and gloved surface tests provided the initial guidance for the development of criteria for allowable waviness as well as for allowable two- and three-dimensional protuberance heights. These criteria provide conservative guidance for the manufacture of NLF surfaces; this conservatism stems from their development origins in wind tunnels where "stream disturbances may exacerbate roughness problems" [27]. In the past, this conservatism may have been partly responsible for the perception that NLF would be difficult to achieve, even on modern production surfaces. This perception was probably heightened by the relatively high unit Reynolds number range, $R_c > 3 \times 10^6$ ft$^{-1}$, for the World War II high-performance fighters on which early NLF applications were attempted (see Figure 2); such freestream conditions make the laminar boundary layer very sensitive to surface imperfections and insect contamination.

In contrast to the difficulties encountered on production airframe surfaces of the 1940's and 1950's, NLF is achievable today because of the small waviness of modern production wings, because of the lower values of unit Reynolds numbers at the higher cruise altitudes of modern airplanes (see Figure 2), and because of the favorable influence of subcritical compressibility on two-dimensional laminar stability at the higher cruise Mach numbers of certain modern airplanes.

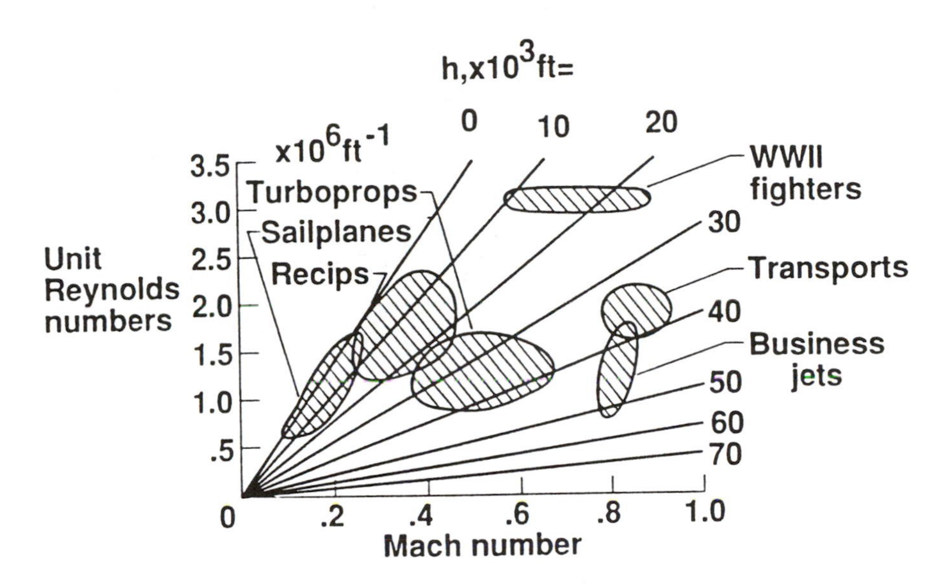


*Figure 2. Unit Reynolds number flight envelopes for several airplane classes.*

## 2. Modern Flight Experiments

As early as 1948, Tani [15] remarked that transition on smooth surfaces in flight typically occurred near or just downstream of the point of minimum pressure. This experience was repeated in the more recent NASA NLF flight experiments [1] on modern airframe surfaces. Physically, these observations mean that transition resulted either from amplified Tollmien-Schlichting (T-S) waves or laminar separation in the adverse pressure gradient. The results also mean that within certain Reynolds number restrictions, atmospheric turbulence and airframe vibration do not

*Table 1. Historical natural laminar flow flight experiments.*

| Principal investigators | References | Airplane | Airfoil | Type of surface |
|---|---|---|---|---|
| Stuper | 3 | Klenm L26Va | | Sanded plywood glove |
| Jones, Stephens, and Haslam | 4,5 | Shark L6103 | t/c=17.5% | Sanded plywood glove |
| | | Hart K1442 | t/c=10% | Metal glove |
| Young and Morris | 6,7 | Anson | NACA 2218 | Metal glove |
| | | Courier | NACA 2219 | Metal glove |
| Young, Serby, and Morris | 8 | Battle | NACA 2417 | Metal glove, production-metal wing surface, camouflage paint |
| Goett and Bicknell | 9 | Fairchild 22 | N-22 | Stiffened metal test panel |
| Bicknell | 10 | Northrup A-17A | NACA 2414.5 | Production metal wing (flush rivets, aft-facing lap joint at x/c=8%), metal glove |
| Wetmore, Zalovcik, and Platt | 11 | Douglas B-18 | NACA 35-215 | Wood glove |
| Zalovcik | 12 | XP-51 | NACA $64_1$2-(1.4), (13.5) | Production metal surface, various surface conditions |

| Chord Reynolds number or speed | Measurements | Results | Comments |
|---|---|---|---|
| $4.88 \times 10^6$ | $C_p$, $u/u_e$ | $(x/c)_t > 30\%$ | First in-flight transition measurements |
| 2.8 to $10.8 \times 10^6$ | $C_d$, $C_p$, $u/u_e$ | $16\% < (x/c)_t < 30\%$ | Waviness measured |
| | | | Effects of steps on transition measured |
| 139 knots | $u/u_e$ | $(x/c)_t = 17\%$ | Measurements inside and outside propeller slipstreams |
| 122 knots | $u/u_e$ | $(x/c)_t = 25\%$, outside propeller slipstream | |
| 12 to $18 \times 10^6$ | $C_d$, $C_p$, $u/u_e$ | $(x/c)_t = 18\%$, on glove | Drag of rivets and lap joints measured<br>No effect of camouflage paint on transition<br>No appreciable NLF on production surface |
| 3.9 to $4.6 \times 10^6$ | $C_d$, $C_p$, $u/u_e$ | $(x/c)_t = 37\%$, downstream of predicted laminar separation | Proximate transition locations for flight and Langley 30- by 60-Foot Tunnel |
| $15 \times 10^6$ | $C_d$, $C_p$, $u/u_e$ | $(x/c)_t = 17.5\%$, on glove | No appreciable NLF on production surface |
| $30 \times 10^6$ | $C_d$, $C_p$, $u/u_e$ | $(x/c)_t = 42.4\%$ | Waviness measured<br>Engine operation effects measured |
| $16 \times 10^6$ | $C_d$, $C_p$, $u/u_e$ | | Waviness measured<br>No appreciable NLF on production surface |

*Table 1. continued.*

| Principal investigators | Refer-ences | Airplane | Airfoil | Type of surface |
|---|---|---|---|---|
| Serby, Morgan, and Cooper | 13 | Hawcon | t/c=14%<br>t/c=25% | Wood glove<br>Metal glove |
| Serby and Morgan | 14 | Hawcon | | |
| | | Heinkel He.70 | t/c=12.5% | Production wood surface |
| Tani | 15 | Japanese biplane | | Wood glove |
| Zalovcik | 16 | Several aircraft | 8 airfoils | Smoothed and gloved surfaces |
| Zalovcik and Skoog | 17 | XP-47F | NACA 66(215)-1 (16.5), a=1.0<br>NACA 67(115)-213, a=0.7 | Production metal surface<br>Smoothed surface |
| Zalovcik | 18 | P-47D | Republic S-3, t/c=11%, t/c=14.6% | Smoothed surface |
| Zalovcik and Daum | 19 | P-47D | Republic S-3 | Production metal surface with camouflage paint |
| Plascott, Higton, Smith, and Bramwell | 20,21 | Hurricane II | NPL t/c=14.8 to 17.9% | Smoothed surface |
| Smith and Higton | 2 | King Cobra | NACA 662x-116<br>NACA 662x-216 | Production metal surface<br>Smoothed surface |
| Britland | 22 | Vampire | NACA 67,1-314, a=1.0 | Metal glove |

| Chord Reynolds Number or speed | Measurements | Results | Comments |
|---|---|---|---|
| 5.7 to 8x10$^6$ | $C_d$, $C_p$, $u/u_e$ | $30\%<(x/c)_t<40\%$ | |
| 5 to 9x10$^6$ | $C_d$ | | |
| 17x10$^6$ | $C_d$ | $C_{d,min}$=.0065 | Measured drag increases with mist deposit on laminar wing<br>Low drag of production wood wing suggested extensive NLF |
| 5 to 10x10$^6$ | $C_d$, $C_p$, $u/u_e$ | $40\%<(x/c)_t<51\%$ | |
| 4 to 32x10$^6$ | $C_d$, $C_p$, $u/u_e$ | Extensive runs of NLF measured | Waviness measured |
| 9 to 18x10$^6$ | $C_d$, $C_p$, $u/u_e$ | $(x/c)_t$=50% | No appreciable NLF on production surface<br>Propeller slipstream effects measured |
| 7.7 to 19.7x10$^6$ | $C_d$, $C_p$, $u/u_e$ | $(x/c)_t$=20%<br>$C_{d,min}$=0.0062 | Waviness measured |
| 0.25<M<0.78<br>8.4 to 23.1x10$^6$ | $C_d$, $C_p$ | $C_{d,min}$=0.0097 compare with reference 17 | No appreciable NLF on production surface<br>Waviness and roughness measured |
| 20x10$^6$ | $C_d$, $C_p$ | $(x/c)_t$=60% | Waviness measured |
| 17x10$^6$ | $C_d$, sublimating chemicals | $(x/c)_t$=65% | Waviness measured<br>No appreciable NLF on production surface |
| M=0.7,<br>30.4x10$^6$ | $C_p$, sublimating chemicals | $(x/c)_t$=50% | Waviness measured |

*Table 1. concluded.*

| Principal investigators | Refer-ences | Airplane | Airfoil | Type of surface |
|---|---|---|---|---|
| Davies | 23 | Several aircraft | | Production surface, smoothed surface |
| Gray and Davies | 24 | King Cobra | NACA 662x-116 NACA 662x-216 | Smoothed surface |
| Montoya, Steers, Christopher, and Trujillo | 25 | F-111 TACT | Super-critical NLF | Glove |
| Banner, McTigue, and Petty | 26 | F-104 | Biconvex t/c=3.4% | Production metal, fiberglass glove |

have significant effects on transition location. Analysis of flight transition data in Reference 28 led to the hypothesis that at moderate values of transition Reynolds numbers ($R_t$ on the order of $6x10^6$) on airfoils with moderately favorable pressure gradients, in dominantly two-dimensional incompressible flows, transition in flight can be expected to occur at laminar separation in the adverse pressure gradient. This hypothesis assumes that sufficient adverse gradient exists to cause laminar separation prior to Tollmien-Schlichting transition. On airfoils with like constraints in subsonic, compressible flows, transition Reynolds numbers as high as $14x10^6$ were observed [1]. These generalizations do not establish upper bounds for NLF transition Reynolds numbers. Supersonically, insufficient experience exists for such generalization.

Transition prediction in these various situations can be accomplished with reasonable confidence using various techniques. The T-S mode of transition is predicted by linear stability theory, otherwise known as the $e^n$ method. The natural

| Chord Reynolds Number or speed | Measurements | Results | Comments |
|---|---|---|---|
| | $C_d$, $C_p$, $u/u_e$, sublimating chemicals | | Waviness measured No appreciable NLF on production surface |
| $17x10^6$ | $C_d$, sublimating chemicals | Skin-joint filler cracks were most serious surface maintenance problem | Insect contamination discussed Laminar flow maintainability studied |
| Up to $30x10^6$ | $u/u_e$ | $(x/c)_t$=56%, at $\Lambda$=10 deg | Sweep effects studied |
| 1.2< M <2 | Hot films, sublimating chemicals | $1.2x10^6$ $<R_t<$ $8x10^6$ | Less laminar flow on production than on gloved surface |

log of the ratio of the T-S wave amplitude (A) to the amplitude at the neutral point ($A_0$, where the instability begins to grow) is defined as n; thus $n=\ln(A/A_0)$. Past analyses of values of n for flight-measured transition [29] have produced values ranging from 12 to 20. With continued flow acceleration downstream to the location of the start of pressure recovery, n=12 may be used as a conservative criterion to predict transition location in flight in dominantly two-dimensional, compressible or incompressible flows. For two-dimensional boundary layers in flight in which the local length Reynolds number at the predicted location of laminar separation is less than about $6x10^6$, transition can be conservatively assumed at that location. Various modern airfoil analysis and design methods incorporate integral boundary layer parameter correlations for transition prediction. Within the flow condition restrictions cited herein these techniques yield conservative transition predictions for flight situations. That is, these methods predict transition earlier than or at the locations observed in flight.

The favorable influence of compressibility on T-S wave damping suggests that transition Reynolds numbers may increase at higher subcritical Mach numbers [29,30]. Figure 3 illustrates this effect. This analysis shows that for a given moderately favorable pressure distribution, the "n-factor" or amplitude ratio does not exceed a value near 12 for predicted transition at 70-percent chord for the highest chord Reynolds numbers at the increasing value of Mach number. This compressible behavior of T-S amplification suggests that at these larger chord Reynolds numbers, transition might still be expected to occur near the minimum pressure location.

The recent NASA and industry flight experiments have demonstrated the achievability of NLF on modern metal and composite airframe surfaces [1] (see Table 2). These experiments, on more than 30 surfaces in total, were conducted over a range of free-stream conditions including Mach numbers up to 0.8,

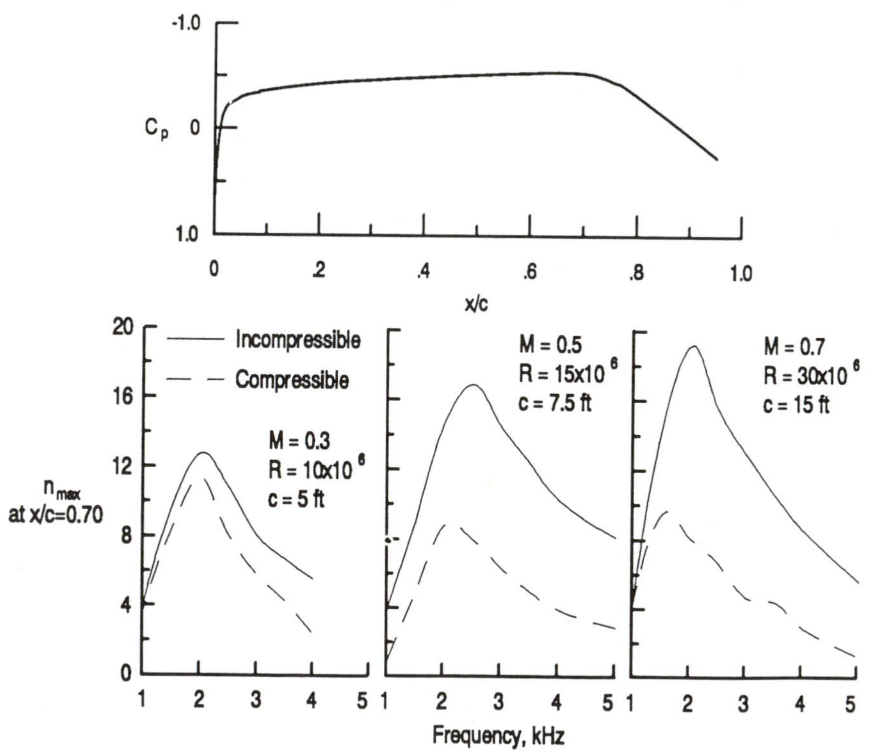


*Figure 3. Favorable compressibility effects on Tollmien-Schlichting wave growth in natural laminar boundary layers.*

transition Reynolds numbers up to $14x10^6$, chord Reynolds numbers up to $30x10^6$, and on wings of relatively small leading-edge sweep angles, typically less than 27 degrees. A selection of aircraft used for recent NLF flight experiments are presented in Figures 4 through 9.

An extensive experimental flight program was conducted using the Bellanca Skyrocket II, a six passenger, 300 mph, all fiberglass, propeller-driven airplane. Prior to testing, the Skyrocket had been unprotected from the natural environment for over 5 years. The wing surfaces were painted only to facilitate flow visualization. No smoothing of the airfoil contours was performed prior to testing. Comprehensive testing included measurements of transition on all lifting surfaces, airfoil pressure distributions, airfoil section drag, total airplane drag, and stall speeds. These tests were conducted both with transition free and transition fixed near the leading edge. Insect debris accumulation studies were also conducted. Transition on the right wing of the airplane is shown near the 50-percent chord location along the wing span in Figure 4 [31]. Extensive runs of more than 50-percent chord length of laminar flow were also recorded on the forward and aft faces of the propeller of this airplane. Laminar flow was documented on the wing in the propeller slipstream as well. Flight-measured drag polars compared very well with original wind-tunnel design validation data for the NACA $63_2$-215 wing section. The results of the Skyrocket tests convincingly demonstrated the reliable achievability of NLF on "production quality" composite airframe surfaces at moderate Reynolds numbers (approximately $6x10^6$).

A flight experiment program [32,33] was conducted to measure the effect of acoustics on the extent of laminar flow on an engine nacelle. A flow-through nacelle designed to achieve extensive runs of natural laminar flow on its external surface, was mounted under the right wing of an OV-1B aircraft (see Figure 5). Acoustic sources were mounted inside the nacelle and outboard of the nacelle with the noise radiating at the nacelle surface. Laminar flow was observed back to 38 percent of the nacelle length with the noise source "off". With the noise source "on" at 1800 Hz and a measured 132 dB on the surface, the transition front moved forward 3-5 percent chord. Within a narrow range of acoustic

***Table 2. Modern natural laminar flow flight experiments.***

| Principle investigators | Refer-ences | Airplane | Airfoil | Type of surface |
|---|---|---|---|---|
| Holmes, Obara, Yip | 1 | Rutan VariEze | Several | Fiberglass/foam core |
| | | Rutan Long-EZ | Several | Fiberglass/foam core |
| | | Rutan Biplane Racer | Several | Fiberglass and graphite/foam core |
| | | Cessna P-210 | NACA 64 series | Dimpled, flush riveted aluminum |
| | | Beech 24R | NACA $63_2$A-415 | Bonded aluminum skins/aluminum honeycomb ribs |
| | | Lear 28/29 | NACA 6 series | Milled aluminum skins, integrally stiffened• |
| Holmes, Obara, Gregorek, Hoffman, Freuler | 31 | Bellanca Skyrocket II | NACA $63_2$-215 | Fiberglass/ aluminum honeycomb core |
| Lee, Wusk, Obara | 74 | Lear 28/29 | | Fiberglass/foam cored glove |
| Ward, Miley, Reininger, Stout | 76 | Mooney 201 | NACA $63_2$-215 | Machined, countersunk flush-riveted aluminum |
| Wentz, Ahmed, Nyenhuis | 77, 78 | Cessna Citation III | NACA supercritical | Bonded, stiffened aluminum |
| Meyer, Trujillo, Bartlett | 79 | F-14 | Modified NACA 6 series | Fiberglass/foam cored glove |
| Boeing Commercial Airplane Co. | 80 | Boeing 757 | | Fiberglass/foam cored glove |
| Befus, Nelson, Latas, Ellis | 81 | Modified Cessna T-210 | NLF-0414 | Aluminum/polyester resin filler |
| Hortsmann, Quast, Redeker | 82 | Experimental prototype | | Glove |
| George-Falvy | 83 | T-33 | | Smoothed wing |

• Leading edge to skin butt joint faired smooth.

| Chord Reynolds number or speed | Measurements | Results | Comments |
|---|---|---|---|
| $0.8\text{-}3.5 \times 10^6$ | Sublimating chemicals | $R_t=1.05 \times 10^{6\dagger}$ | Fixed transition effects measured, $\Lambda=27$ deg |
| $1.2\text{-}4.0 \times 10^6$ | Sublimating chemicals | $R_t=1.51 \times 10^{6\dagger}$ | Fixed transition effects measured, $\Lambda=23$ deg |
| $1.5\text{-}3.5 \times 10^6$ | Sublimating chemicals | $R_t=2.24 \times 10^{6\dagger}$ | Propeller slipstream effects observed |
| $5\text{-}8 \times 10^6$ | Sublimating chemicals | $R_t=3.31 \times 10^{6\dagger}$ | Flush rivets caused premature transition |
| $4 \times 10^6$ | Sublimating chemicals | $R_t=2.73 \times 10^{6\dagger}$ | NLF on propeller blades measured |
| $1.7\text{-}30 \times 10^6$ | Sublimating chemicals | $R_t=11.5 \times 10^{6\dagger}$ | $\Lambda=17$ deg |
| $7.5\text{-}12 \times 10^6$ | Sublimating chemicals, $C_p$, $C_d$, $C_l$ | $R_t=11.5 \times 10^{6\dagger}$ | Fixed transition effects measured. Insect contamination effects measured. |
| $10\text{-}20 \times 10^6$ | Hot films, temperature distribution, $C_p$, liquid crystals | $R_t=6.5 \times 10^6$ at $R=17 \times 10^6$ | Detailed laminar boundary layer measurements made using hot films. |
| $5\text{-}7 \times 10^6$ | Sublimating chemicals, $C_p$, $C_d$, $C_l$ | $(x/c)_t=50\%^\dagger$ | Flush rivets do not cause transition |
| $6\text{-}19 \times 10^6$ | Sublimating chemicals, hot films | $(x/c)_t=.5\text{-}15\%^\dagger$ | Sweep effects documented. Transition at pressure peak at M=0.70, $\Lambda=25$ deg |
| M=0.60-0.80 | $C_p$, hot films | $R_t=5\text{-}17.6 \times 10^6$ | Sweep effects studied |
| M=0.63-0.82 | hot films, microphones | $(x/c)_{t,max}=30\%$ | Effect of engine noise measured. |
| $4.5\text{-}6.0 \times 10^6$ | Sublimating chemicals, hot films, $C_p$, $C_d$, $C_l$ | $(x/c)_t=76\%$ | 14 knot improvement on cruise performance due to laminar flow wing |
| $3\text{-}10 \times 10^6$ | Infrared, $C_p$ | $1\text{-}7 \times 10^6$ | "n" factors on the order of 13.5 |
| M=0.35-0.70 | $C_p$, Pitot probes, hot films, liquid crystals, sublimating chemicals | $R_t=11.5 \times 10^{6\dagger}$ | Waviness documented |

† Transition occured near or downstream of pressure peak, near predicted laminar separation.

*Figure 4. Bellanca Skyrocket II airplane used in natural laminar flow flight experiments.*

disturbance levels and Reynolds numbers, these tests validated a semi-empirical method [34] for predicting the effect of noise on transition. The sharp inlet design required for NLF nacelles may be precluded by the need to have a larger radius at the nacelle leading edge to maintain proper engine mass flows and avoid internal separation. It is then possible to design an LFC nacelle to avoid this problem [35].

*Figure 5. OV-1B Mohawk airplane used in acoustic effects on natural laminar flow flight experiments.*

An industry evaluation of the effects of loss of natural laminar flow on performance and handling qualities of a high performance, single-engine, high-wing business airplane was conducted using a Cessna 210 with an NLF wing (Figure 6). These tests provided an opportunity to evaluate these effects for an NLF airfoil designed for minimum loss of lift with loss of laminar flow. The airfoil was the NASA designed, NLF 0414F [36,37]. Transition was measured at 70-percent chord on both the upper and lower surfaces of the wing. The cruise speed of this aircraft was increased by 15

***Figure 6. Cessna P210 airplane with natural laminar flow wing.***

knots over the previous aircraft configuration as a result of the reduced drag from the new NLF wing [38]. The results of the evaluation showed no significant impact of the loss of laminar flow on airplane handling qualities related to airworthiness certification; the effects of loss of laminar flow on performance were also measured [39].

A study of transition on a very thick laminar airfoil was conducted using the Dragonfly, "homebuilt," single-engine, tandem-wing airplane shown in Figure 7. This study was conducted to evaluate the effects of loss of laminar flow for the case in which early transition caused loss of lift due to increased flow separation on a lifting surface. The aircraft incorporated a foreplane airfoil section of 20-percent thickness to chord ratio which is shown in Figure 8. The Figure also depicts the surface velocity distribution for a typical climb or approach speed of 80 mph. The locations of boundary-layer transition and turbulent separation are indicated for free and fixed (x/c=0.075) transition.

***Figure 7. Dragonfly tandem-wing airplane used in natural laminar flow flight experiments.***

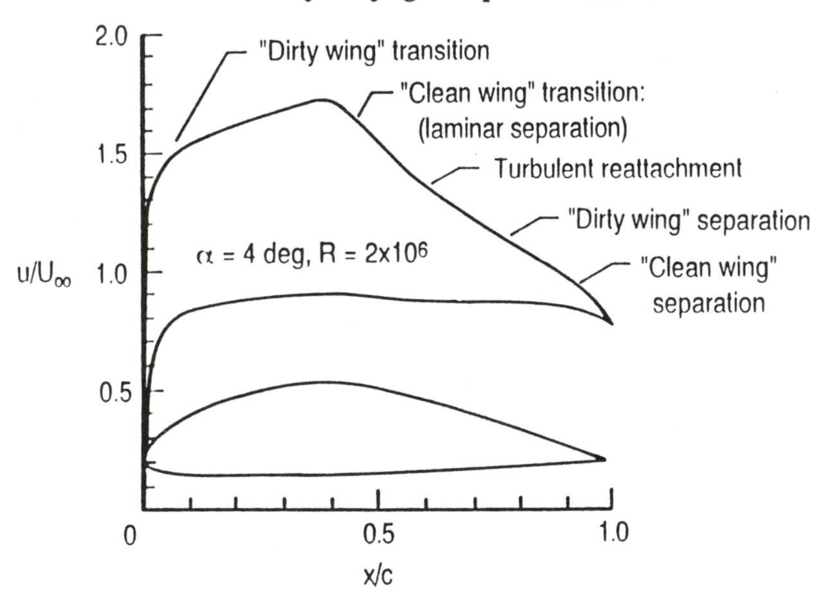


***Figure 8. Inviscid velocity distribution and calculated transition and separation locations for Dragonfly foreplane section shape ($\alpha$=4 degrees, R=2x10^6^).***

In the case of free transition, laminar separation is predicted near x/c=0.50. Separation of the turbulent boundary layer then is predicted near x/c=0.95. With transition fixed near the leading edge, the resulting turbulent boundary layer is unable to remain attached during the pressure recovery on the aft part of the airfoil, and it separates near x/c=0.75, much further forward than for the free transition case [40]. The results of the Dragonfly flight experiments provided an appreciation for the potential severity of the effects of loss of lift associated with loss of laminar flow on certain kinds of NLF airfoils; the results included means by which such effects could be ameliorated.

To investigate transition behavior at larger values of unit and length Reynolds numbers and Mach number on NLF, a series of tests were conducted on a Lear 28/29 twin engine, business jet (Figure 9). The airplane cruises at Mach numbers up to 0.81 and altitudes up to 51,000 feet. The test surface was a modified airfoil contour, constructed with body filler, which covered the critical surface discontinuity at the leading edge joint with the wing upper surface skin. The test surface had "production quality" in that measured waviness did not exceed that typical for production wings for the airplane. Transition was measured with several techniques near 45-percent chord for a flight condition of M = 0.7, h = 48,000 ft, and R = $1.8 \times 10^6$ ft$^{-1}$. These results illustrated the achievability and maintainability of NLF at transonic speeds, and with relatively small wing leading edge sweep (17 degrees) at moderate Reynolds numbers.

***Figure 9. Lear 28/29 airplane used in natural laminar flow flight experiments.***

This limited selection of results illustrates the wide variety of aircraft surfaces and flight conditions for which NLF achievability and maintainability on modern production quality airframe surfaces has been studied. The significant implications of the past research are the following:

1. Achievability: NLF is a practical drag reduction technology on modern metal and composite airframe surfaces for Mach numbers as high as 0.8, chord Reynolds numbers as large as $30x10^6$, and wing sweep angles as high as 17 to 27 degrees, depending on length and unit Reynolds numbers and Mach number.
2. Maintainability: NLF is more persistent and durable at high-speed subsonic conditions than previously expected, sufficiently so that aircraft designed to utilize NLF can be expected to sustain laminar flow with reasonable (day to day) attention to cleanliness of the laminar surfaces.

## 3. Lessons Learned

Many of the lessons learned from these NLF flight experiments have significance for current efforts to design, flight test, and operate NLF airplanes. In particular, these lessons relate to using NLF on practical airplanes in typical operating environments. This section summarizes the lessons learned concerning the following topics:

1. Effects of laminar flow on total airplane drag
2. Manufacturing tolerances for laminar surfaces
3. Effects of loss of laminar flow on stability and control
4. Effects of loss of laminar flow on maximum lift
5. Insect accretion behavior and effects on laminar flow
6. Effects of clouds and precipitation on laminar flow
7. Laminar flow behavior in propeller slipstreams
8. Fixed transition flight testing

While the lessons of the past have been very instructive for current efforts to apply NLF to aircraft designs, research efforts continue to explore the limits of practical applications for NLF. These limits may be thought of as combinations of maximum angles of sweep, Reynolds numbers, and Mach numbers for which NLF can be achieved and maintained on practical wings in typical operating environments. Beyond these limits for NLF, laminar flow control (LFC) by suction is a promising means of achieving laminar viscous drag reduction benefits. This chapter concentrates on NLF subjects.

## 3.1. Effects of Laminar Flow on Total Airplane Drag

The reduction of airplane drag with laminar flow results directly from changes in skin friction and any ensuing changes in pressure drag. Practical boundary-layer considerations on pressure recovery limit the maximum lengths of NLF runs to be between 50 and 70 percent of the total length of a surface (depending on Reynolds number, sweep, and airfoil pressure gradients). For these lengths of laminar runs, the potential drag reduction ranges between about 30- and 60-percent compared to the drag of a "good" turbulent airfoil (NACA 23-015), as illustrated in Figure 10. The Figure also illustrates the nearly 100 percent increase in airfoil section cruise drag with turbulent compared to laminar conditions for the NACA $63_2$-215 NLF airfoil measured in flight on the Bellanca Skyrocket II airplane [31].

Flight measured increases in total airplane cruise drag of 25 percent caused by loss of laminar flow were reported in Reference 1 for three airplanes. These three airplanes were the Rutan VariEze, the Rutan Long-EZ, and the Bellanca Skyrocket II. The drag increases on the first two airplanes were aggravated by flow separation associated with loss of laminar flow on the thick canard airfoil. The Skyrocket's NACA 6-series airfoil did not experience significant flow separation with loss of laminar flow; for this airplane, the drag change was dominated by the change in skin friction caused by early transition.

For a high-performance business jet, the potential drag reduction with NLF ranges between about 12 percent (for NLF on the wing only) to about 24 percent (for NLF on the wing, fuselage nose, empennage, and engine nacelle's) (see Figure 11). These drag reductions are calculated for NLF added to an existing con figuration; larger benefits would accrue for integrated design calculations.

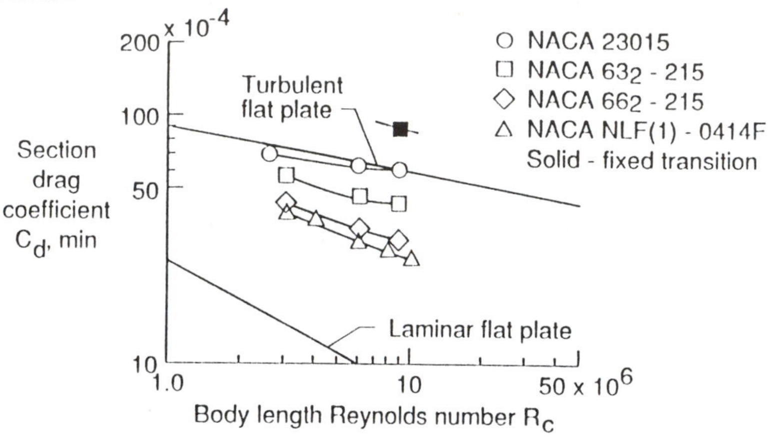


*Figure 10. Effect of natural laminar flow on section drag coefficients.*

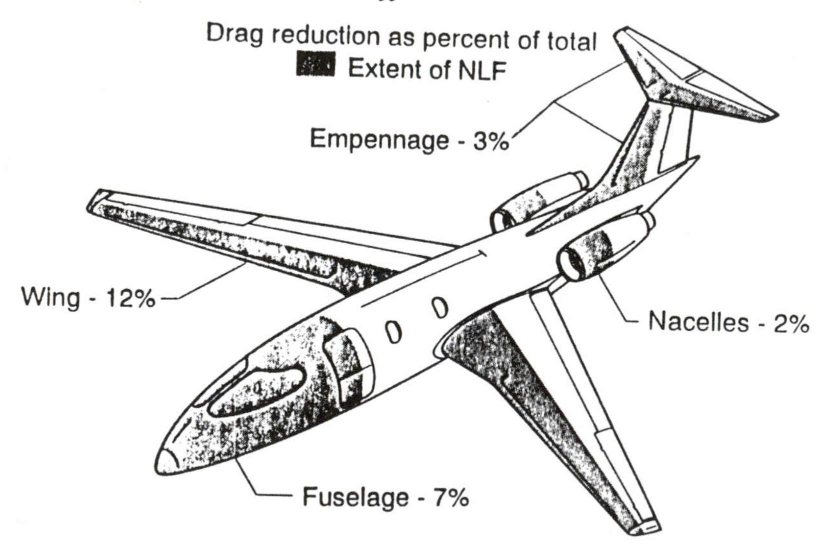


*Figure 11. Predicted drag reduction benefits of laminar flow for a typical subsonic business jet airplane.*

## 3.2. Manufacturing Tolerances for Laminar Surfaces

Many modern metal and composite airframe manufacturing techniques can provide surface smoothness which is compatible with NLF requirements. Specifically, this has been shown in flight investigations over a range of free-stream conditions including Mach numbers up to 0.8, chord Reynolds numbers up to about 30 million, and transition Reynolds numbers up to about 14 million. Surface smoothness requirements relate to waviness, to two-dimensional steps and gaps, and to three-dimensional roughness elements. The laminar flow flight experiments reported in Reference 1 included careful documentation of the surface quality for all the numerous test surfaces on which transition locations were measured. The airplanes for the experiments were selected to provide transition behavior information for a broad range of modern, production-quality airframe surfaces. The flight experiments were conducted on various surfaces including flush-riveted, thin aluminum skins; integrally-stiffened, milled, thick aluminum skins; bonded, thin aluminum skins; and composite surfaces. The most important conclusion concerning manufacturing to be drawn from these experiences is that the waviness of the production-quality surfaces in the tests met the NLF criterion for the free-stream conditions flown.

In addition to waviness however, there are other important surface imperfections to be considered for NLF. Perhaps the most important consideration is the effect of surface imperfections in the form of steps and gaps. The principal challenge to the design and manufacture of laminar flow surfaces today appears to be at the joints which occur in the installation of leading-edge panels on wings, nacelles, and empennage surfaces. Another similar challenge is in the installation of access panels, doors, windows, and the like on fuselage noses. While much work has been done in the past, many unknowns still exist concerning the influences of wing sweep, compressibility, and shapes of steps or gaps on manufacturing tolerances for laminar flow surfaces. Even less information is available concerning NLF requirements related to practical three-dimensional roughness elements such as flush screw head slots and protruding rivets.

Existing criteria for NLF surface quality deal with waviness and with both two- and three-dimensional roughness. Each of these types of surface imperfections can cause transition by different mechanisms in the boundary layer. The mechanisms of most practical interest include amplification of free shear layer instabilities associated with laminar separation, amplification of Tollmien-Schlichting (T-S) waves, amplification of crossflow vorticity, and interactions between any of these mechanisms. In addition, freestream turbulence and acoustic disturbances can interact with these mechanisms to influence critical waviness and roughness heights. Criteria exist only for critical waviness and roughness which cause either laminar separation or amplification of T-S waves. No comprehensive criteria exist which fully address surface-imperfection-induced transition related to crossflow amplification on swept wings or interactions between the various transition mechanisms and free-stream disturbances.

The following definitions appear in the literature and are useful for the present discussion. Critical waviness height to length ratio $(h/\ell)$ and critical step height or gap width can be defined as those which produce transition forward of the location where it would occur in the absence of the surface imperfection. Figure 12 illustrates possible effects of a two-dimensional surface imperfection on transition. A subcritical condition exists when transition is unaffected by the disturbance (top of figure). The middle of Figure 12 depicts the critical condition at which the disturbance just begins to affect transition. In the extreme, a surface imperfection could cause sufficiently rapid disturbance amplification for transition to occur very near the wave itself, as illustrated at the bottom of Figure 12. Experimentally, premature transition was identified in past work as the first appearance of turbulent bursts downstream of either a waviness or roughness surface imperfection. This is the definition used in References 41 to 45 to establish critical conditions for surface imperfections.

Another limiting condition of practical interest is the occurrence of transition related to the disturbance amplification associated with a large wave in the laminar region. Using flight data [28], Figure 13 illustrates the predicted local increase in growth rate of T-S instability caused by a surface wave. The surface wave

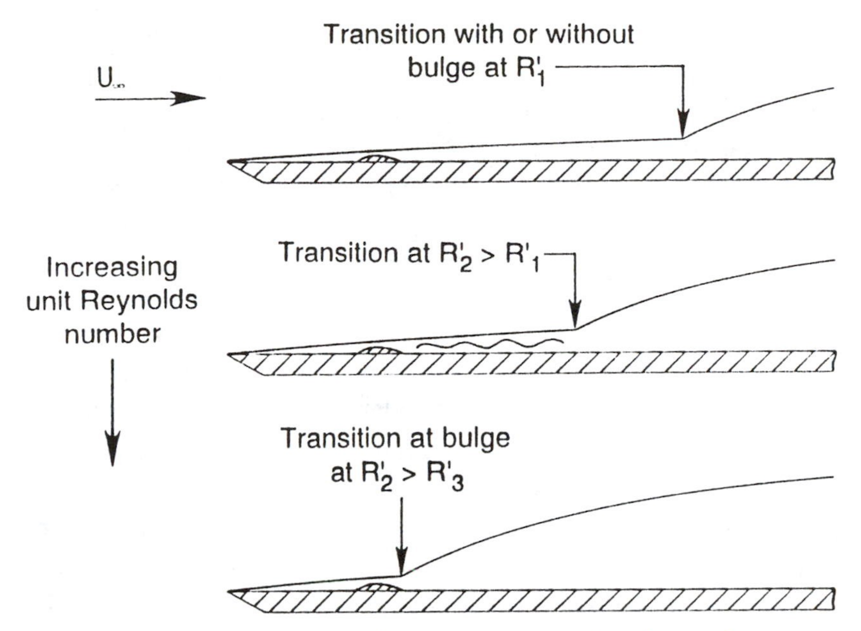


*Figure 12. Effects of two-dimensional surface imperfections on laminar flow.*

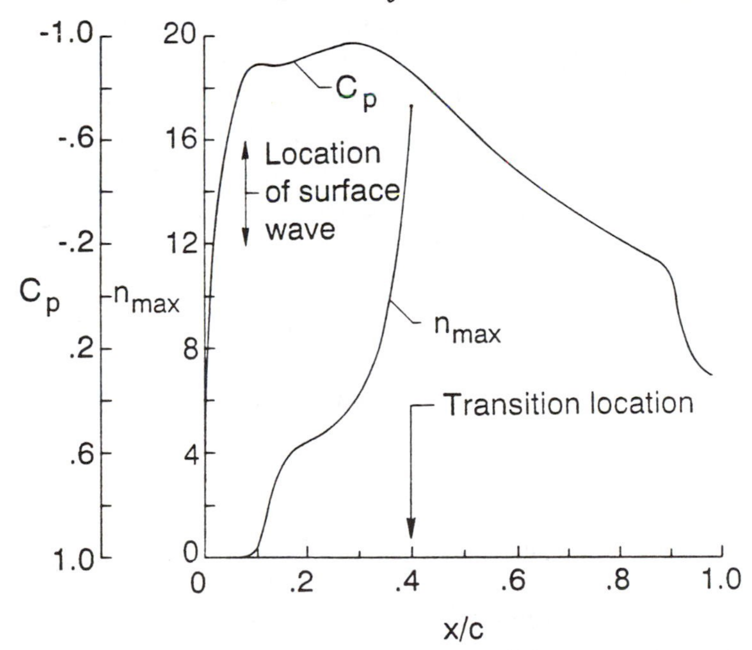


*Figure 13. Tollmien-Schlichting instability growth in the presence of a surface wave.*

tested was of height $h$=0.010 in. and length $\lambda$=2.5 in.; the effects of this wave on the pressure distribution between $0.10<x/c<0.13$ and on maximum T-S amplitude ratios are apparent in the Figure. In the adverse pressure gradient of the wave, $n_{max}$ is seen to grow from less than 1 to near 4. Elsewhere, in favorable pressure gradients, the rate of growth of the T-S disturbance is damped. In this example the surface wave had no effect on transition for the range of Reynolds numbers tested.

Surface waves may or may not have laminar separation zones associated with them, depending on several factors. From Schlichting [46], the laminar boundary layer will separate for $(q^2/\nu)\ (du_e/dx) < -0.1567$ where q is the boundary-layer momentum thickness, $\nu$ is the local kinematic viscosity, and $u_e$ is the local potential flow velocity. Calculation of values of $(q^2/\nu)\ (du_e/dx)$ for both Fage's [42] and Carmichael's [43] surface imperfections indicates that the critical value for laminar separation was exceeded at most of the test conditions for those studies. For example, at the conditions shown in Figure 14 (from Fage), $(q^2/\nu)\ (du_e/dx) = -0.19$. Similar results occur for analysis of Carmichael's data from Reference 45. It appears, then, that for many of the critical surface imperfections tested by Fage and Carmichael, laminar separation at the imperfection would have been present. Thus, the mechanism for forward movement of transition due to a surface imperfection could involve both the effect of local adverse pressure gradient on T-S amplification and the effect of the free shear layer inflectional instability.

In practice, more complete modeling of the parameters involved in surface imperfection effects is needed to provide more fully useful manufacturing tolerances criteria. The enormous variety of flow conditions and surface geometries makes experimental determination of complete criteria virtually impossible. Therefore, computational modeling approaches are being pursued for criteria development. Some effort in this direction has been made by Nayfeh and others [47-50]. These efforts have resulted in improved understanding of the mechanisms of instability amplification in flows over surface imperfections; however, further work is needed to validate computational approaches before practical applications are feasible.

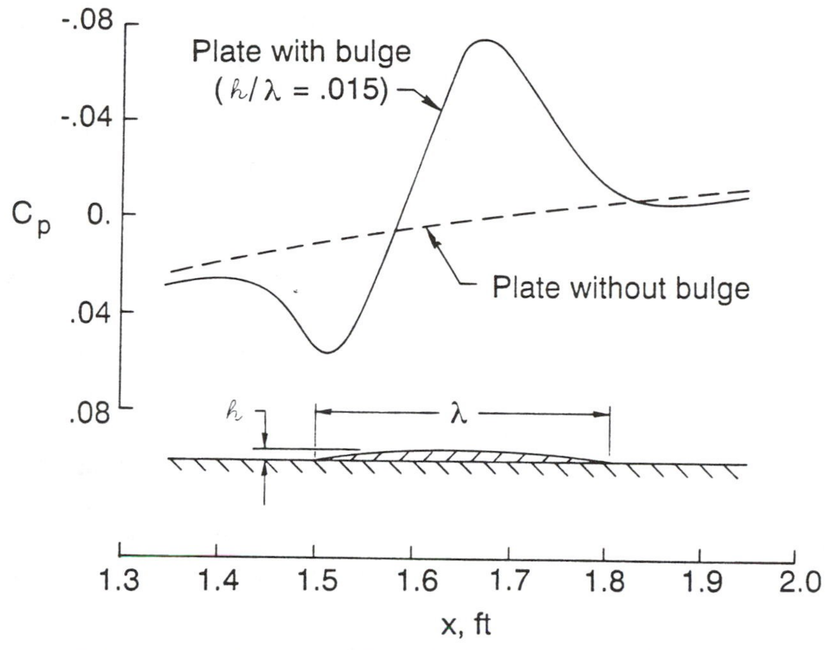


*Figure 14. Pressure distribution over a bulge [42].*

### 3.3. Effects of Loss of Laminar Flow on Stability and Control

For several NLF flight experiments, changes in stability and control characteristics caused by the loss of laminar flow have been observed [1,40]. These changes were brought on by the behavior of the particular airfoils selected for use on the forward control surfaces for several canard-configured airplanes. These particular airfoils experienced boundary-layer separation near the trailing edge if transition occurred near the leading edge. This design feature is not typical of NLF airfoils. In general, NLF airfoils are designed or selected which do not experience flow separation and lift loss upon loss of laminar flow. The Dragonfly shown previously in Figure 7 experienced significant changes in stability and control characteristics with the loss of laminar flow on the forward wing. Difficulties were encountered in elevator effectiveness, climb performance, and handling qualities on approach and landing. Correction of this behavior was achieved with the application of vortex generators at mid-chord on the upper surface of the airfoil [40].

Figure 8 previously illustrated the predicted velocity distributions and transition locations for the forward wing on this airplane. Free transition (clean wing) is predicted near the 45-percent chord location. In flight, transition occurred at this location where a laminar separation bubble was observed with a length of about 10-percent chord. When transition occurs near the leading edge (dirty wing), the thick turbulent boundary layer is unable to remain attached during the pressure recovery on the aft part of this airfoil, and turbulent separation is predicted near the 75-percent chord location. Excellent agreement was observed between these predictions and the flight-measured separation location with transition fixed near the leading edge. Figure 15 illustrates the differences in airfoil performance (lift and drag) which result from these changes in transition. A very large, approximately 100-percent, increase in drag results from the combination of laminar flow loss and the increase in form drag caused by separation near $C_l = 1.0$. The effect of loss of laminar flow on the forward wing also causes a 15-percent reduction of airplane lift curve slope.

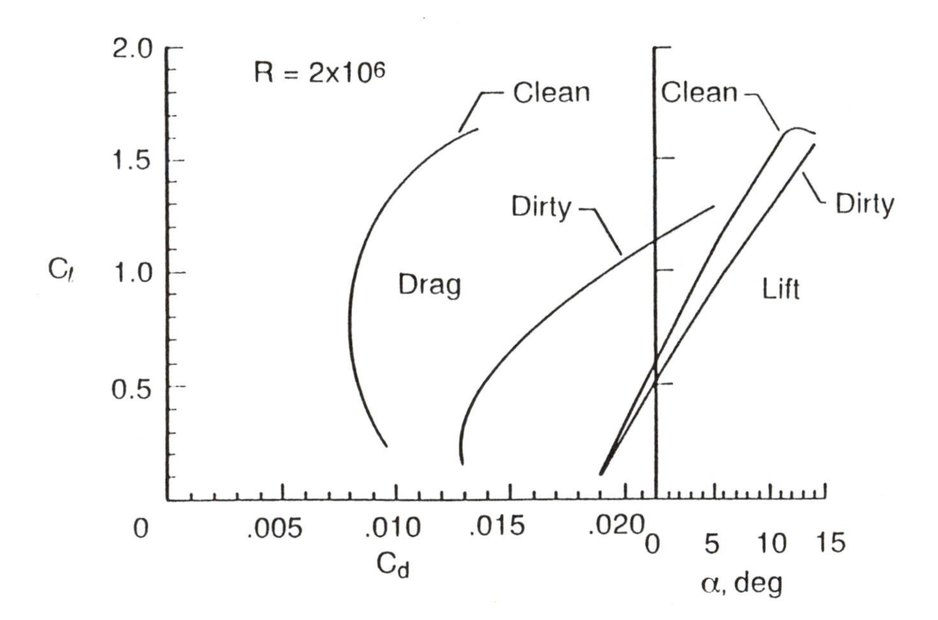


*Figure 15. Effect of loss of laminar flow on airfoil performance for the Dragonfly foreplane.*

This behavior is precisely the cause of pitch-trim changes observed in flight with loss of laminar flow in this airplane. Figure 16 shows the configuration of small vortex generators installed at the 45-percent chord location to energize the turbulent boundary layer and alleviate the effects of loss of laminar flow. In addition, these devices increased the top speed of the airplane in the clean wing (laminar) condition by about 10 mph and decreased the minimum trim speed by about 8 mph. This improvement resulted from the elimination of the relatively large laminar separation bubble on this airfoil and from the ensuing reduction in turbulent separation. Small improvements were observed for maximum and minimum speeds with transition fixed near the leading edge. Climb performance was improved by the vortex generators as well. Thus, the devices were very effective in alleviating the flow separation present for this laminar flow airfoil in both the laminar and turbulent conditions. In doing so, the stability and control of the airplane were greatly improved.

On airplanes for which winglets provide substantial levels of directional stability, loss of laminar flow can affect lateral-directional stability and control characteristics. Reference 40 explores the potential consequences of loss of laminar flow on stability and control in greater detail.

***Figure 16. Vortex generators installed on the Dragonfly foreplane.***

### 3.4. Effects of Loss of Laminar Flow on Maximum Lift

Careful selection of NLF airfoils can preclude difficulties related to maximum lift changes with loss of laminar flow. Two examples given here illustrate two possible outcomes depending on airfoil section behavior. The flight data presented in the previous section for the Dragonfly airplane illustrated the effect of loss of laminar flow on minimum trim speed. This effect was caused by the flow separation which resulted from early transition, thus affecting section lifting behavior. For the Dragonfly airplane, loss of laminar flow caused an estimated increase in minimum trim speed of 18 mph. This speed change corresponds to a 40-percent reduction in maximum trimmed lift coefficient. Reductions in maximum trimmed lift coefficient between 20 and 27 percent were reported in Reference 1 for the VariEze and Long-EZ airplanes using canard airfoils on which lift was sensitive to loss of laminar flow. By proper airfoil design, the dramatic effects of loss of laminar flow on lifting behavior can be avoided.

The NACA 6-series airfoil on the Skyrocket wing, for example, [31] actually experienced a slight increase in maximum lift in flight with transition artificially fixed near the wing leading edge. This effect is explained by the elimination of an upper-surface, leading-edge laminar-separation bubble at high angles of attack, by the transition strip.

These observations reinforce the need for selection of NLF airfoils which do not experience significant flow separation and lift or moment loss associated with the loss of laminar flow. These examples also show that care must be taken during testing of NLF airplanes to account for the effects of transition location.

### 3.5. Insect Accretion Behavior and Effects on Laminar Flow

In spite of the long history of NLF flight research, little quantitative information is in the literature concerning the seriousness of insect contamination on laminar flow airplanes in practical operating environments. Specifically, no data are available which establish the increase in drag which can be expected to occur on laminar flow airplanes flying in representative insect population densities. In practice, the seriousness of insect debris contamination will likely be dependent on airplane characteristics and mission.

The occurrence of insect accumulation on aircraft surfaces varies widely in terms of frequency, location of impact, and resulting debris height. The population density of insects is affected by local terrain, vegetation, temperature, moisture, humidity, wind, and height above ground level [51]. The insect impact pattern, as shown in recent analytical studies by Bragg [52], is affected by airfoil section geometry. Insect accumulation on aircraft occurs predominantly at low altitudes (less than 500 ft above ground level), mostly on the takeoff roll and initial climb and on final approach and landing [53]. Under many conditions (very cool or very warm temperatures, for example), very small rates of insect accumulation will occur, even at low altitudes. Maximum rates of insect accumulation will occur for an ambient temperature of 77 degrees F under light wind conditions and high humidity [54]. During recent NASA flight experiments by Croom [55] of an insect contamination protection system, the ambient conditions noted above were observed to produce maximum rates of insect accumulation. Figures 17 and 18 illustrate the sensitivity of rates of insect accumulation to ambient temperature and wind conditions, respectively. The results of these flight experiments indicate that below temperatures of about 60 degrees F or above about 100 degrees F, and for wind speeds above 20 mph insect accumulation rates will be much less significant. Altitude is another factor which affects population density of insects. Figure 19 shows the effect of altitude on insect population as reported by Coleman [53] and Croom [55]. For altitudes above ground level of 500 ft the insect population appears insignificant.

Flight-measured insect debris patterns on the Skyrocket airplane provide data illustrating the relative insensitivity to insect contamination of its particular airfoil at the conditions of the test. Figure 20 illustrates an insect debris pattern accumulated during a 2.2-hour flight at low altitudes. Sublimating chemicals were used in flight at sea level at 178 knots to determine which insect strikes caused transition. As shown, only about 25 percent of the insects

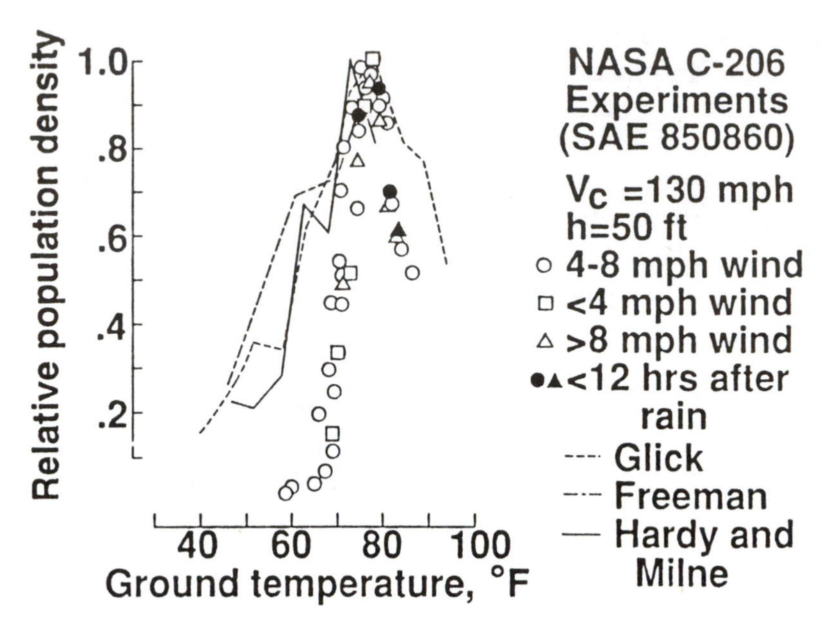


*Figure 17. Effect of temperature on normalized insect population density.*

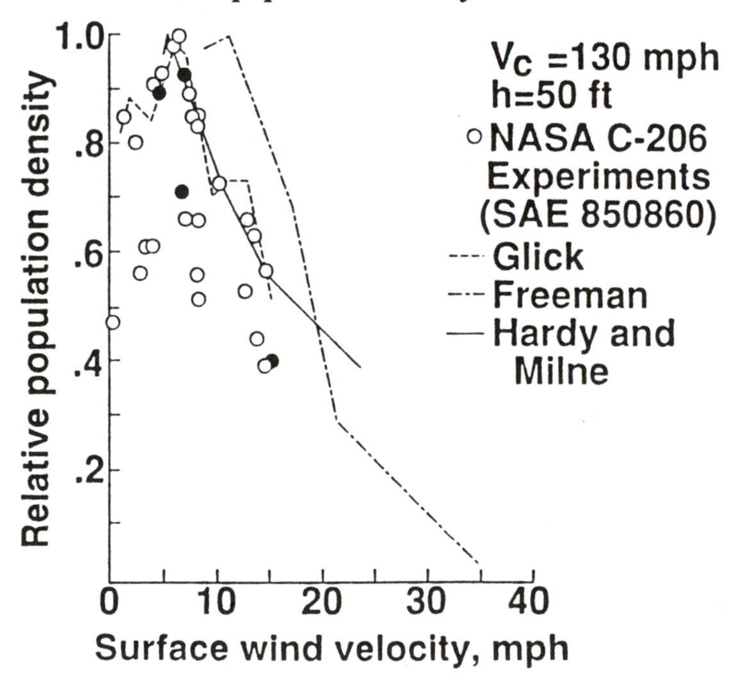


*Figure 18. Effect of wind velocity on normalized insect population density.*

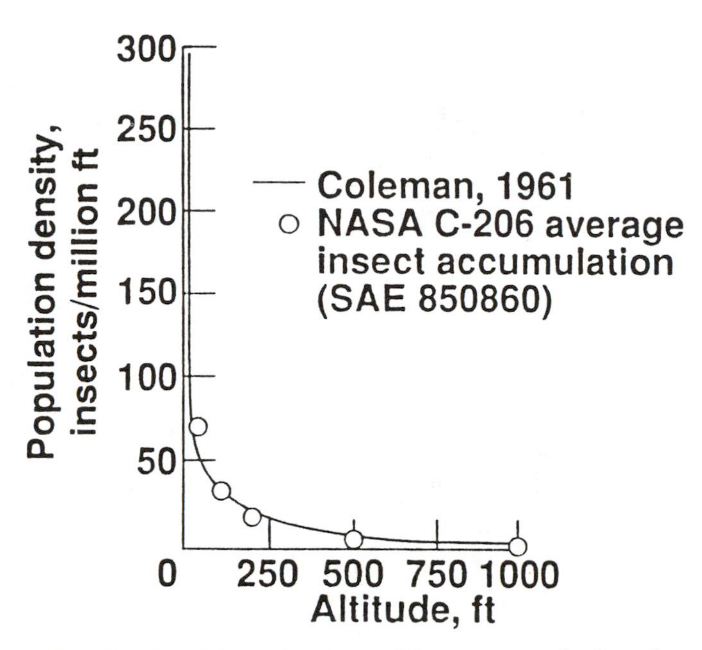


*Figure 19. Vertical distribution of insect population density.*

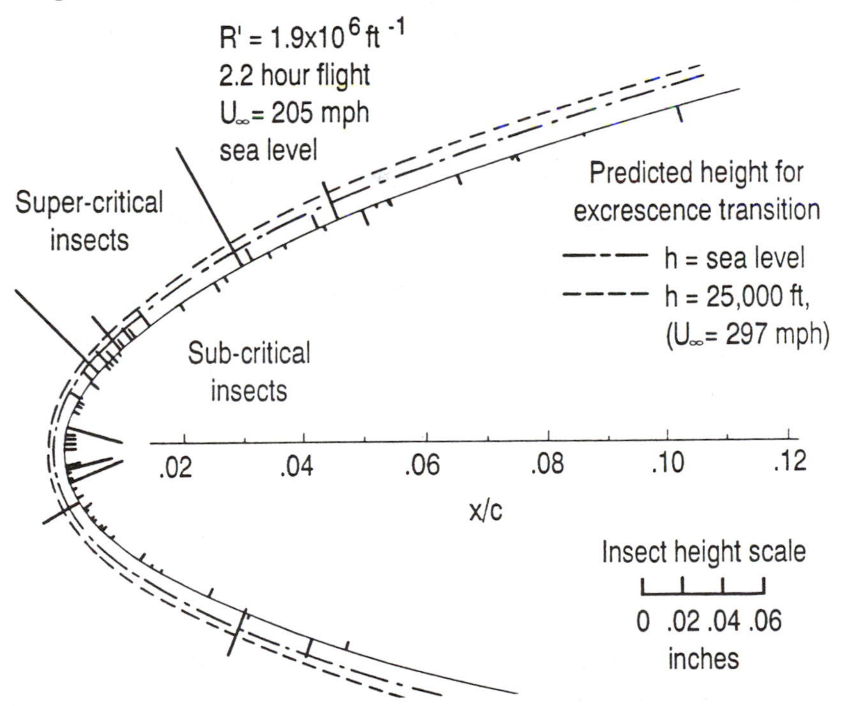


*Figure 20. Insect contamination pattern on the Bellanca Skyrocket II natural laminar flow wing, accumulated in flight.*

collected exceeded the critical height at the particular airfoil location to cause transition. For illustrative purposes in the Figure, supercritical insects are shown as protruding outward from the airfoil surface and subcritical ones protruding inward. Very near the stagnation point, rather large insect remains were recorded which did not cause transition. These insects were located forward of the location where disturbances can begin to amplify in the laminar boundary layer. An analysis using a value of critical-roughness-height Reynolds number of 600 was conducted to predict which insects would cause transition at a more typical cruise altitude of 25,000 ft. The dashed line in the Figure depicts the height of roughness required to cause transition at this altitude. It shows that only about 9 percent of the insects collected would have caused transition. Thus, even though large numbers of insects might be collected on a wing leading edge, relatively few of them can be expected to cause transition at high cruise altitudes.

## 3.6. Effects of Flight Through Clouds and Precipitation on Laminar Flow

Under certain conditions, the operation of a laminar flow airplane can be affected by either precipitation onto the NLF surface or by the flux of free-stream cloud particles through the laminar boundary layer. Precipitation can cause loss of laminar flow by creating three-dimensional roughness elements on the airfoil surface which, in sufficient quantity and size, act as a boundary-layer trip near the leading edge. Cloud particles (e.g., ice crystals at higher altitudes and liquid-phase particles at lower altitudes) can cause loss of laminar flow by the shedding of turbulent wakes from the particles as they traverse the laminar boundary layer. At sufficient flux (particles per unit area per unit time) and sufficient particle Reynolds number, partial or total loss of laminar flow can occur.

The VariEze wind-tunnel experiments of References 1 and 57 provided limited data on the effects of precipitation on NLF. In those experiments the effects of rain were studied by spraying water on the canard and wing. (see Figure 21). Comparison of the

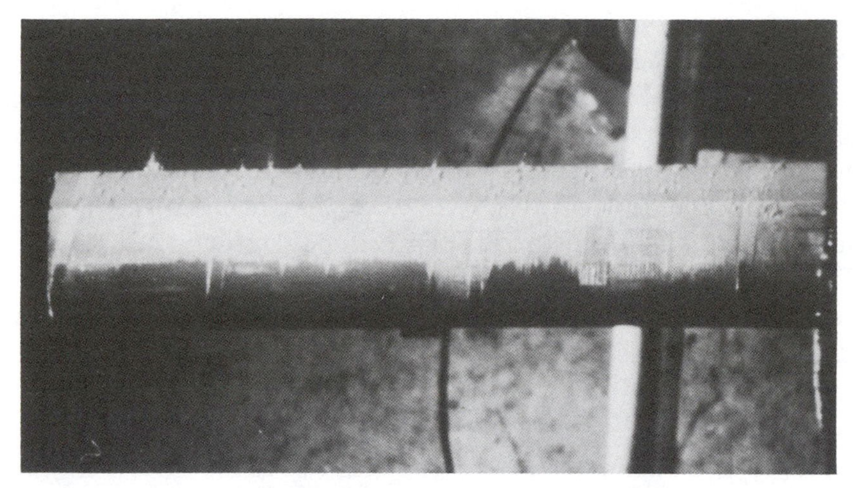

*Figure 21. Water spray pattern on the VariEze canard.*

aerodynamic characteristics of the canard in a heavy water spray and with transition fixed by artificial roughness showed that the effect of the water drops on the airfoil was to increase the drag to the same level as for the case with transition fixed near the leading edge. Figure 22 illustrates the effect of water spray on the VariEze canard in the Langley 30- by 60-Foot Tunnel. For these condi-

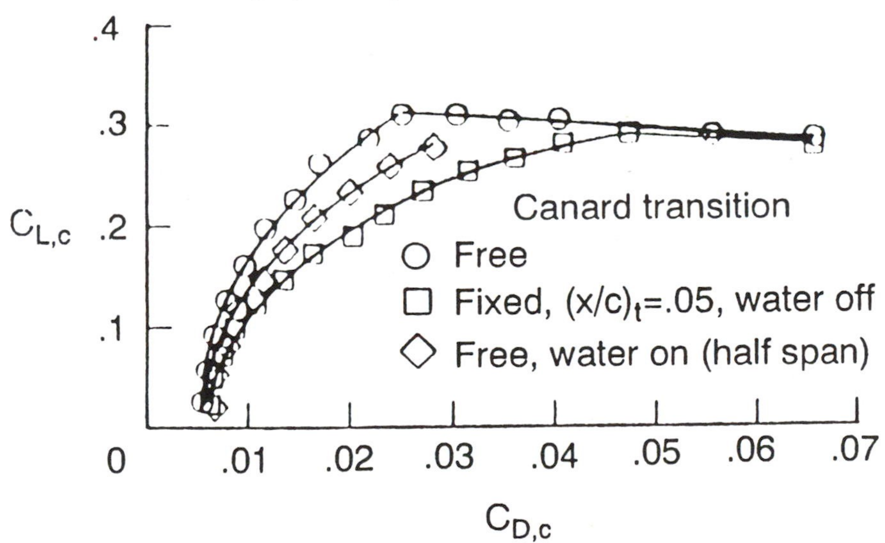


*Figure 22. Effect of water spray on the performance characteristics of the VariEze canard.*

tions, transition is suspected to occur near the leading edge, with separation of the turbulent boundary layer near the 55-percent chord location. Results of two flight experiments have shown that when a mist deposit occurs on a laminar flow surface during flight through clouds, the boundary layer becomes turbulent. During the early Hawcon flights [14], wake-rake drag measurements were made with a mist deposit from flight through clouds on the wing. The Heinkel measurements included in the paper [14] showed a 42-percent increase in section drag (i.e., loss of laminar flow) caused by the mist deposit at chord Reynolds numbers between $6.5 \times 10^6$ and $8.5 \times 10^6$.

For the more recent NASA T-34C NLF glove flight experiments [28], transition location was measured using hot films during flight through cloud particles for which no mist deposit occurred on the wing. For these tests, laminar flow was unaffected by the cloud particles in the free stream. By using Hall's criterion [58,59] for a critical Reynolds number of 400 (based on particle diameter), the speed required for an average-sized cloud particle of 20 microns to cause transition is estimated as 587 knots at a unit Reynolds number of $1.4 \times 10^6$.

In the X-21 LFC flight experiments [60], laminar flow was lost as a result of flight through ice-crystal clouds. For these tests, the critical particle Reynolds number was exceeded for the flight conditions involved. This occurred because of the much lower value of critical particle Reynolds number for the larger and prism-shaped ice crystals encountered in the stratosphere.

For both the X-21 and the T-34C flights, laminar flow was restored immediately upon exiting from a cloud. These results indicate the insensitivity of the laminar boundary layer to flight through clouds at low altitudes where the particles do not deposit on the surface and where the critical particle Reynolds number is not exceeded. The mechanism for loss of laminar flow in clouds at lower altitudes involves deposit of mist which creates supercritical roughness in the boundary layer.

### 3.7. Laminar Flow Behavior in Propeller Slipstreams

Early observations of the effect of the propeller slipstream on boundary-layer transition have resulted in different conclusions. Young and Morris [6,7], and Hood and Gaydos [61] concluded that in the propeller slipstream the boundary-layer transition location moved forward to near the leading edge. Results of Zalovcik and Skoog [17,18] showed little effect of the slipstream on boundary-layer transition for an NACA 230 section, whereas for an NACA 66-series wing section, transition moved from 50-percent chord to 20-percent chord. A drawback to these early experiments, which may explain the inconsistencies, is that the use of total pressure probes in combination with large volume pressure transducers provided only time-averaged information. Time-dependent behavior, characteristic of blade passing frequencies, was not measurable by the techniques employed in these early experiments.

Recent experiments [1,62] to measure the time-dependent characteristics of the propeller slipstream have shown the existence of a cyclic turbulent behavior resulting in convected regions of turbulent packets between which the boundary layer remains laminar (see Figure 23). The flight experiments of Holmes [1] using surface hot films in a laminar boundary layer in the propeller slipstream on the T34C airplane (Figure 24) illustrate this cyclic nature. The hot films located on the "mini glove" show a cycle of turbulent bursts in the otherwise laminar signals. The NLF glove hot film outputs are shown for comparison. Analysis of some data from Wenzinger [63] indicates that the drag increase of laminar airfoils in propeller slipstreams is significantly less than that due to total loss of laminar flow (Figure 25).

The implication of these recent observations is that the section drag increase associated with the transition changes in propeller slipstreams may not be as large as that for fixed leading-edge transition. Thus, NLF airfoils may provide drag reduction benefits, even on multi-engine configurations with wing-mounted tractor engines.

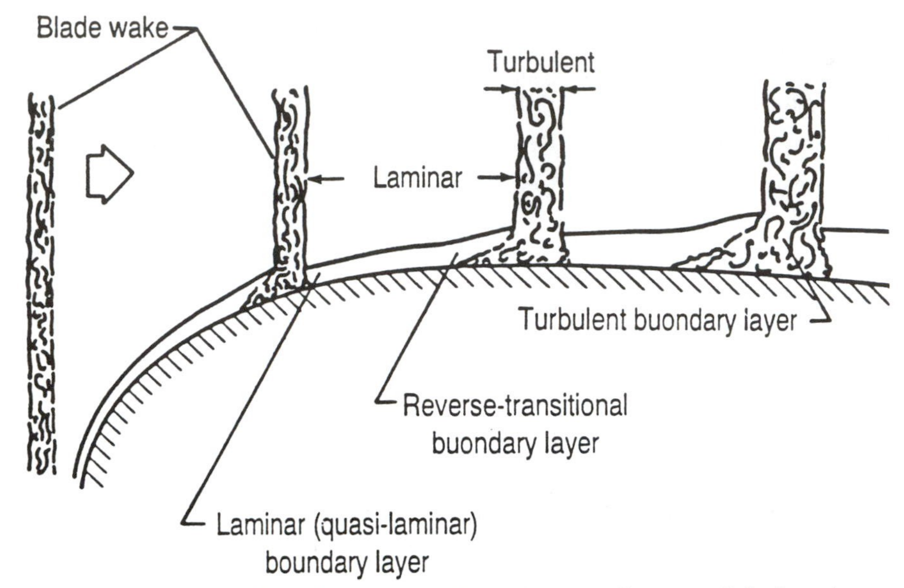


***Figure 23. Propeller slipstream disturbance flow model showing turbulent response in a laminar boundary layer.***

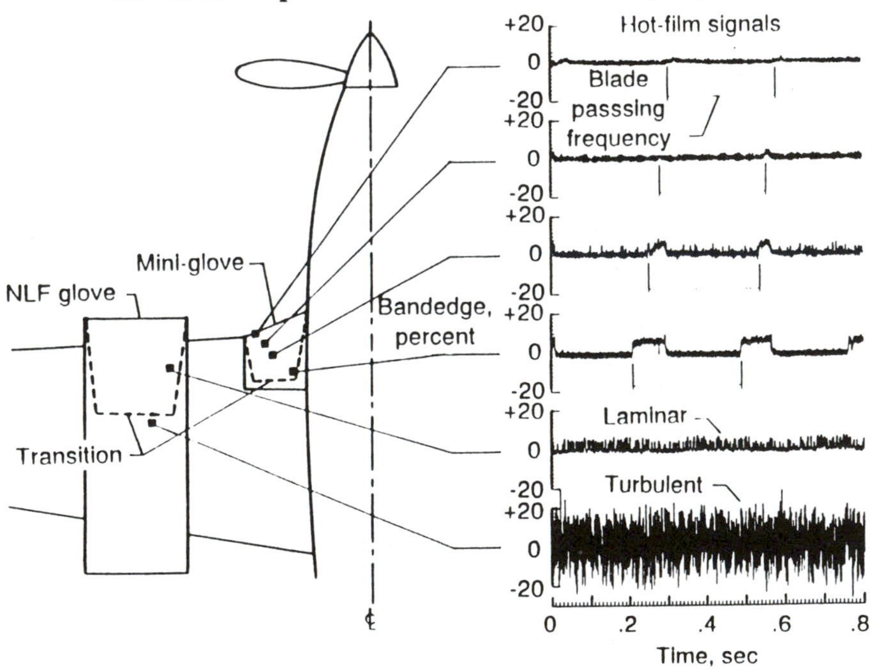


***Figure 24. Time dependent behavior of the laminar boundary layer in a propeller slipstream (V=165 Knots, 175 RPM, 3-bladed propeller).***

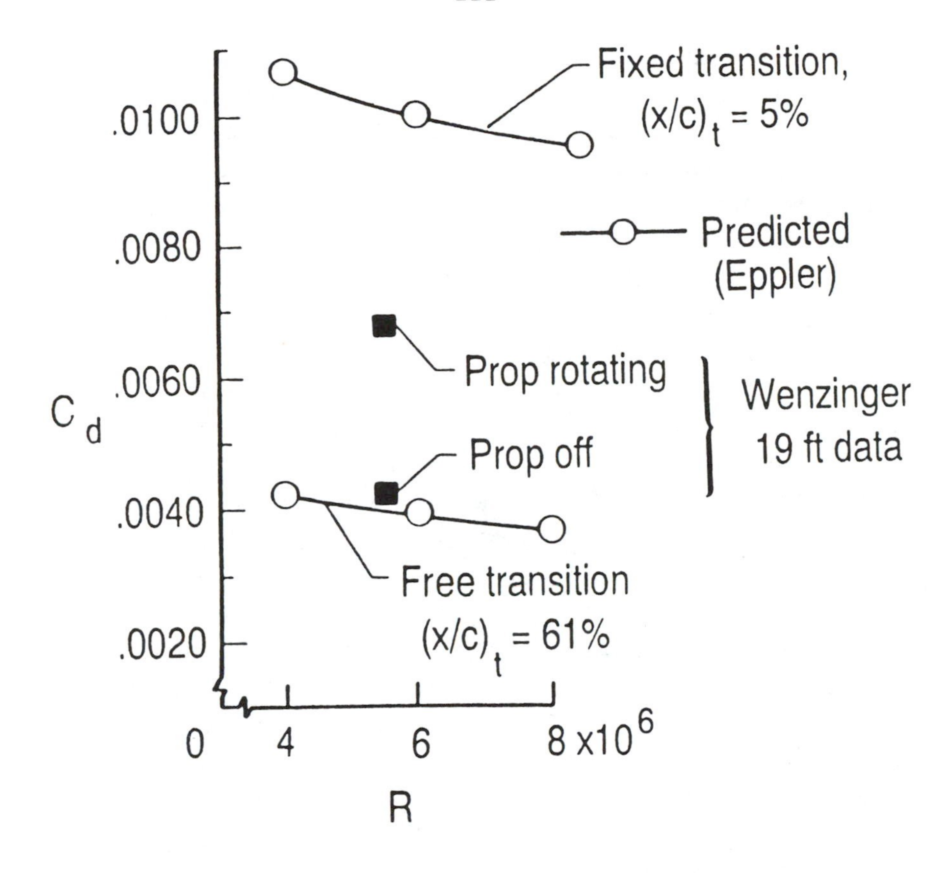


*Figure 25. Effect of propeller slipstream on the section drag of a natural laminar flow airfoil.*

## 3.8. Fixed Transition Flight Testing

In certain respects, the design, testing, and operation of NLF airplanes differ from those considerations for turbulent airplanes. Laminar flow airplane designs must include the consideration that for certain environmental conditions, laminar flow will be lost. Analysis and testing of these airplanes must include the evaluation of the potential effects of fixing transition near the leading edges of the laminar surfaces. Operators of laminar flow airplanes must have information concerning the differences in airplane characteristics with and without laminar flow.

One important conclusion from the recent NASA NLF flight experiments is that fixed transition tests are an important inclusion in flight research or in certification flight testing on airplanes on which significant differences in airplane behavior are anticipated to result from loss of laminar flow. Fixed transition testing will be increasingly important for correlation of wind-tunnel, analytical, and flight-test characteristics for laminar-flow airplanes. Furthermore, since several propeller surfaces have been observed to support significant runs of NLF, there is additional value in conducting tests with transition fixed on the propeller as well. Standard wind-tunnel transition-fixing procedures are directly applicable to flight testing. Braslow's critical roughness criteria for both two-dimensional and three-dimensional boundary layers [64] can be used for sizing of grit to produce transition without excessive grit drag. Very thin (0.001 in.) double-back tape is available from large manufacturers of industrial tapes and is very useful for applying grit in a fashion which makes removal easy after testing. Two-dimensional transition strips (e.g., tape or wire) can be used as an alternative to grit. Sizing of two-dimensional trip strips can be accomplished using Reference 15 for a tape trip and Reference 45 for a wire trip. The trip disk method has also been very successfully adapted to flight test [39,65]. This technique offers greater control of trip sizing, while at the same time being somewhat simpler to use than grit.

## 4. Laminar Flow Experimental Methods

Much of the development of experimental test techniques is done as a consequence of the "need to know" philosophy. With progress in the understanding of the usefulness of laminar flow came the "need to know" more about detailed transition behavior over larger surface areas and over wider ranges of flight conditions than earlier techniques were capable of. While there are many techniques available, this chapter will discuss the relatively recent developments in flow visualization and hot-film anemometry used in flight applications.

Flow visualization complements hot-film anemometry techniques for determining aerodynamic characteristics such as boundary-layer state (laminar, transitional, turbulent, or separated), shock wave location, and surface flow direction in flight at subsonic and supersonic speeds. The most popular boundary-layer visualization methods for flight applications include the sublimating chemical technique [66], the liquid crystal technique [67,68], and infrared imaging [69,70]. Each method has certain advantages and disadvantages which constrain the applications. In addition to flight testing, all three methods are applicable to general use in high- and low-speed wind tunnel and water-tunnel testing.

A refinement of the sublimating chemical technique has been developed to define both the boundary-layer transition location and the transition mode; however, the method is restricted (subsonically) to altitudes below approximately 20,000 ft. In response to the need for flow visualization at subsonic and transonic speeds and altitudes between 20,000 ft and 50,000 ft, the liquid crystal technique has been developed. A third flow visualization technique that has been developed recently is infrared imaging, which offers non-intrusive transition detection over a wide range of test conditions.

More detailed measurements of the characteristics of the boundary layer can be made with hot-film anemometry. Flight applications of surface-mounted, thin hot-film devices were introduced in the late 1950's to measure transition location [26]. Recent developments have extended the use of hot films to measure the details of the boundary layer. Hot film sensors have been developed to make time-dependent and streamwise measurements of the frequency content [71] in the boundary layer as well as detecting regions of separation [72] and crossflow [73]. Even more recent development and refinement of arrayed hot-film devices is reported in Reference 74. This work has proven the usefulness of the method for detection of stagnation, separation, and reattachment locations.

General descriptions of the use of these flow visualization and hot-film techniques are given in the following sections. Detailed procedures for the techniques should be obtained from the cited literature.

## 4.1. Sublimating Chemicals

The sublimation method involves coating the surface to be observed with a very thin film of a volatile chemical solid. During exposure to a free-stream airflow, areas develop in which the chemical film sublimates more rapidly due to greater local shear stress and attendant mass transfer within the boundary layer. Greater rates of sublimation will therefore occur in turbulent flow, lesser rates in laminar flow.

In order for the chemical coatings to remain solid, opaque, and durable at temperatures for which transition indications are obtained, they must have high melting points, have no adverse effects on surface finishes, have low vapor pressures for aerodynamic use, and be soluble in a fast evaporating carrier. Another consideration for selecting appropriate chemicals is safety from health hazards associated with the use of such compounds. Four useful compounds which meet these requirements are naphthalene, biphenyl, acenaphthene, and fluorene. An added feature of fluorene is its fluorescent properties, which makes it possible to obtain high quality photographic transition pattern data by using ultraviolet lighting. The solvents found to be most suitable to carry the solids are acetone and light petroleum fractions such as 1,1,1 trichloroethane and trichlorotrifluoroethane (Freon TF). However, since the chlorinated fluorocarbon, trichlorotrifluoroethane, poses an environmental hazard, and does not provide significant advantages over 1,1,1 trichloroethane, the latter is most suitable for the majority of applications. A more detailed description on how to select and apply sublimating chemicals is given in Reference 65.

Figure 26 depicts characteristics of sublimating chemical pattern development in response to various modes of transition based on numerous experiments. In the subsonic flight environment, the transition modes of practical interest include Tollmien-Schlichting instability in two-dimensional boundary layers, laminar separation-induced inflectional instability, crossflow instability in three-dimensional boundary layers, and instabilities which may bypass these "natural" modes (caused by surface imperfections or acoustics, for example). The Figure illustrates these common transition modes, as they appear in sublimating chemical coatings.

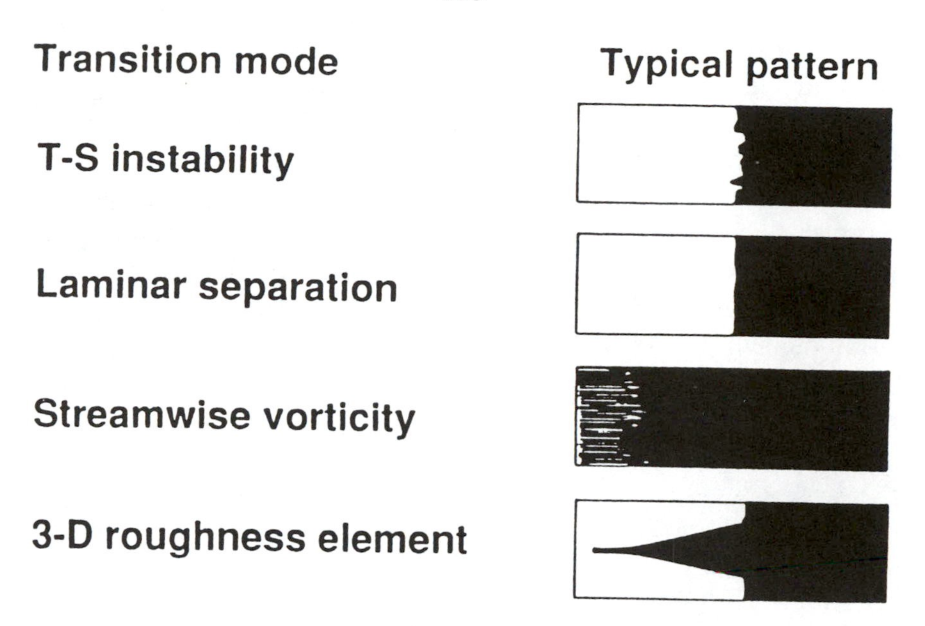


*Figure 26. Boundary-layer transition mode characteristics in sublimating chemical patterns.*

Flight evaluations of various flow visualization methods were conducted using a Lear Model 28/29 business jet airplane (Figure 9). The flow visualization on the winglet, shown in Figure 27, was conducted using the sublimating chemical acenaphthene at an altitude of 17,500 ft, a Mach number of 0.6, and a unit Reynolds number of $2.1 \times 10^6$ $ft^{-1}$. On the upper span of the winglet, with 33 degrees of leading-edge sweep, the crisp, straight transition front at x/c=0.65 is characteristic of the laminar separation mode of transition. A roughness particle that prematurely tripped the boundary layer is evident in the mid-span region. Another mode of transition appears in the lower-span region, where between 5- and 10-percent chord, there is a manufacturing joint with a large aft facing step and several screw slots. The screw slots were not smoothed over, and consequently caused premature transition. In the lowest region of the photo the chemical pattern reveals that an aft-facing step was the cause of premature transition.

While the results shown in Figure 27 provide some useful data, the test conditions at which the data were acquired were not that of cruise flight. Ultimately it is very important that transition studies be conducted at the speeds and altitudes at which the laminar-flow technology is intended to be applied. This is important because on laminar airframe surfaces which are practical to build and maintain, the transition process will be influenced by unit Reynolds number as well as length Reynolds number. The simultaneous scaling of both of these parameters is not possible; large-scale flight conditions are required for complete understanding of transition behavior. The flight conditions for the data shown in Figure 27 are near the upper limit of altitude for transonic use of sublimating chemicals.

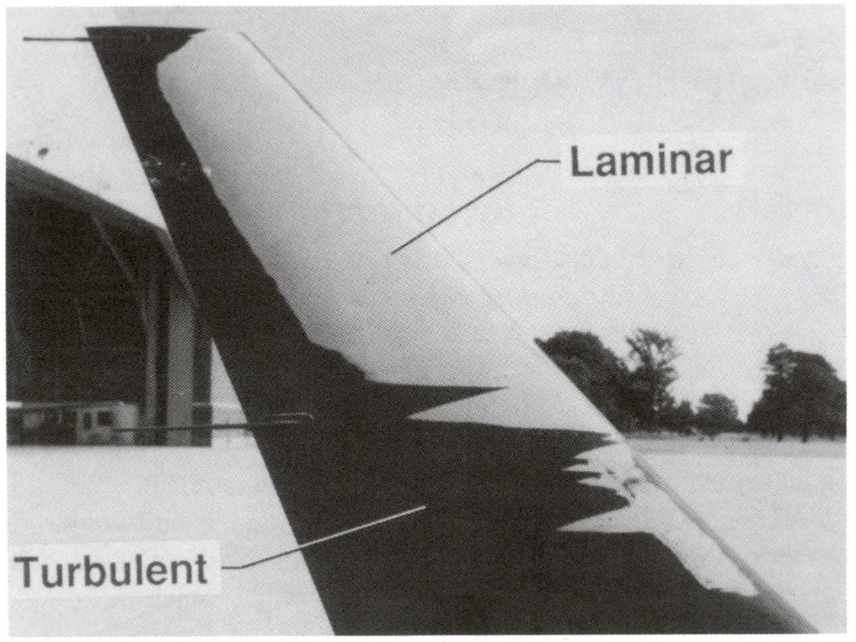


***Figure 27. Boundary-layer transition on a Lear 28/29 winglet indicated by sublimating chemicals ($M=0.6$, $h=17{,}500$ ft, $R'=2.1 \times 10^6$ ft$^{-1}$).***

## 4.2. Liquid Crystals

Liquid crystals are a peculiar substance with properties of both liquid- and solid-phase materials. Although they appear as oily liquids, they are unique in their ability to exhibit the optical properties of solid crystals. In particular, liquid crystals scatter incident light very selectively, in ways which can be related to aerodynamic properties such as shear stress at the wall. The optical characteristics exhibited by liquid crystals are a result of the behavior of their molecular structure. Within a liquid crystal film, the axis of alignment of molecules is rotated in a helical fashion with a characteristic pitch length. Fortuitously, the helix pitch lengths are in the range of wavelengths of visible light. The pitch length of the helix can be altered by external stimuli. In this fashion, liquid crystal coatings selectively reflect discrete wavelengths (color) of light in response to physical phenomena. Typically, temperature and shear stress are the primary stimuli which can alter the pitch length, although other phenomena such as ferromagnetism, electrical potential, and chemical vapors also work. Since the fundamental chemical structure is unaffected by these changes, a liquid crystal coating can respond repeatedly to the same physical changes; thus, the color changes can be reversible and rapid. This feature allows several test conditions to be observed with one application of the liquid crystal material, as compared to the sublimating chemical technique, which is limited to one test condition per application. The particular color observed depends on viewing angle and on the local shear and temperature to which the liquid crystal is subjected. The vividness and brilliance of the reflected colors also depend on the amount and angle of incident light.

The application of the liquid-crystal transition visualization technique requires care in the preparation of the test surface and in the preparation of the proper liquid-crystal formulation for the specific test conditions. Arrangement of adequate lighting to provide graphic, clear, visual data is also important. It is important that the surface be painted with a flat-black paint which is resistant to the solvents used to thin the liquid crystals. This matte texture provides a surface to which a very thin liquid crystal coating can best adhere, while the black color allows for destructive interaction of the light reflected from the liquid crystal coating with the light reflected from the test surface. Reference 68 describes the development and applications of the liquid crystal flow visualization method in more depth.

Figure 28 shows a liquid crystal transition pattern obtained in flight on the Lear Model 28/29 wing. The transition front is shown to be near the 30- to 35-percent chord location at a Mach number of 0.80 and an altitude of 48,000 ft. The unit Reynolds number for this condition was $R'=1.8x10^6$ $ft^{-1}$. The regions of laminar flow appear blue in color, and no color appears in the turbulent areas; a few turbulent wedges can be seen emanating from tape edges near the leading edge. To reduce the chance of debris accumulation during these experiments, the leading edge was not coated with the liquid crystal material.

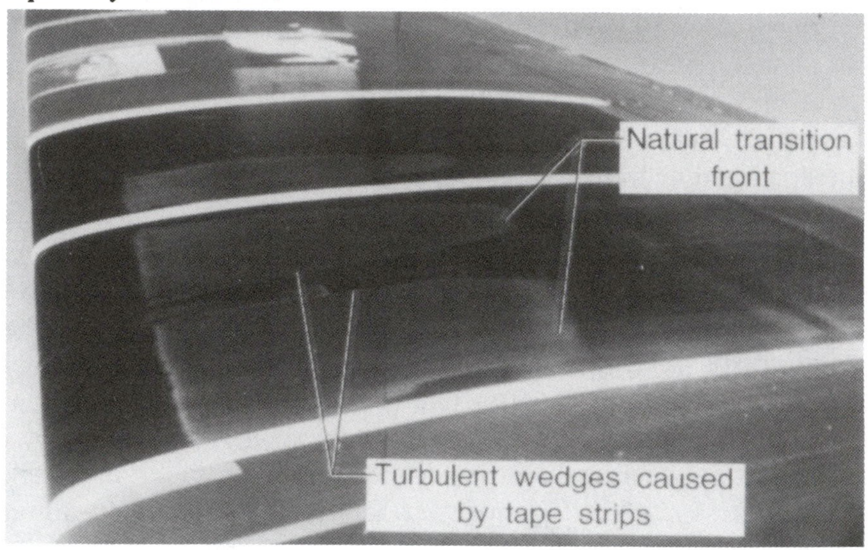


***Figure 28. Boundary-layer transition on a Lear 28/29 wing indicated by liquid crystals ($M=0.80$, $h=48,000$ ft, $R'=1.8x10^6$ $ft^{-1}$).***

An additional objective of the liquid crystal flight experiments on the Lear Model 28/29 was to qualitatively evaluate the time response of liquid crystal coatings to changes in transition location. Figure 29 shows photographs made from a video recording of the transition front motions on the winglet. Liquid crystals were brushed on the left winglet and the airplane was flown at an altitude of 17,150 ft, a Mach number of 0.53, and a unit Reynolds number of $2.1x10^6$ $ft^{-1}$. During the flight the airplane was yawed to sideslip angles of +3.9 to -3.5 degrees with a period of 0.56 seconds. The resulting maneuver provided variations in local angles of attack and pressure distributions on the winglet, causing transition to move from the leading edge to the 70-percent chord location. On this particular airfoil, pressure recovery at small angles of attack begins at about 65-percent chord. The observed movement of transition was in phase with the sideslip oscillation, indicating the ability of the coating to respond rapidly to varying conditions.

The liquid crystal transition visualization technique development efforts have led to the capability to visualize transition virtually throughout the flight envelopes (altitudes and speeds) of all modern subsonic aircraft. The method clearly has value in ground facilities as well.

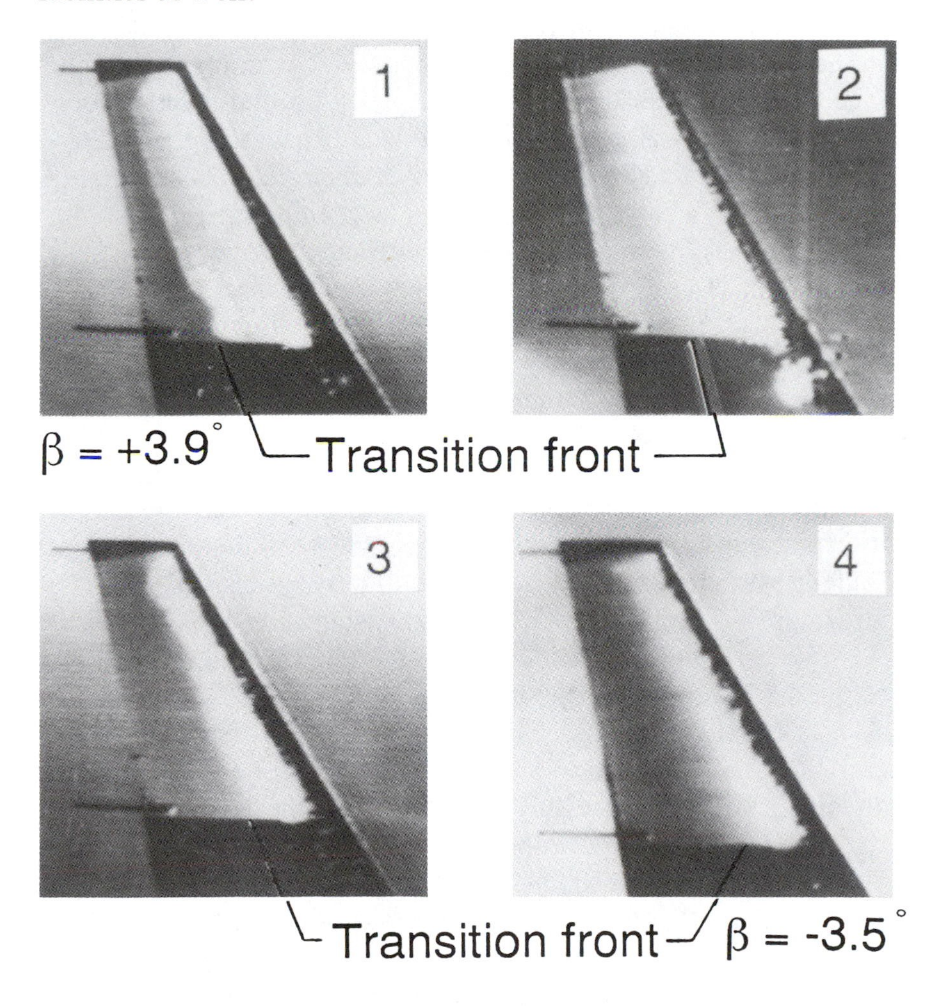


*Figure 29. Dynamic response of liquid crystals to transition motion on a Lear 28/29 winglet ($M=0.53$, $h=17{,}150$ ft, $R'=2.1\times10^{6}$ ft$^{-1}$).*

## 4.3. Infrared Imaging

Infrared boundary-layer transition visualization relies on heat transfer properties of laminar and turbulent flow. In laminar flow, the heat transfer rate between the fluid and surface is low due to low rates of convection. Conversely, the turbulent boundary-layer flows exhibit higher rates of heat transfer due to increased rates of convection. This difference in heat transfer rate results in a surface temperature difference between laminar and turbulent regions that can be detected by an IR imager. Factors which affect transition detection using IR include wing surface emittance in the imager passband, the imager spectral passband, the solar absorption, the skin thermal conductivity and heat capacity, optical filtering, intervening gas absorption and radiation, Mach number (compressibility), and rate of change of ambient temperature.

A flight test was conducted to evaluate infrared (IR) flow imaging techniques for boundary-layer flow visualization [70]. The flight tests used a single-engine, turboprop Beechcraft T-34C airplane with an onboard infrared imaging system that displayed and recorded real-time boundary-layer transition visualization. The actual test surface was an NLF glove mounted on the left wing of the aircraft as shown in Figure 30. The low thermal conductivity of the black fiberglass NLF glove with an emittance value of 0.9 helped reduce the conduction of heat between the laminar and turbulent portions of the glove, allowing for nearly steady-state temperature differences to be maintained between the two flow-field areas on the glove. Additionally, the low thermal mass of the fiberglass allowed the temperature differences to occur rapidly for changing flight conditions. The 92-inch chord, 3-foot span glove incorporates a NASA NLF(1)-0215F airfoil section. This airfoil maintains a favorable pressure gradient from the leading edge to near 45-percent chord location on the upper surface at low angles of attack. Tests were conducted over an indicated speed range from 80 to 210 knots and a pressure altitude range of 3500 to 18,000 ft during day and night lighting conditions. All data runs were conducted with the canopy open so that no loss of imager sensitivity occurred from infrared absorption by the plexiglass canopy.

***Figure 30. T34C airplane fitted with a natural laminar flow glove for infrared imaging flight experiments.***

Figure 31 shows the IR visualization in sunlight at an indicated velocity of 187 knots and an altitude of 6500 ft. Note that this figure and the following figure are photographs taken from a video image and hence some clarity has been lost in the process. The lighter color in the laminar region was due to a decreased convection coefficient of the laminar boundary layer and hence a warmer temperature than in the turbulent region. The outside air temperature was lower than the NLF glove surface temperature due to solar heating, and the turbulent flow area was cooled at a faster rate than the laminar area. Figure 32 shows the IR results for a slower airplane speed of 100 knots. The boundary-layer transition front has moved forward to the 20-percent chord location as a result of an increase in angle of attack for the lower speed.
Other testing was performed to investigate various effects of using the IR imager. Rapid ascents and descents were made to assess the effects of ambient temperature and to determine the transient characteristics of the IR images. Also, day and night flights were performed with different glove colors. This testing

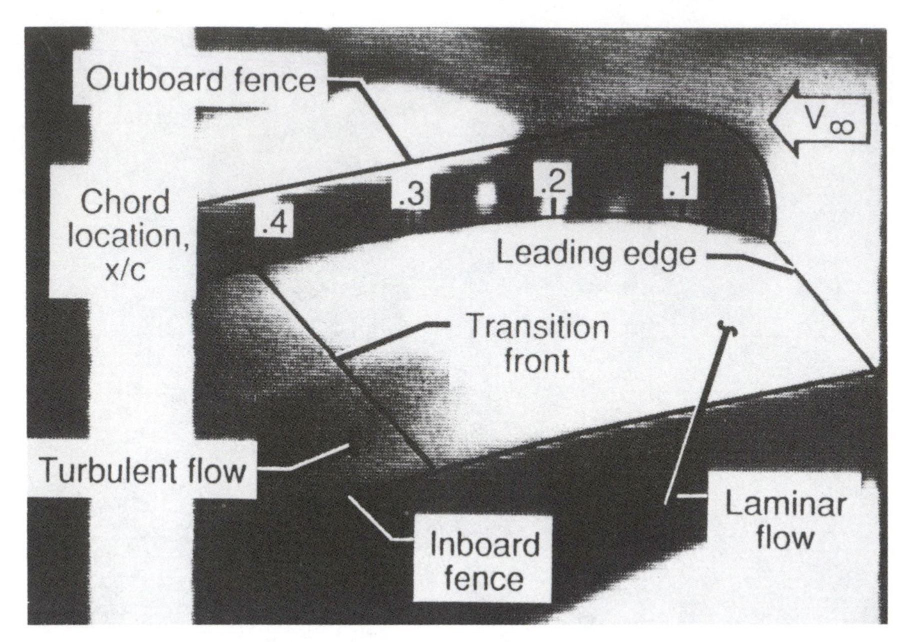


*Figure 31. Infrared image from the T34C natural laminar flow glove ($V_i$=187 knots).*

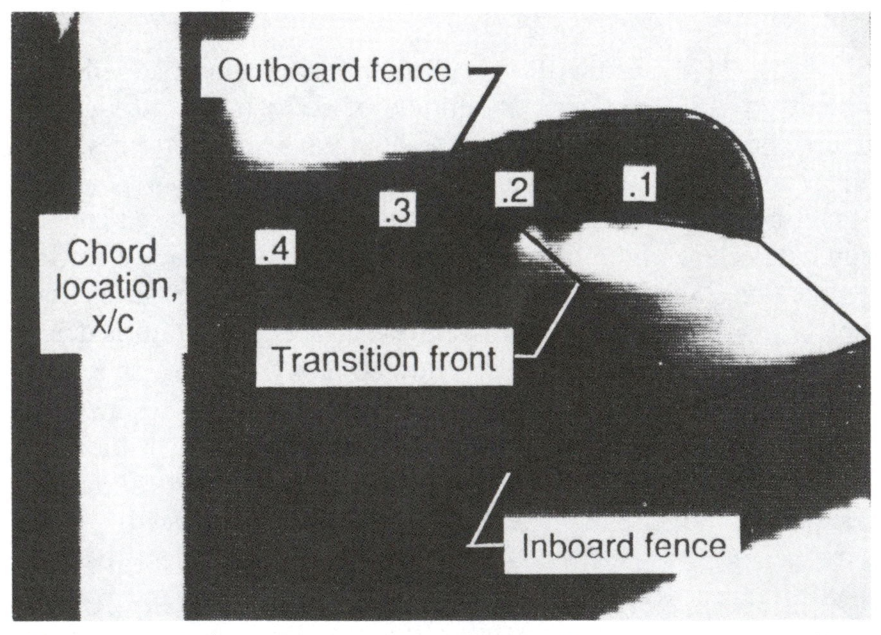


*Figure 32. Infrared image from the T34C natural laminar flow glove ($V_i$=100 knots).*

showed that night flights gave good results but required higher speed and/or altitude changes to produce a steady temperature difference between the two boundary layer areas on the glove. The black glove provided the clearest IR result while orange and white glove colors reflected the solar radiation, and the imager could not clearly identify the very small temperature gradients on the glove due to flow differences. In addition, the results were compared to those obtained using the sublimating chemical technique with good agreement throughout the flight envelope.

## 4.4. Arrayed Hot-Film Transition Sensors

Many flight research efforts require surface mounted, minimum installed thickness (less than about 0.007 inches for most subsonic applications), hot-film sensors. In many cases these sensors must make time-dependent and streamwise measurements. Three surface-mounted, arrayed hot-film instrument and analysis systems have been developed to meet these needs: the multi-element transition sensor, the multi-element laminar separation sensor, and the crossflow vorticity sensor. All the sensors consist of nickel-film elements deposited on a substrate of polyimide film. Each hot-film is operated in a constant temperature mode. The anemometer electronics consist of each hot-film sensor as one side of a wheatstone bridge circuit and include an amplifier with internal programmable gain and filtering. The multi-element sensor and a prototype laminar separation sensor were evaluated in flight on the Lear 28/29 described earlier. The arrayed multi-element laminar separation sensor and the crossflow vorticity sensor were evaluated in the NASA Langley 14- by 22-foot Wind Tunnel and the NASA Langley 17- by 22-inch Small Calibration Facility, respectively.

## 4.5. Multi-Element Transition Sensor

The continuous, multi-element, hot-film transition sensor overcomes the disadvantages of the staggered individual hot-film sensors by integrating the required number and spatial distribution of hot-film sensing elements into a long, continuous thin sheet. The length of the sheet covers the streamwise area of interest for transition measurements, beginning at the leading edge and continuing to downstream of the transition region. Figure 33 shows the geometric detail of a typical 25-element sensor design.

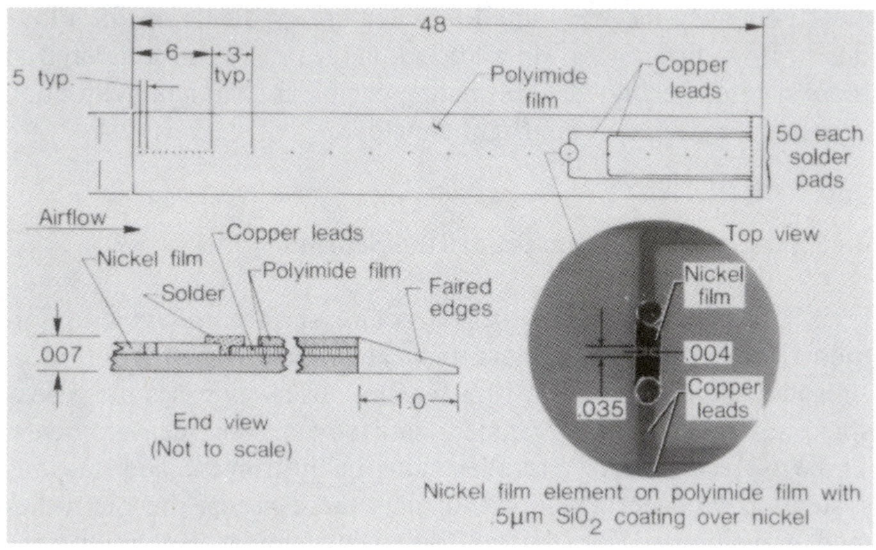


***Figure 33. Multi-element transition sensor geometric detail (dimensions in inches).***

A flight experiment was conducted on the Lear 28/29 to evaluate the multi-element transition sensor. For this flight test, the 25 sensors were controlled through a six channel switching system, due to the limited number of onboard anemometers (six). Figure 34 illustrates typical data which can be acquired with the multi-element transition sensor. The data were taken during a steady, level-flight condition at an altitude of 39,000 ft and a Mach number of 0.77. The time segments that show no data is a result of the sensors being switched around due to the limited number of anemometers. Transition onset is located at 41-percent chord as shown in the figure. The length of the transition region extends over about 8-percent of the chord. This transition region length, the character of the signal, and the flat pressure gradient that was measured are indicative of T-S initiated transition. To further analyze the hot-film signals, a follow on experiment was conducted to

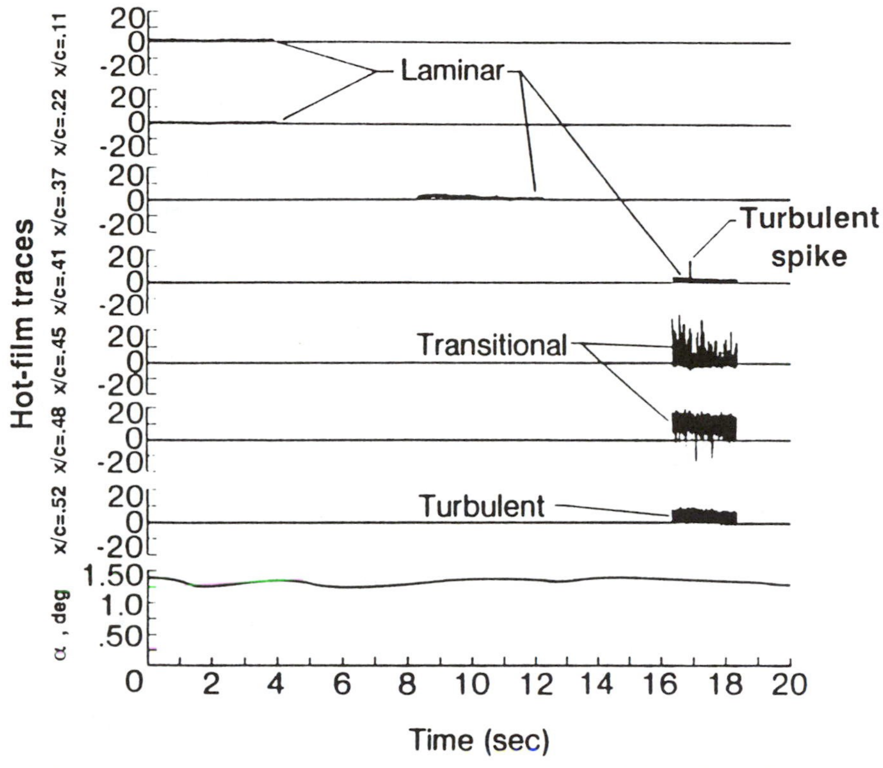


***Figure 34. Lear 28/29 flight-measured transition from laminar to fully turbulent flow.***

study the frequency characteristics of the boundary layer [74]. In addition to the standard signal filtering and amplification incorporated in the anemometer electronics, for these experiments, each signal was externally amplified with a gain of 56. This amplification provided more detailed information on the laminar signal characteristics while the turbulent signal was completely offscale. These laminar signals were then processed through a power spectral density (PSD) analyzer. This technique is particularly useful because system noise that may prevail throughout the time domain tends to appear only at discreet frequencies in the PSD plots, avoiding saturation of the T-S data. Figure 35 shows a PSD plot of two hot-film traces obtained at 40-percent chord. A shift in the relative amplitudes of the two traces shown is due to the difference in the overheat ratio of the sensors from which these data were obtained. Both traces show a range of frequencies from 3900-

5300 Hz over which the T-S instability was observed. A range of frequencies would be expected instead of a single peak because of the disturbance growth process that occurs in the boundary layer. The breakdown from laminar to turbulent flow may begin with the linear amplification of selective T-S waves, but as these disturbances grow to appreciable magnitudes, three-dimensional effects occur.

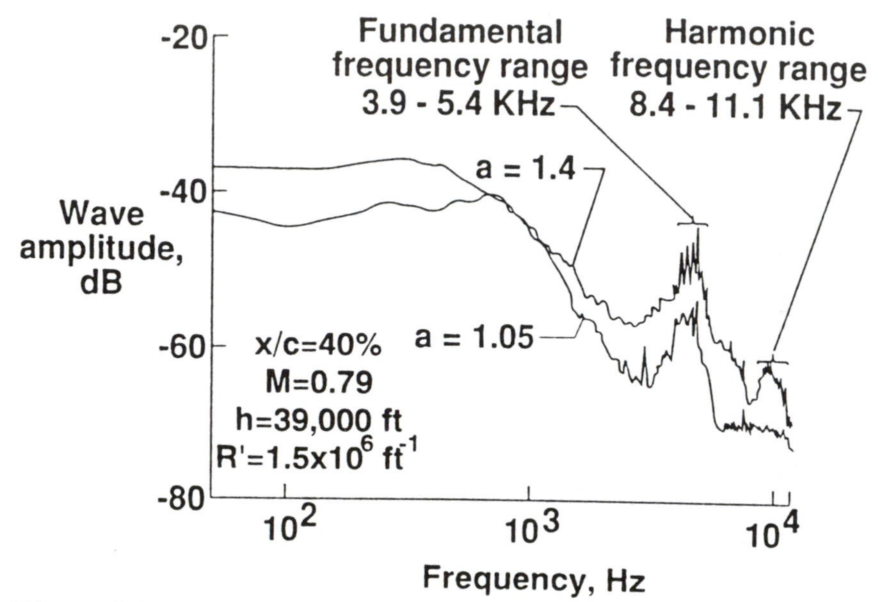


***Figure 35. Power spectral density analysis of hot-film signals at 40% chord (M=0.79, h=39,000 ft, R'=1.5x10⁶ ft⁻¹).***

Another important observation of the data in figure 35 was the indication of higher harmonics from 8350 to 11,000 Hz, following the growth and decay of the fundamental frequencies. The harmonics could only be detected by the more sensitive, higher overheat ratio sensor. Saric [75], in his vibrating ribbon experiment, showed that a higher harmonic, twice the frequency of the critical disturbance, could be generated but not predicted by linear stability theory. The ability to detect this higher harmonic from flight data is readily accomplished with PSD analysis.

Detecting T-S instability frequencies and their higher harmonics associated with transition from flight experiment data is of vital importance. The multi-element transition sensor has demonstrated the capabilities of detecting T-S disturbances in the laminar boundary layer prior to transition. Without this capability, other causes of transition such as roughness or waviness cannot be ruled out and a complete understanding of the transition phenomena is not possible.

## 4.6. Laminar Separation Sensor

The laminar separation sensor design is based on the fact that the airflow recirculates in the separated region, with flow reversal occurring in the laminar bubble that forms. The device consists of a flush array of three or more parallel films oriented perpendicular to the freestream air flow ($U_\infty$) as sketched in figure 36. The center film is electronically heated by a constant temperature anemometer (CTA) which acts as a hot-film transition sensor, showing a typical laminar or turbulent signal. The upstream and downstream films on either side of the CTA are incorporated into two legs of a differential bridge amplifier circuit for use as resistance thermometers. When the laminar separation sensor is

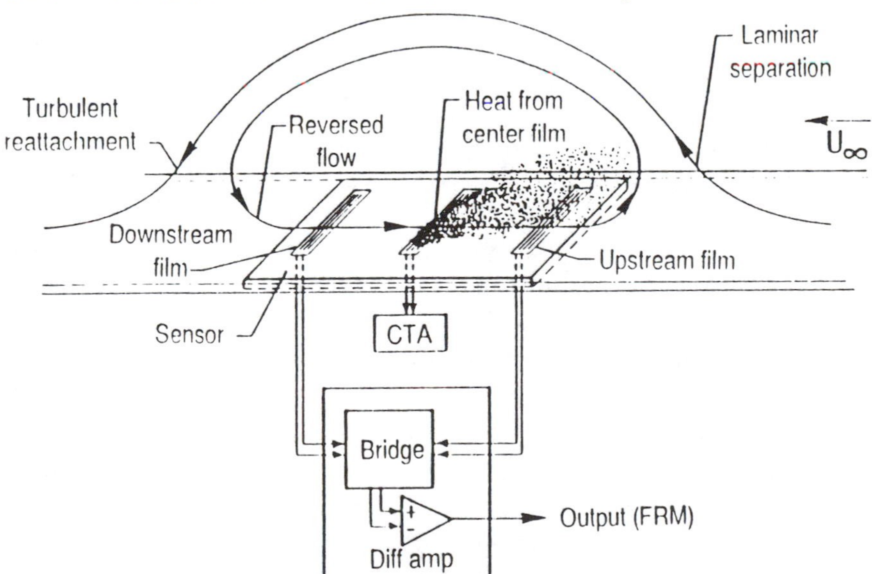


*Figure 36. Conceptual operating principle of the laminar separation sensor.*

exposed to the airflow, heat is transferred to either the upstream or downstream film depending on the direction of the air flow over the center element. The temperature difference between the upstream and downstream films, which results in film resistance changes, is measured by the bridge and high-gain differential amplifier. The direction of the air flow is then determined by the polarity of the output signal from the amplifier which is monitored using a flow reversal meter (FRM) as shown in figure 36.

The multi-element separation sensor was mounted on a six-inch chord natural laminar flow airfoil in the NASA Langley 14- by 22-Foot Wind Tunnel. The sensor, which has a total thickness of less than 0.003 inches, was applied with a spray adhesive. The polyimide substrate was extended around the airfoil leading edge so that the upper-surface boundary-layer would not be disturbed. The sensor was placed on the airfoil model such that the 48 elements were located along the airfoil surface from 22- to 70-percent chord. The elements in the sensor were selected for 0.125 inch sub-array spacing (using every other element for a given sub-array) instead of the 0.0625 inch center-to-center spacing because the closer spacing caused substrate heating contamination at very low speed test conditions. Figure 37 shows a data sample for the experiment for an airspeed of $U_{\infty}$=91 ft/sec, chord Reynolds number of 292,000, and an angle of attack of 6 degrees. The figure shows CTA and FRM signals from five sub-arrays to depict the transition process from laminar to turbulent flow and corresponding local flow directions. Note, for example, that by using the 0.125 inch spacing, element 28 was heated and elements 26 and 30 were used as the flow direction detecting elements. CTA elements numbered 1 through 28 indicated that laminar flow existed through the 50-percent chord location. FRM elements surrounding CTA elements 30 and 32 indicated reversed flow, while FRM elements surrounding element 34, located at the 56-percent chord location, indicated downstream airflow at the point of turbulent reattachment. Data obtained at slower velocities produced similar results, but with increased bubble length.

The laminar separation sensor has been shown to be a viable technique for detecting transition as well as indicating reversed flow in a laminar separation bubble. This sensor overcomes the limitations of other intrusive techniques that disturb the boundary layer, especially at high Reynolds number conditions where bubbles are small and easily disturbed.

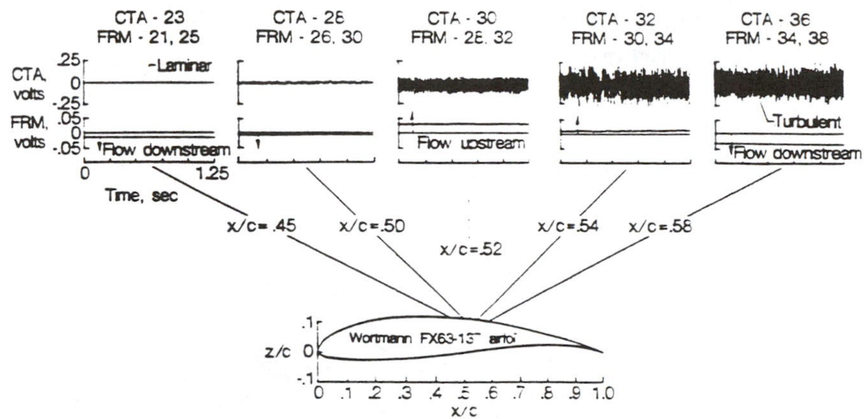


***Figure 37. Multi-element laminar separation sensor data sample ($\alpha$=6 degrees, $U_{oo}$=91 ft/sec).***

## 4.7. Crossflow Vorticity Sensor

The arrayed crossflow vorticity sensor was designed to detect the spanwise variations in local flow velocity or heat transfer that occur when streamwise vortices develop in the laminar boundary layer. The prototype sensor, shown in figure 38, is comprised of an array of closely spaced hot-film elements vapor deposited on a thin polyimide substrate film. With a closely spaced array of hot-film elements, the spanwise distribution in local flow velocity can provide information on the vortex spacing and/or frequency of crossflow disturbances. A center-to-center element spacing of 0.03 inches was chosen to provide as many as 5 elements per wavelength for future in-flight streamwise vorticity studies. Each nickel element is V-shaped in order to minimize the dependence of the sensor sensitivity to local flow direction. The geometric arrangement of the elements is intended to capture the periodic nature of the variations in heat transfer, in addition to giving such information as whether or not the vortices are moving or stationary. For moving crossflow vortices, a time series analysis of the sensor's signals would be required to provide the frequency of the most amplified crossflow disturbance. In the experiment reported herein, only steady-state sensor signals were recorded in order to show that the sensor could discriminate different convective flow regimes and thereby distinguish between disturbed and undisturbed laminar boundary layers. The experiment was designed to validate the hot-film vorticity sensor concept by determining the spatial resolution capability of the sensor in measuring laminar boundary-layer disturbances. This task was accomplished by creating lateral variations in the local flow velocity over the sensor using surface imperfections.

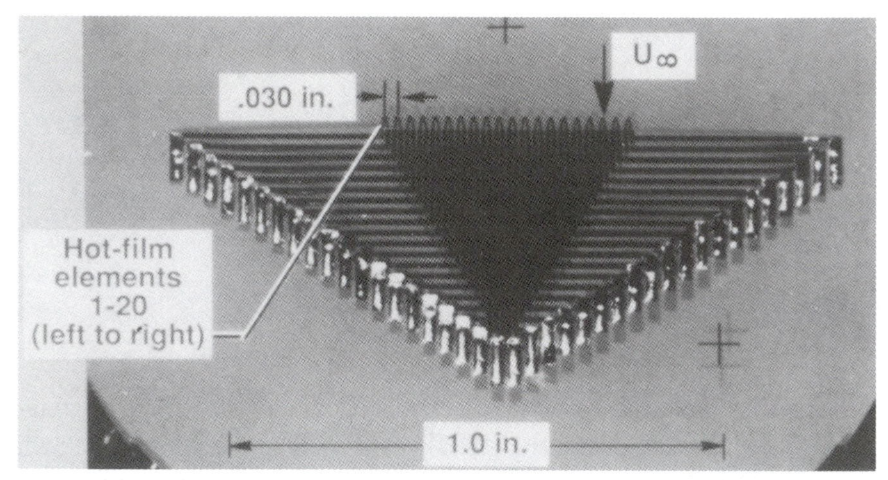


***Figure 38. Prototype arrayed crossflow hot-film sensor.***

The experiment was conducted in the 17- by 22-inch test section of the Instrument Research Division's Small Calibration Facility. The airfoil model upon which the hot-film sensor was mounted was a one foot chord section of the NASA NLF(1)-0414 natural laminar flow airfoil The vorticity experiment was conducted by using glass microspheres to generate streamwise vorticity disturbances in the laminar boundary layer, thus creating spanwise variations in the local flow velocity for the sensor to detect. Theoretical sizing criteria by Braslow [64] were used to determine that a glass microsphere 0.033 inches in diameter would create a critical roughness height suitable for investigating laminar, vortical disturbance, and turbulent boundary-layer conditions over the tunnel speed range capabilities. The glass spheres were glued to the model surface, upstream of the sensor location. A horseshoe vortex system propagated downstream from each sphere and passed over the sensor. At sufficient speed, each of these horseshoe vortex systems caused a local transition to turbulent flow, and through spanwise contamination, the turbulence would spread laterally forming a turbulent wedge as it propagated downstream. Thus, discreet placement of microspheres allowed for spacing and number of disturbances encountered by the sensor to be easily altered to create the aerodynamic simulation required for the steady-state sensor calibration. Disturbances were created using from one to four spheres placed approximately one inch upstream of the sensor. Measurements were made at tunnel speeds up to 130 ft/sec. In addition, laminar or no disturbance cases were also run for each sensor element for the tunnel speed range of interest. Flow visualization using sublimating chemicals provided detailed visual indications of the boundary-layer transition patterns for the experiment.

Figure 39 shows flow visualization and normalized RMS voltage data, respectively, for all of the odd numbered sensor elements (3, 5, 7...19) except #1, which was not operational. The data in Figure 39 are for a test condition of $U_{\infty}$=127 ft/sec and a boundary-layer roughness trip configuration of three upstream spheres. For this configuration microspheres were positioned directly upstream from elements 1, 11, and 20. The flow visualization and normalized RMS data show that elements 9 and 11 were in a fully turbulent region and elements 3, 13, and 19 were in a predominantly turbulent region, while elements 5, 7, 15 and 17 were in a region of undisturbed flow. The fully developed turbulent wedges shown by the flow visualization and detected by the hot-film sensor were as expected for this speed condition, where the boundary layer was sufficiently thin (approximately 0.03 inches thick) to allow the 0.033 inch diameter spheres to induce transition.

The results from this experiment show the arrayed crossflow sensor's ability to detect spanwise variations in a flow field, and thus detect vorticity disturbances in laminar boundary layers and to detect transition and turbulent flow. The steady-state calibration of this sensor will offer the basis for further tests of this sensor.

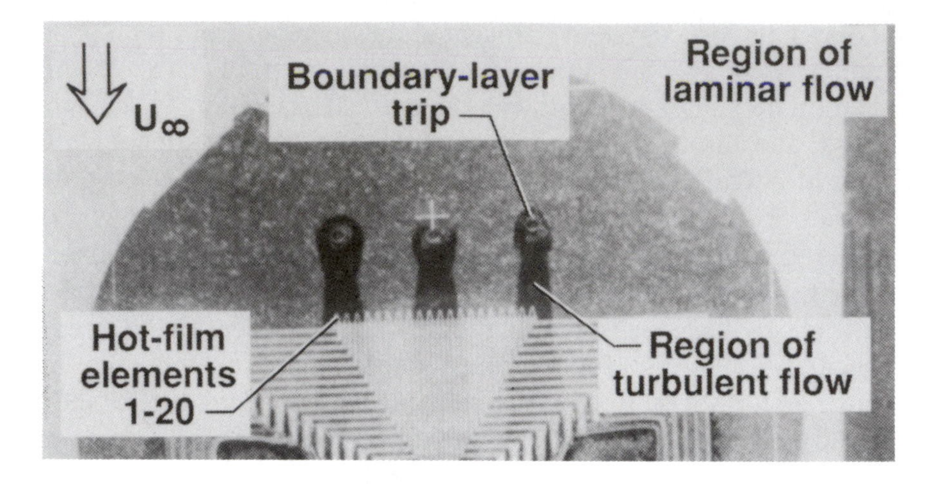


*a. Flow visualization*

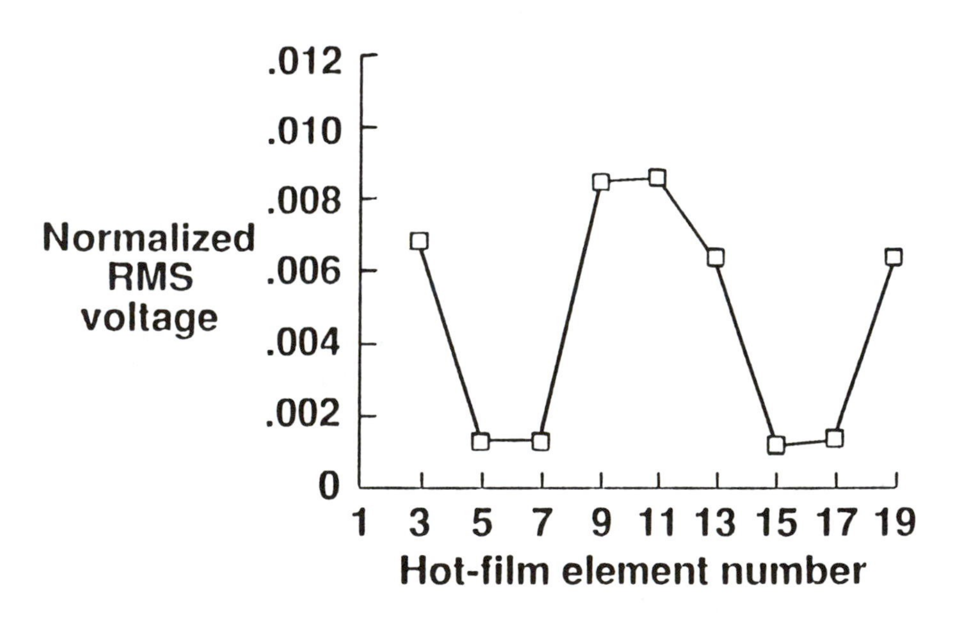


*b. Hot-film RMS output*

*Figure 39. RMS signal output for the crossflow vorticity sensor ($U_{\infty}$=127 ft/sec).*

## 5. Concluding Remarks

NLF flight experiences from the 1930's to the present have been reviewed to provide information on the achievability and maintainability of NLF in typical airplane operating environments. Significant effects of loss of laminar flow on airplane performance have been observed for several airplanes, indicating the importance of providing information on these changes to laminar flow airplane operators. Significant changes in airplane stability and control and maximum lift were observed in flight experiments with the loss of laminar flow. However, these effects can be avoided by proper selection of airfoils. Conservative laminar flow airfoil designs should be employed which do not experience loss of lift (caused by flow separation) or change in pitching moment upon the loss of laminar flow. Mechanisms have been observed for the effects of insect accumulation, flight through clouds and precipitation, and propeller slipstreams on laminar flow behavior. Fixed-transition testing, in addition to free-transition testing, is recommended as a new standard procedure for airplanes with surfaces designed to support laminar flow. With proper airfoil selection for the design flight conditions and current fabrication techniques enable the safe and practical use of NLF for viscous drag reduction.

## References

[1] Holmes, Bruce J.; Obara, Clifford J.; and Yip, Long, P.: *Natural Laminar Flow Flight Experiments on Modern Airplane Surfaces.* NASA TP-2256, 1984.

[2] Smith, F. and Higton, D.J.: *Flight Tests on "King Cobra" FZ 440 To Investigate the Practical Requirements for the Achievement of Low Profile Drag Coefficients on a "Low Drag" Aerofoil.* R. & M. No. 2375, British A.R.C., August 1945.

[3] Stuper, J.: *Investigation of Boundary Layers on an Airplane Wing in Free Flight.* NACA TM 751, 1934.

[4] Jones, M.: *Flight Experiments on the Boundary Layer.* J. Aeronaut. Sci., vol. 5, no. 3, January 1938, pp. 81-94.

[5] Stephens, A.V. and Haslam, J.A.G.: *Flight Experiments on Boundary Layer Transition in Relation to Profile Drag.* R. & M. No. 1800, British A.R.C., 1938.

[6] Young, A.D. and Morris, D.E.: *Note on Flight Tests on the Effect of Slipstream on Boundary Layer Flow.* R. & M. No. 1957, British A.R.C., 1939.

[7] Young, A.D. and Morris, D.E.: *Further Note on Flight Tests on the Effect of Slipstream on Boundary Layer Flow.* Report No. B.A. 1404b, British R.A.E., September 1939.

[8] Young, A.D.; Serby, J.E.; and Morris, D.E.: *Flight Tests on the Effect of Surface Finish on Wing Drag.* R. & M. No. 2258, British A.R.C., 1939.

[9] Goett, H.J. and Bicknell, J.: *Comparison of Profile-Drag and Boundary- Layer Measurements Obtained in Flight and in the Full-Scale Wind Tunnel.* NACA TN 693, 1939.

[10] Bicknell, Joseph: *Determination of the Profile Drag of an Airplane Wing in Flight at High Reynolds Numbers.* NACA Rep. 667, 1939.

[11] Wetmore, J.W.; Zalovcik, J.A.; and Platt, R.C.: *A Flight Investigation of the Boundary-Layer Characteristics and Profile Drag of the NACA 35-215 Laminar Flow Airfoil at High Reynolds Numbers.* NACA WR L-532, 1941. (Formerly NACA MR.)

[12] Zalovcik, J.A.: *A Profile-Drag Investigation in Flight on an Experimental Fighter-Type Airplane - The North American XP-15 (Air Corps Serial No. 41-38).* NACA ACR, November 1942.

[13] Serby, J.E.; Morgan, M.B.; and Cooper, E.R.: *Flight Tests on the Profile Drag of 14% and 25% Thick Wings.* R. & M. No. 1826, British A.R.C., 1937.

[14] Serby, J.E. and Morgan, M.B.: *Note on the Progress of Flight Experiments on Wing Drag.* Rep. No. B.A. 1360, British R.A.E., December 1936.

[15] Tani, I.: *On the Design of Airfoils in Which the Transition of the Boundary-Layer is Delayed.* NACA TM-1351, 1952.

[16] Zalovcik, J.A.: *Profile-Drag Coefficients of Conventional and Low-Drag Airfoils as Obtained in Flight.* NACA WR L-139, 1944. (Formerly NACA ACR L4E31.)

[17] Zalovcik, J.A. and Skoog, R.B.: *Flight Investigation of Boundary-Layer Transition and Profile Drag of an Experimental Low-Drag Wing Installed on a Fighter-Type Airplane.* NACA WR L-94, 1945. (Formerly NACA ACR L5C08a.)

[18] Zalovcik, J.A.: *Flight Investigation of Boundary-Layer and Profile-Drag Characteristics of Smooth Wing Sections of a P-47D Airplane.* NACA WR L-86, 1945. (Formerly NACA ACR L5H11a.)

[19] Zalovcik, J.A. and Daum, F.L.: *Flight Investigation at High Speeds of Profile Drag of Wing of a P-47D Airplane Having Production Surfaces Covered with Camouflage Paint.* NACA WR l-98, 1946. (Formerly NACA ACR L6B21.)

[20] Plascott, R.H.: *Profile Drag Measurements on Hurricane II Z.3687 Fitted With "Low Drag" Section Wings.* Rep. No. Aero. 2153, British R.A.E., September 1946.

[21] Plascott, R.H.; Higton, D.J.; Smith, F.; and Bramwell, A.R.: *Flight Tests on Hurricane II, Z.3687 Fitted With Special Wings of "Low Drag" Design.* R. & M. No. 2546, British A.R.C., August 1945.

[22] Britland, C.M.: *Determination of the Position of Boundary Layer Transition on a Specially-Prepared Section of Wing in Flight at Moderate Reynolds Number and Mach Number.* Tech. Memo. No. Aero 193, British R.A.E., September 1951.

[23] Davies, Handel: *Some Aspects of Flight Research.* J. R. Aeronaut. Soc., vol. 55, June 1951, pp. 325-361.

[24] Gray, W.E. and Davies, H.: *Note on the Maintenance of Laminar Flow Wings.* British ARC R&M No. 2485, 1952.

[25] Montoya, L.C.; Steers, L.L.; Christopher, D.; and Trujillo, B.: *F-111 TACT Natural Laminar Flow Glove Flight Results.* Advanced Aerodynamics - Selected NASA Research, NASA CP-2208, 1981, pp. 11-20.

[26] Banner, R.D.; McTigue, J.G.; and Petty, G., Jr.: *Boundary-Layer-Transition Measurements in Full-Scale Flight.* NACA RM H58E28, 1958.

[27] Bushnell, D.M. and Tuttle, M.H.: *Survey and Bibliography on Attainment of Laminar Flow Control in Air Using Pressure Gradient and Suction.* Volume I, NASA RP 1035, 1979.

[28] Obara, Clifford J.; and Holmes, Bruce J.: *Flight Measured Laminar Boundary-Layer Transition Phenomena Including Stability Theory Analysis.* NASA TP-2417, April 1985.

[29] Holmes, B.J.: *Progress in Natural Laminar Flow Research.* AIAA Paper 84-2222CP, July 1984.

[30] Vijgen, P.M.H.W.; Dodbele, S.S.; Holmes, B.J.; and van Dam, C.P.: *Effects of Compressibility on Design of Subsonic Natural Laminar Flow Fuselages.* AIAA Paper 86-1825CP, June 1986.

[31] Holmes, Bruce J.; Obara, Clifford J.; Gregorek, Gerald M.; Hoffman, Michael J.; and Freuler, Rick J.: *Flight Investigation of Natural Laminar Flow on the Bellanca Skyrocket II*. SAE Paper 830717, 1983.

[32] Hastings, E.C.; Schoenster, J.A.; Obara, C.J.; and Dodbele, S.S.: *Flight Research on Natural Laminar Flow Nacelles: A Progress Report*. AIAA Paper 86-1629, June 1986.

[33] Hastings, E.C.; Faust, G.K.; Mungur, P.; Obara, C.J.; Dodbele, S.S.; Schoenster, J.A.; and Jones, M.G.: *Status Report on a Natural Laminar Flow Nacelle Flight Experiment*. NASA CP-2487, pp. 887-921, March 1987.

[34] Swift, G. and Mungur, P.: *A Study of the Prediction of Cruise Noise and Laminar Flow Control Noise Criteria for Subsonic Air Transports*. NASA CR-159104, October 1978.

[35] Mount, J.S. and Millman, V.: *Development of an Active Laminar Flow Nacelle*. AIAA Paper 85-1116, July 1985.

[36] Viken, J.K.: *Aerodynamic Design Considerations and Theoretical Results for a High Reynolds Number Natural Laminar Flow Airfoil*. M.S. Thesis, George Washington University, January 1983.

[37] McGhee, R.J.; Viken, J.K.; Pfenninger, W.; Beasley, W.D.; and Harvey, W.D.: *Experimental Results for a Flapped Natural-Laminar-Flow Airfoil with High Lift/Drag Ratio*. NASA TM-85788, 1984.

[38] Befus, J.; Nelson, R.; Latas, J., Sr.; Ellis, D.: *Flight Test Investigations of a Wing Designed for Natural Laminar Flow*. SAE Paper 871044, April 1987.

[39] Manuel, G.S. and Doty, W.A.: *A Flight Test Investigation of Certification Requirements for Laminar-Flow General Aviation Airplanes*. AIAA Paper 90-1310, May 1990.

[40] van Dam, C.P. and Holmes B.J.: *Boundary-Layer Transition Effects on Airplane Stability and Control.* AIAA Journal of Aircraft, Volume 25, Number 8, August 1988, pp. 702-709.

[41] Holmes, B.J.; Obara, C.J.; Martin, G.L.; and Domack, C.S.: *Manufacturing Tolerances for Natural Laminar Flow Airframe Surfaces.* SAE Paper 850863, April 1985.

[42] Fage, A.: *The Smallest Size of Spanwise Surface Corrugation Which Affects Boundary Layer Transition on an Airfoil.* British ARC R&M No. 2120, 1943.

[43] Carmichael, B.H.; Whites, R.C.; and Pfenninger, W.: *Low Drag Boundary Layer Suction Experiments in Flight on the Wing Glove of an F-94A Airplane.* Northrup Aircraft Inc. Report No. NAI-57-1163 (BLC-101), 1957.

[44] Carmichael, B.H.: *Surface Waviness Criteria for Swept and Unswept Laminar Suction Wings.* Norair Report No. NOR-59-438 (BLC-123), 1959.

[45] Carmichael, B.H. and Pfenninger, W.: *Surface Imperfection Experiments on a Swept Laminar Suction Wing.* Norair Report No. NOR-59-454 (BLC-124), 1959.

[46] Schlichting, H.: Boundary Layer Theory. McGraw-Hill, 7th ed., 1979, p. 539.

[47] Nayfeh, A.H. and Ragab, S.A.: *Effect of a Bulge on the Secondary Instability of Boundary Layers.* AIAA Paper No. 87-0045, January 1987.

[48] Nayfeh, A.H.; Ragab, S.A.; and Al-Maaitah, A.A.: *Effect of Bulges on the Stability of Boundary Layers.* The Physics of Fluids, Vol. 31, No. 4, April 1988, pp. 796-806.

[49] Ragab, S.A.; Nayfeh, A.H.; and Krishna, R.C.: *Stability of Compressible Boundary Layers over Smooth Backward- and Forward-Facing Steps*, submitted for publication, The Physics of Fluids.

[50] Nayfeh, A.H.: *Stability of Compressible Boundary Layers. Transonic Symposium: Theory, Application, and Experiment*, NASA CP-3020, pp. 629-689.

[51] Atkins, P.B.: *Wing Leading Edge Contamination by Insects*. Flight Note 17, Aeronautical Research Laboratories, October. 1951.

[52] Bragg, M.B.; and Maresh, J.L.: *The Role of Airfoil Geometry in Minimizing the Effect of Insect Contamination of Laminar Flow Sections*. AIAA Paper 84-2170, 1984.

[53] Coleman, W.S.: *Roughness Due to Insects*. Boundary-Layer and Flow Control, vol. 2, Pergamon Press, 1961.

[54] Coleman, W.S.: *Wind Tunnel Experiments on the prevention of Insect Contamination by Means of Soluble Films and Liquids Over the Surface*. Report to the Boundary-Layer Control Committee, BLCC Note 39, 1952.

[55] Croom, C.C.; and Holmes, B.J.: *Flight Evaluation of an Insect Contamination Protection System for Laminar Flow Wings*. SAE Paper No. 850860, 1985,

[57] Yip, L.P.: *Wind-Tunnel Investigation of a Full Scale Canard-Configured General Aviation Airplane*. NASA TP-2382, March 1985.

[58] Hall, G. R.: *On the Mechanics of Transition Produced by Particles Passing Through an Initially Laminar Boundary Layer and the Estimated Effect on the LFC Performance of the X-21 Aircraft*. Northrup Corp., October. 1964.

[59] Hall, G. R.: *Interaction of the Wake From Bluff Bodies With an Initially Laminar Boundary Layer.* AIAA J., vol. 5, no. 8, August. 1967, pp. 1386-1392.

[60] Anon: *Final Report on LFC Aircraft Design Data Laminar Flow Control Demonstration Program.* Norair Report No. NOR-67-136, June 1967.

[61] Hood, M.J. and Gaydos, M.E.: *Effects of Propellers and of Vibration on the Extent of Laminar Flow on the NACA 27-212 Airfoil.* NACA WR L-784, October 1939. (Formerly NACA ACR)

[62] Howard, R. M.; Miley, S. J.; and Holmes, B. J.: *An Investigation of the Effects of the Propeller Slipstream on the Laminar Boundary Layer.* SAE Paper 850859, 1985.

[63] Wenzinger, C.J.: *Wind-Tunnel Investigation of Several Factors Affecting the Performance of a High-Speed Pursuit Airplane with Air-Cooled Radial Engine.* NACA ACR, November 1941.

[64] Braslow, Albert L.; and Knox, Eugene C.: *Simplified Method for Determination of Critical Height of Distributed Roughness Particles for Boundary-Layer Transition at Mach Numbers from 0 to 5.* NACA TN 4363, 1958.

[65] Holmes, J.D.: *Transition Trip Technique Study in the McAir Polysonic Wind Tunnel.* McDonnell Aircraft Company TM-4426, November 1984.

[66] Obara, C.J.: *Sublimating Chemical Technique for Boundary-Layer Flow Visualization in Flight Testing. AIAA Journal of Aircraft,* Volume 25, Number 6, June 1988, pp. 493-498.

[67] Gall, P. D.; and Holmes, B.J.: *Liquid Crystals for High-Altitude in Flight Boundary Layer Flow Visualization.* AIAA Paper No. 86-2592, 1986.

[68] Holmes, B.J.; and Obara, C.J.: *Advances in Flow Visualization Using Liquid Crystal Coatings*. SAE Paper No. 871017, 1987.

[69] Quast, A.: *Detection of Transition by Infrared Image Techniques*. Schriftliche Fassung des Vortrags auf dem XX OSTIV Kongress, Banalla, Australia, Jan. 1987.

[70] Brandon, J.M.; Manual, G.S.; Wright, R.E.; and Holmes, B.J.: *In-Flight Flow Visualization Using Infrared Imaging*. AIAA Paper No. 88-2111, 1988.

[71] Holmes, B.J.; Croom, C.C.; Gall, P.D.; Manuel, G.S.; and Carraway, D.L.: *Advanced Boundary Layer Transition Measurement Methods for Flight Applications*. AIAA Paper No. 86-9786, 1986.

[72] Manuel, G.S.; Carraway, D.L.; and Croom, C.C.: *The Laminar Separation Sensor: An Advanced Transition Measurement Method for Use in Wind Tunnels and Flight*. SAE Paper 871018, 1987.

[73] Wusk, M.S.; Carraway, D.L.; and Holmes, B.J.: *An Arrayed Hot-Film Sensor for Detection of Laminar Boundary-Layer Flow Disturbance Spatial Characteristics*. AIAA Paper 88-4677CP, September 1988.

[74] Lee, C.C.; Obara, C.J.; and Wusk, M.S.: *Flight-Measured Streamwise Disturbance Instabilities in Laminar Flow*. AIAA Paper 90-1283, 1990.

[75] Saric, W.S. and Reynolds, G.A.: *Experiments on the Stability of Nonlinear Waves in a Boundary Layer*. Laminar-Turbulent Transition Symposium, Stuttgard, Germany, 1979.

[76] Ward, D.T.; Miley, S.J.; Reininger, T.L.; and Stout, L.J.: *Flight Test Techniques for Obtaining Airfoil Pressure Distributions and Boundary Layer Transition*. AIAA Paper 83-2689, November 1983.

[77] Wentz, W.H., Jr.; Ahmed, A.; and Nyenhuis, R.: *Natural Laminar Flow Flight Experiments on a swept-Wing Business Jet.* AIAA Paper 84-2189, August 1984.

[78] Wentz, W.H., Jr.; Ahmed, A.; and Nyenhuis, R.: *Further Results of Natural Laminar Flow Flight Test Experiments.* SAE Paper 850862, 1985.

[79] Meyer, R.R.; Trujillo, B.M.; and Bartlett, D.W.: *F-14 VSTFE and Results of the Cleanup Flight Test Program.* NASA CP-2487, pp. 819-844, 1987.

[80] Boeing Commercial Aircraft Co.: *Flight Survey of the 757 Wing Noise Field and its Effect on Laminar Boundary Layer Transition.* NASA CR-178419, Vol. 3 - Extended Data Analysis, May 1988.

[81] Befus, J.; Nelson, E.R.; Ellis, D.R.; and Latas, J.: *Flight Test Investigations of a Wing Designed for Natural Laminar Flow.* SAE Paper 871044, April 1987.

[82] Horstmann, K.H.; Quast, A.; and Redeker, G.: *Flight and Wind-Tunnel Investigations on Boundary-Layer Transition.* AIAA Journal of Aircraft, Volume 27, Number 2, February 1990, pp. 146-150.

[83] George-Falvy, D.: *In Quest of the Laminar Flow Airliner: Flight Experiments on a T-33 Jet Trainer.* Ninth Hungarian Aeronautical Science Conference, November 1988.

# SUBSONIC NATURAL-LAMINAR-FLOW AIRFOILS

*Dan M. Somers*[1]

NASA Langley Research Center
Hampton, VA 23665

## 1. Historical Background

From its very beginning, the National Advisory Committee for Aeronautics (NACA) recognized the importance of airfoils as a cornerstone of aeronautical research and development. In its first Annual Report to the Congress of the United States, the NACA called for "the evolution of more efficient wing sections of practical form, embodying suitable dimensions for an economical structure, with moderate travel of the center of pressure and still affording a large range of angle of attack combined with efficient action" (Ref. 1). By 1920, the Committee had published a compendium of experimental results from various sources (Ref. 2). Shortly thereafter, the development of airfoils by the NACA was initiated at the Langley Memorial Aeronautical Laboratory (Ref. 3). The first series of airfoils, designated "M sections" for Max M. Munk, was tested in the Langley Variable-Density Tunnel (Ref. 4). This series was significant because it represented a systematic approach to airfoil development as opposed to earlier, random, cut-and-try approaches. This empirical approach, which involved modifying the geometry of an existing airfoil, culminated in the development of the four- and five-digit-series airfoils in the mid 1930's (Refs. 5–7).

Concurrently, Eastman N. Jacobs began work on laminar-flow airfoils. Inspired by discussions with B. Melvill Jones and G. I. Taylor in England, Jacobs inverted the airfoil analysis method of Theodore Theodorsen (Ref. 8) in order to determine the airfoil shape which would produce the pressure distribution he desired (decreasing pressure with distance from the leading edge over the forward portion of the airfoil). This pressure distribution, it was felt, would sustain laminar flow.

Thus, the basic idea behind modern airfoil design was conceived: the desired boundary-layer characteristics result from the pressure distribution which results from the airfoil shape. The inverse method

[1]Currently with Airfoils, Incorporated, State College, PA 16803.

mathematically transforms the pressure distribution into an airfoil shape whereas the designer intuitively/empirically transforms the boundary-layer characteristics into the pressure distribution.

The resulting 2- through 7-series airfoils, the most notable of which are the 6-series, were tested in the Langley Low-Turbulence Tunnel and the Langley Low-Turbulence Pressure Tunnel (LTPT) in the late 1930's and early 1940's (Refs. 9 and 10). In order to concentrate on high-speed aerodynamics, the NACA got out of the airfoil business in the 1950's, leaving the world with a large number of systematically designed and experimentally tested airfoils (Ref. 11). The four- and five-digit-series turbulent-flow airfoils produced relatively high maximum lift coefficients although their drag coefficients were not particularly low whereas the 6-series laminar-flow airfoils offered the possibility of low drag coefficients although their maximum lift coefficients were not especially high. The quandary faced by the aircraft designers of the day over the type of airfoil to select, laminar- or turbulent-flow, was solved by the available construction techniques which produced surfaces that were insufficiently smooth and rigid to support extensive laminar flow.

The airfoil scene then shifted to Germany where F. X. Wortmann and Richard Eppler were engaged in laminar-flow airfoil design. Wortmann employed singularity and integral boundary-layer methods (Refs. 12 and 13) to develop a catalog of airfoils intended primarily for sailplanes (Ref. 14). Because the theoretical methods he used were relatively crude, however, final evaluation of the airfoils was performed in a low-turbulence wind tunnel. Eppler, on the other hand, pursued the development of more accurate theoretical methods (Refs. 15 and 16).

The successor to the NACA, the National Aeronautics and Space Administration (NASA), reentered the airfoil field in the 1960's with the design of the supercritical airfoils by Richard T. Whitcomb (Ref. 17). The lessons learned during the development of these transonic airfoils were transferred to the design of a series of turbulent-flow airfoils for low-speed aircraft. The basic objective of this series of airfoils was to achieve higher maximum lift coefficients than the earlier NACA airfoils. It was assumed that the flow over these airfoils would be turbulent because of the construction techniques then in use by general-aviation manufacturers. While these NASA turbulent-flow airfoils (Ref. 18) did achieve higher maximum lift coefficients, the cruise drag coefficients were no lower than those of the NACA four- and five-digit-series airfoils. Emphasis was therefore shifted toward natural-laminar-flow (NLF) airfoils in an attempt to combine the low-drag characteristics of the NACA 6-series airfoils with the high-lift characteristics of the NASA low-speed airfoils. In this context, the term 'natural-laminar-flow airfoil' refers to an airfoil which can achieve significant extents of laminar flow ($\geq$ 30-percent chord) on both the upper and lower surfaces simultaneously, solely through favorable pressure gradients (no boundary-layer suction or cooling).

Finally, the advent of composite structures (Ref. 19) has also fueled the resurgence in NLF research. This construction technique allows NLF airfoils to achieve, in flight, the low-drag characteristics measured in low-turbulence wind tunnels (Ref. 20).

## 2. Philosophy

At this point, it is important to emphasize that the goal of our effort was *not* the design of a series of natural-laminar-flow airfoils for low-speed applications. Rather we adopted the philosophy of Richard Eppler which is to develop a (theoretical) method and verify it such that others can use the method to design airfoils for their own specific applications. The key to the success of this philosophy is the verification of the method. Our objective then was to accomplish this verification through selective testing of various airfoil concepts in the Langley Low-Turbulence Pressure Tunnel.

Note that this philosophy is contrary to the approach taken by the NACA and F. X. Wortmann. They developed catalogs from which aircraft designers could select airfoils for their proposed vehicles. This approach was necessary for the NACA because the theoretical methods of the day were too primitive to predict accurately the aerodynamic characteristics of an airfoil. The use of catalogs has been successful, however, because the applications for which the airfoils were used were indeed those for which the airfoils were intended. In addition, the section characteristics were painstakingly measured in good, low-turbulence wind tunnels at the appropriate Reynolds numbers. Thus, the airfoils, their measured characteristics, and their applications coincided well. But, as applications have become more diverse, the older airfoils and the measured characteristics have become less appropriate. Today, with applications ranging from high-altitude vehicles to commuter aircraft, the use of airfoils designed for aircraft having Reynolds numbers of 3 to 9 $\times$ $10^6$ and relatively low lift coefficients is unacceptable.

The catalogs also suffer from a lack of coverage. Each aircraft design requires a specific performance from the airfoil. If this performance falls within the range of characteristics contained in a catalog, an airfoil can be selected from that catalog for the given aircraft. More than likely, however, this airfoil will still represent a compromise because its characteristics do not match exactly those of the aircraft.

A related advantage of the theoretical airfoil design method is that it allows many different concepts to be explored economically. Such efforts are generally impractical in wind tunnels because of time and money constraints.

Thus, the need for a theoretical airfoil design method is threefold: first, for the design of airfoils which fall outside the range of applicability of existing catalogs; second, for the design of airfoils which more exactly match the requirements of the intended application; and third, for the economic exploration of many airfoil concepts.

The ultimate acceptance of this philosophy faces one final hurdle which can be summed up by the following saying:

> No one believes the theory except the one who developed it.
>
> Everyone believes the experiment except the one who ran it.

This hurdle can be overcome by a rigorous verification of the method.

## 3. Theory

In 1975, we began working with the Eppler Airfoil Design and Analysis Program (Refs. 21 and 22). This program contains a conformal-mapping method for the design of airfoils with prescribed velocity-distribution characteristics, a panel method for the analysis of the potential flow about given airfoils, and a boundary-layer method. With this program, airfoils with prescribed boundary-layer characteristics can be designed and airfoils with prescribed shapes can be analyzed.

In all other inverse methods, the velocity distribution is specified at one angle of attack and the airfoil shape which will produce that velocity distribution is computed. Thus, the airfoil is designed at a single point. All other conditions are considered 'off-design' and must be taken into account intuitively and analyzed later to determine acceptability.

The conformal-mapping method in the Eppler program is unique because it allows the velocity distribution to be specified along different segments of the airfoil *at different angles of attack.* This is an extremely powerful capability because it allows the important features of many velocity distributions to be incorporated into the airfoil design from the outset. Thus, the airfoil is designed at several points simultaneously and the off-design conditions can be taken into account in the initial specification.

The following sketch helps to illustrate, in a very simplified way, the use of this capability.

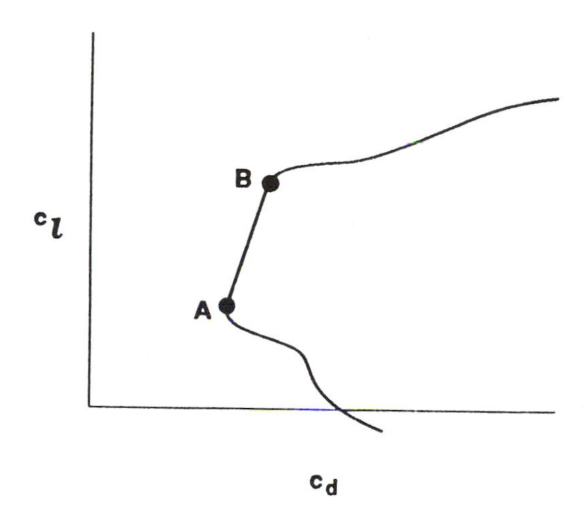


As determined from the desired aircraft performance, the airfoil must produce low drag over the range of lift coefficients from points A to B. Point A corresponds to the lift coefficient below which the transition point moves rapidly forward along the lower surface. Thus, for this point, the design of the lower surface is critical. Point B corresponds to the lift coefficient above which the transition point moves rapidly forward along the upper surface. For this point, the upper surface is critical. Using conventional design methods, the velocity distribution must be specified at point A or point B or, possibly, some intermediate point. With the Eppler program, however, the velocity distribution along the lower surface at point A is specified as is the distribution along the upper surface at point B.

It should be noted that the actual, absolute velocity distributions are not specified in the method, only the velocity gradients. For details, see Reference 21.

The panel method allows the velocity distribution about a given airfoil to be computed. This is obviously required for the analysis of specified airfoil shapes but, with respect to airfoil design, it is also necessary for determining the effect of a simple flap deflection on the velocity distribution. This method employs third-order panels with vorticity distributed parabolically along each panel.

An integral method is used for the prediction of the boundary-layer development for each velocity distribution. The method can predict laminar and turbulent boundary layers, transition, and separation, both laminar and turbulent. The occurrence of significant (drag-producing) laminar separation bubbles is also indicated. The method is semi-empirical and contains a boundary-layer displacement iteration.

An important feature of the Eppler program is the connection between the boundary-layer method and the conformal-mapping method. This connection allows the boundary-layer characteristics to be controlled directly during the airfoil design process. This is a particularly significant capability for the design of laminar-flow airfoils and represents a major step forward from the procedure used for the design of the NACA laminar-flow airfoils. Now, instead of intuitively or empirically transforming the desired boundary-layer characteristics into a velocity distribution, the designer can determine directly the modifications to the velocity distribution which will produce the desired boundary-layer development at any given angle of attack. See Reference 23.

## 4. Verification

To verify the theory, several airfoils have been designed using the Eppler program and the majority have been tested in the Langley Low-Turbulence Pressure Tunnel (Refs. 24 and 25). Four of these airfoils will be used to illustrate the process of verification.

### NLF(1)-0416 Airfoil

In 1977, the NASA NLF(1)-0416 airfoil (Ref. 26) was designed for advanced, light, single-engine, general-aviation aircraft. For this application, low profile-drag coefficients at a Reynolds number of about $4 \times 10^6$ are desirable for the cruise lift coefficient ($c_l = 0.4$) as well as for the climb lift coefficients ($c_l = 0.5$ to $1.0$).

Two specific objectives were identified for this airfoil. First, the airfoil should produce a maximum lift coefficient at a Reynolds number of $3 \times 10^6$, which is at least as high as that of the best available NASA turbulent-flow airfoil (i.e., $c_{l,\mathrm{max}} \geq 1.76$). An important requirement related to this objective was that $c_{l,\mathrm{max}}$ not decrease with transition fixed near the leading edge. In other words, the maximum lift coefficient cannot depend on the achievement of laminar flow. Thus, if the leading edge of the wing is contaminated by insect remains or other matter, the maximum lift coefficient should not decrease. This requirement stems from safety concerns relating to stall and, accordingly, landing speeds. This characteristic was also desirable to counter the commonly-held belief that the maximum lift coefficients of all laminar-flow airfoils are significantly decreased by roughness. The second objective for this airfoil was to obtain a cruise profile-drag coefficient similar to those achieved by comparable NACA 6-series airfoils. In addition, a profile-drag coefficient lower than those typical of comparable laminar- or turbulent-flow airfoils was desired for $c_l = 1.0$.

To further define the airfoil, four constraints were placed on the design. First, the extent of the favorable pressure gradient on the upper surface at the cruise lift coefficient was limited to 30-percent chord. This constraint grew out of skepticism concerning the achievement of extensive laminar flow in practice. Second, the airfoil thickness must be greater than 12-percent chord for structural reasons. Third, the zero-lift, quarter-chord pitching-moment coefficient should be no more negative than that of the NASA LS(1)-0413 airfoil (i.e., $c_{m,0} \geq -0.10$). This constraint arose from criticism by the general-aviation industry of the large pitching-moment coefficients of the NASA low-speed, turbulent-flow airfoils. Finally, the airfoil should not have a flap in deference to simplicity.

The design specifications for this first test case for the theory can be summarized as follows: limited laminar flow, moderate Reynolds numbers, incompressible, and unflapped or, in other words, generally conservative.

The NLF(1)-0416 airfoil shape and selected theoretical (inviscid and incompressible) velocity distributions are shown in Figure 1. The most notable features of the airfoil shape are its relatively blunt leading edge and aft camber. Both these features result from the quest for a high maximum lift coefficient. Their influence on the velocity distributions is evidenced by the relatively low suction peaks at the leading edge at high angles of attack and the aft loading.

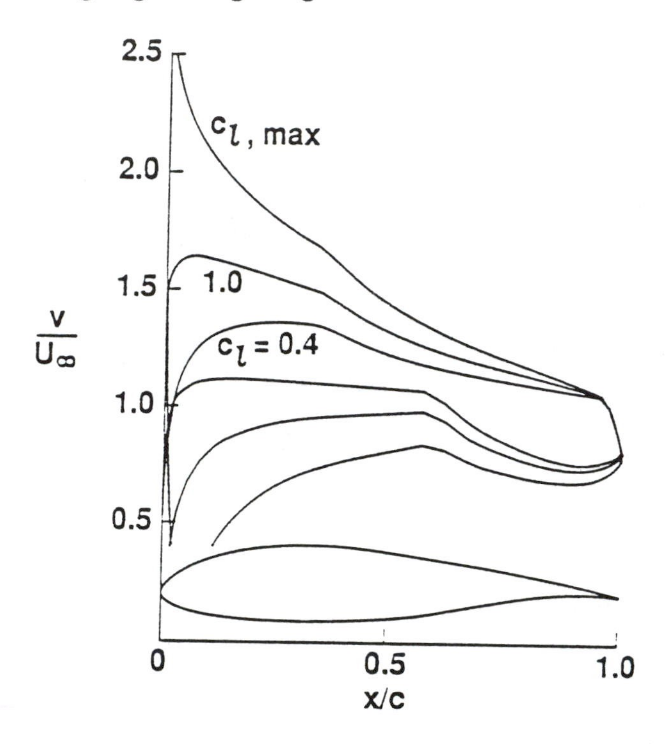


Figure 1.- NLF(1)-0416 airfoil shape and inviscid, incompressible velocity distributions.

The effect of roughness on the measured section characteristics for a Reynolds number of $2 \times 10^6$ is shown in Figure 2. Of note is the fact that the maximum lift coefficient is completely unaffected by roughness, as intended. This result can be traced to the movement of the transition point on the upper surface as shown in Figure 3. As $c_{l,max}$ is approached, the transition point moves steadily toward the leading edge. At $c_{l,max}$, the flow along the entire upper surface is turbulent. Thus, the addition of leading-edge roughness has essentially no effect on the boundary-layer development along the upper surface.

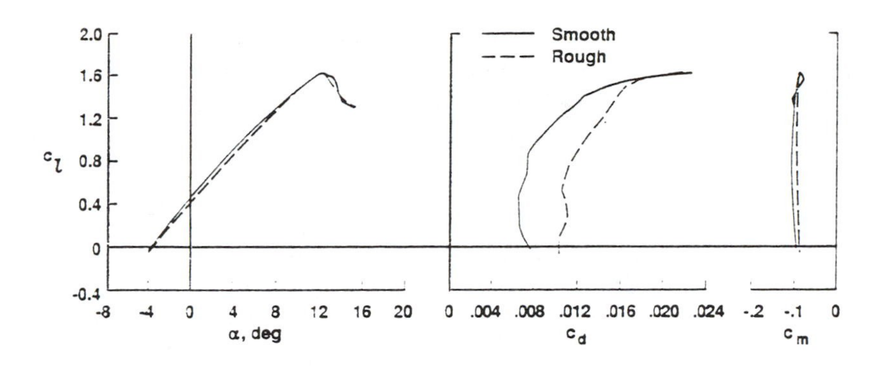


Figure 2.- Effect of roughness on section characteristics of NLF(1)-0416 airfoil for $R = 2 \times 10^6$.

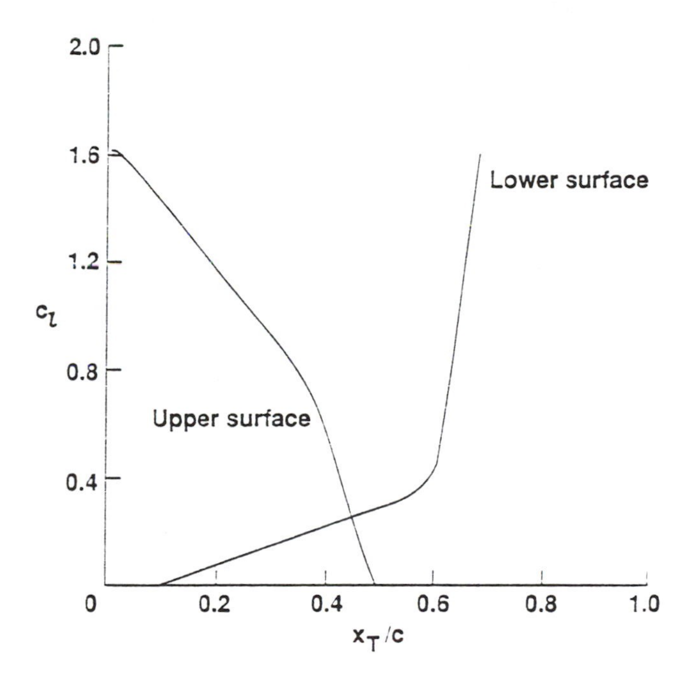


Figure 3.- Transition location on NLF(1)-0416 airfoil for $R = 2 \times 10^6$.

The theoretical and experimental pressure distributions are compared in Figure 4. The theoretical pressure distributions are inviscid and incompressible whereas the experimental pressure distributions were obtained at $R = 4 \times 10^6$ and $M = 0.10$. Although the magnitudes of the pressure coefficients at $c_l = 0.45$ (roughly the cruise $c_l$) do not match exactly, the pressure gradients agree well. This is important because the accurate prediction of the boundary-layer development depends primarily on the pressure gradients. At $c_l = 1.00$, the decambering viscous effects are more apparent and the disparities include small differences in the pressure gradients as well as larger differences in the magnitudes of the pressure coefficients. At $c_{l,\mathrm{max}}$, the agreement is poor primarily because of the upper-surface, trailing-edge separation which is not modeled in the pressure distributions predicted by the theory.

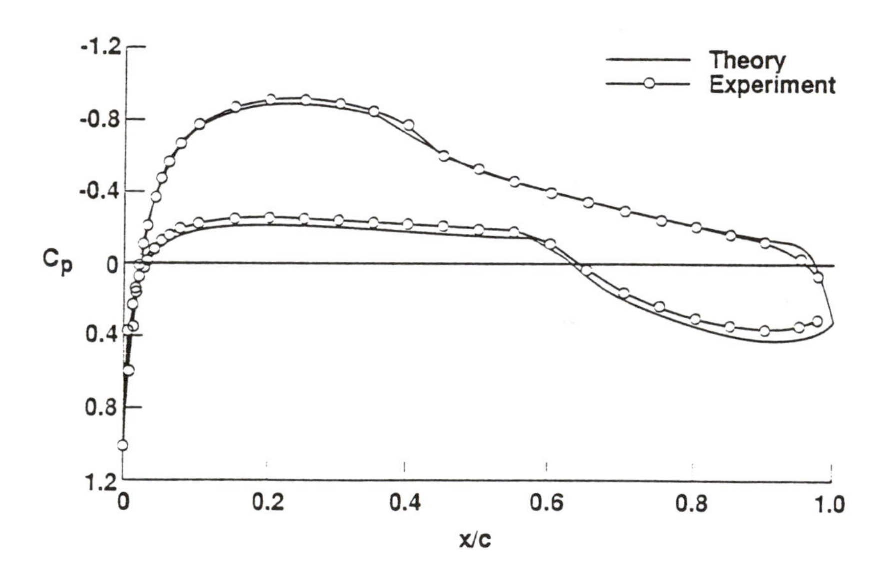


(a) $c_l = 0.45$.

Figure 4.- Comparisons of theoretical and experimental pressure distributions of NLF(1)-0416 airfoil.

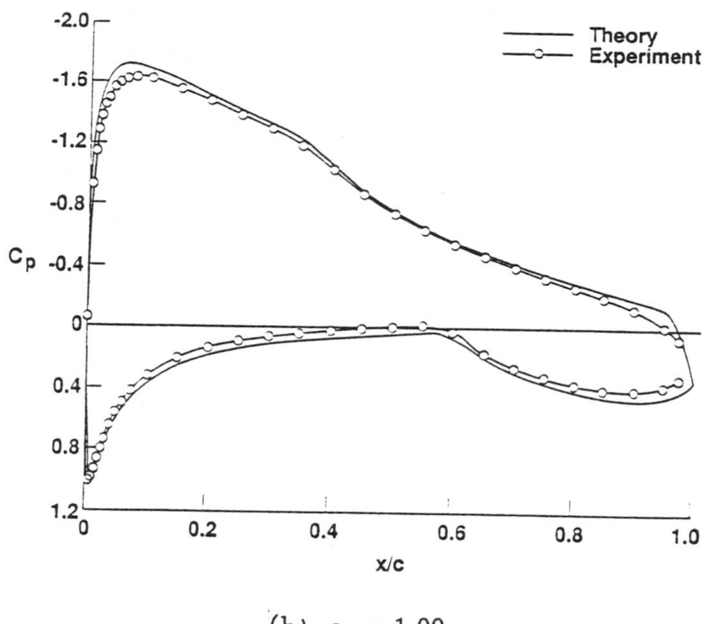


(b) $c_l = 1.00$.

Figure 4.-Continued.

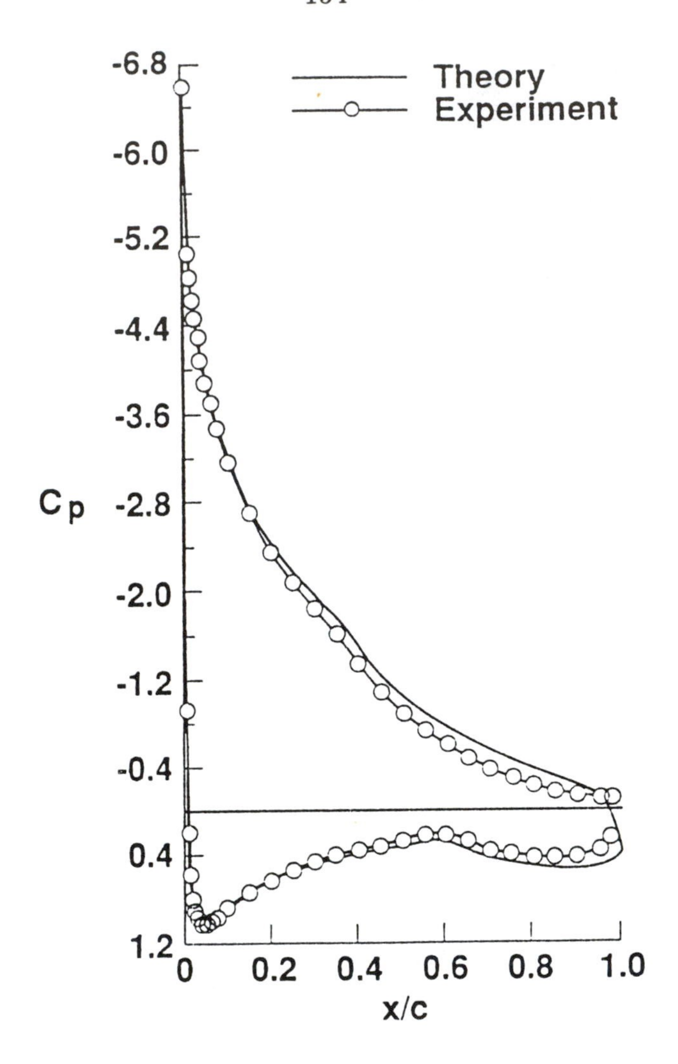


(c) $c_{l,\max}$.

Figure 4.- Concluded.

The predicted and measured transition locations for $R = 2 \times 10^6$ are compared in Figure 5. The theory consistently predicts transition upstream of the locations measured in the wind tunnel. This result is obtained because the method 'defines' the transition location as the end of the laminar boundary layer whether due to natural transition or laminar separation. In the wind tunnel, transition can normally be confirmed only by the observation of attached turbulent flow. Thus, as conditions change to produce shorter laminar separation bubbles (higher lift coefficients for the upper surface and lower lift coefficients for the lower surface and/or higher Reynolds numbers), the agreement between theory and experiment improves.

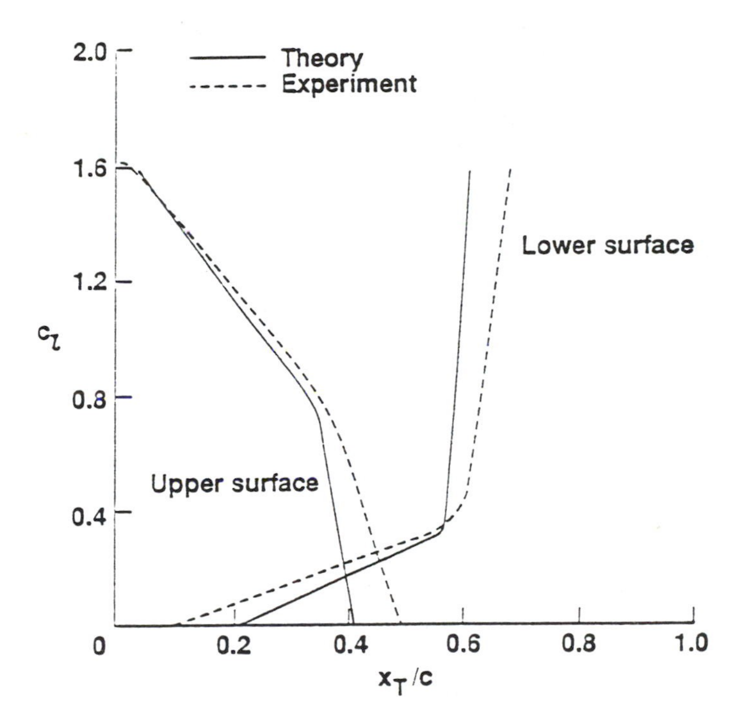


Figure 5.- Comparison of theoretical and experimental transition locations on NLF(1)-0416 airfoil for $R = 2 \times 10^6$.

The comparison of theoretical and experimental section characteristics for $R = 2 \times 10^6$ is shown in Figure 6. The lift-curve slope and maximum lift coefficient are predicted quite accurately. The zero-lift angle of attack and the pitching-moment coefficients are, however, overpredicted because the method did not contain a boundary-layer displacement iteration at the time. The agreement between the predicted and measured drag coefficients is very good.

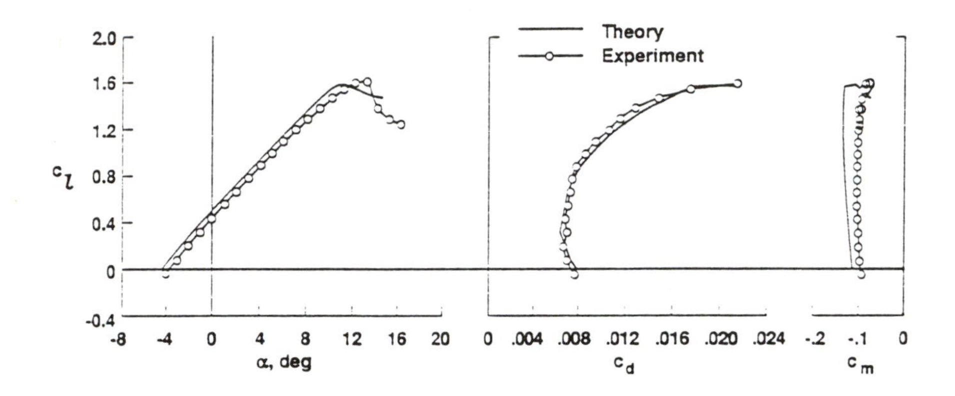


Figure 6.- Comparison of theoretical and experimental section characteristics of NLF(1)-0416 airfoil for $R = 2 \times 10^6$.

In summary, the first test case showed that the method was reasonably accurate. The most significant discrepancies resulted from the boundary-layer displacement effect.

The overall success of the first test case was somewhat fortuitous because we had selected a relatively benign application, where the Reynolds number was not so high that laminar flow was extremely difficult to achieve or so low that laminar separation bubbles had a dominant influence on the flow. Based on this success, the next obvious step was to explore more extreme operating conditions—and more extensive laminar flow.

NLF(1)-0215F Airfoil

In 1979, the second NASA natural-laminar-flow airfoil, the NLF(1)-0215F (Ref. 27), was designed for a high-performance, single-engine, general-aviation airplane—specifically, the Bellanca Skyrocket II (Ref. 28). The overall goal was to design and fabricate an advanced-composite (graphite- and aramid-fiber) wing to replace the fiberglass, 6-series wing of the Skyrocket for a flight test program. This project never came to fruition and the NLF(1)-0215F airfoil remains the sole legacy of the effort. This application requires low profile-drag coefficients at a Reynolds number of about $9 \times 10^6$ for the cruise lift coefficient ($c_l = 0.2$) as well as for the climb lift coefficients ($c_l = 0.5$ to $1.0$).

Two primary objectives were identified for this airfoil. First, the airfoil should produce a maximum lift coefficient at $R = 3 \times 10^6$, which is comparable to those of the NASA low-speed, turbulent-flow airfoils. The same requirement that $c_{l,\mathrm{max}}$ not decrease with transition fixed near the leading edge applies to this airfoil as it did to the NLF(1)-0416 airfoil. The second objective was to obtain low profile-drag coefficients from the cruise lift coefficient of 0.2 to about 1.0.

Three constraints were placed on this airfoil design in order to make it compatible with the Skyrocket. First, the airfoil thickness must be 15-percent chord. Second, the pitching-moment coefficient should be no more negative than $-0.05$ at the cruise lift coefficient ($c_l = 0.2$). Third, the airfoil must incorporate a simple flap having a chord equal to 25 percent of the airfoil chord.

The above objectives and constraints are not sufficient to design the airfoil, however, primarily because of the variables introduced by the flap and the unconstrained extents of laminar flow on the upper and lower surfaces. To evaluate the effects of these variables, it is necessary to examine the design goals with respect to overall aircraft performance. For this design, the primary goal is the reduction of wing parasite drag. This goal can be achieved in a number of ways, two of which seemed appropriate to this application. First, if a higher maximum lift coefficient can be achieved, the wing area can be reduced relative to that of a wing with a lower $c_{l,\mathrm{max}}$. This conclusion is based on the assumption that both aircraft must achieve the same minimum speed. Second, if the extent of laminar flow on one or both surfaces can be increased, the profile-drag coefficient will be reduced.

Further analysis indicated that by maximizing $c_{l,max}/c_{d,cruise}$, the wing parasite drag would be minimized. Unfortunately, a high maximum lift coefficient and a low profile-drag coefficient are generally conflicting goals because, as the extent of laminar flow on the upper surface increases, the pressure gradient over the aft portion of the upper surface steepens thereby decreasing the maximum lift coefficient. By trial and error, it was determined that $c_{l,max}/c_{d,cruise}$ would be maximized *for this application* if the extent of laminar flow was about 40-percent chord on the upper surface and about 60-percent chord on the lower surface.

The effect of the flap on the design can be evaluated by examining the constraint on the pitching-moment coefficient ($c_{m,cruise} \geq -0.05$). The objective of a high maximum lift coefficient is in conflict with the pitching-moment constraint. For this design, the flap can be used to alleviate this conflict by employing negative (up) flap deflections. This concept allows an airfoil to be designed which has a fairly large amount of camber (conducive to a high $c_{l,max}$) but retains the ability to achieve a low pitching-moment coefficient at the cruise lift coefficient. This concept has the important additional advantage that, by deflecting the flap up and down, the low-drag range is shifted to lower and higher lift coefficients, respectively. (See Ref. 29.) Based on experience with other airfoils, the negative flap deflection was limited to $-10°$.

The design specifications for this second test case for the theory can be summarized, relative to those for the NLF(1)-0416 airfoil, as follows: more extensive laminar flow, higher Reynolds numbers, incompressible, and flapped or, in other words, less conservative.

The NLF(1)-0215F airfoil shape with various flap deflections and selected theoretical (inviscid and incompressible) velocity distributions are shown in Figure 7. Of particular interest is the velocity distribution with a flap deflection of $-10°$ at the cruise lift coefficient ($c_l = 0.2$). Along the lower surface, the velocity gradient is initially adverse, then zero, and then increasingly favorable. Thus, transition is imminent over the entire forward portion of the lower surface. This concept was originated by Richard Eppler (Ref. 30).

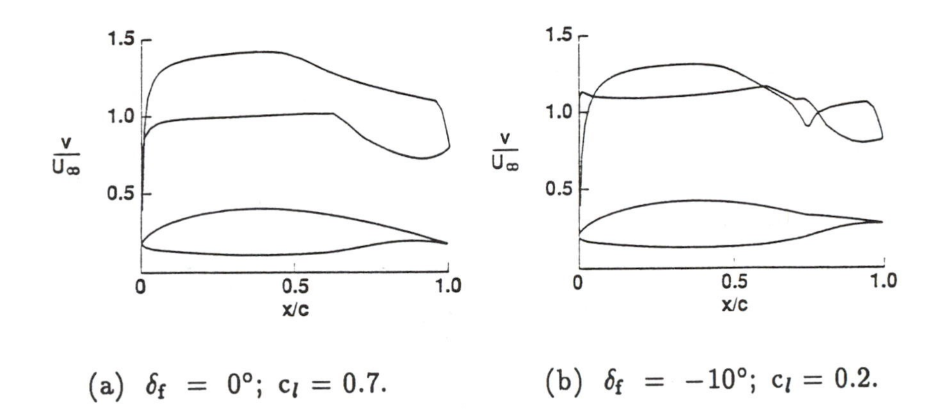


(a) $\delta_f = 0°$; $c_l = 0.7$.  (b) $\delta_f = -10°$; $c_l = 0.2$.

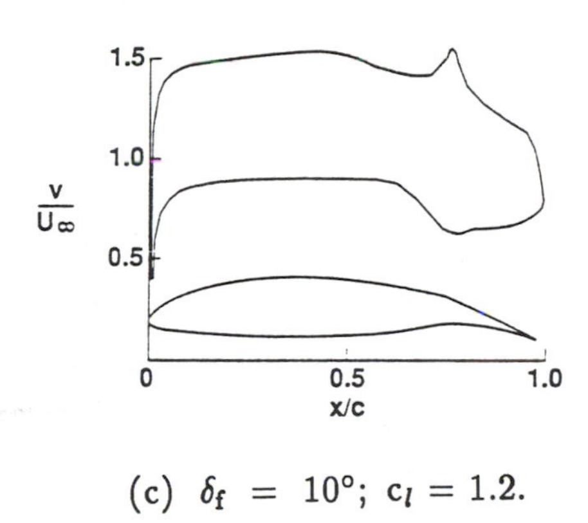


(c) $\delta_f = 10°$; $c_l = 1.2$.

Figure 7.- NLF(1)-0215F airfoil shape and inviscid, incompressible velocity distributions.

The comparisons of theoretical and experimental section characteristics for R = 3, 6, and 9 $\times$ $10^6$ are shown in Figure 8. The lift-curve slopes are again predicted well. The agreement between the predicted and measured drag coefficients becomes poorer as Reynolds number increases probably because the transition criterion is too conservative. This is corroborated by the fact that the transition criterion is based on numerous wind-tunnel experiments which were probably conducted with turbulence levels higher than those found in the Low-Turbulence Pressure Tunnel or in flight.

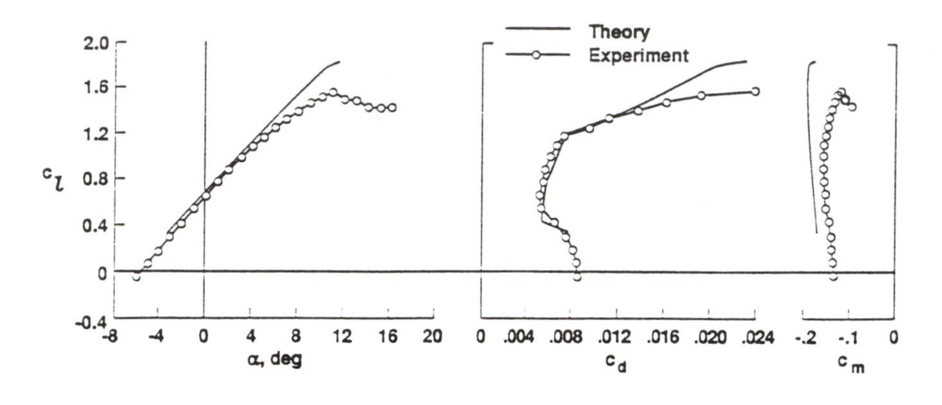


(a) $R = 3 \times 10^6$.

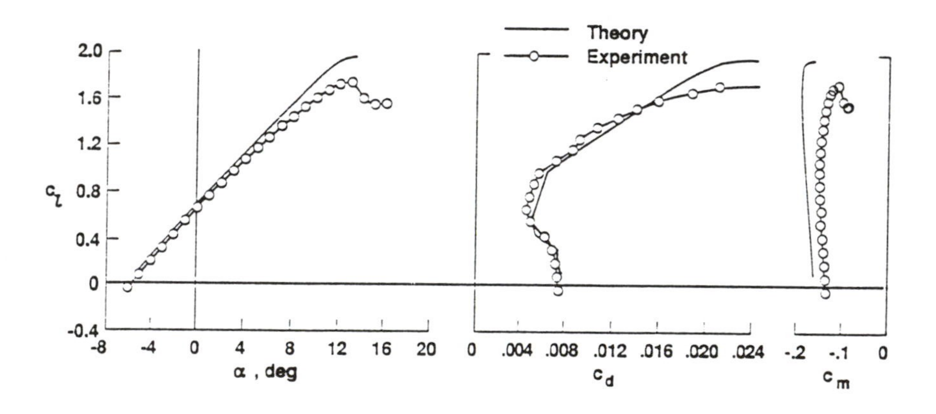


(b) $R = 6 \times 10^6$.

Figure 8.- Comparisons of theoretical and experimental section characteristics of NLF(1)-0215F airfoil with $\delta_f = 0°$.

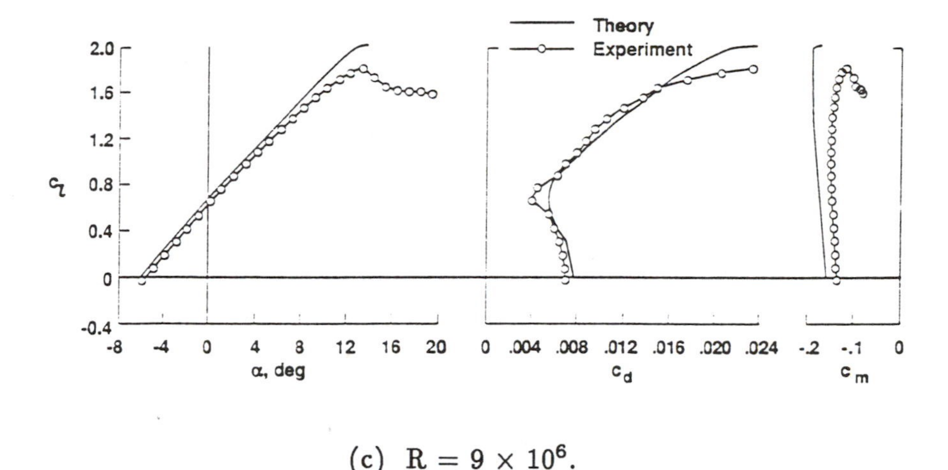


(c) $R = 9 \times 10^6$.

Figure 8.- Concluded.

In summary, the second test case confirmed that the method was relatively accurate and expanded the range of applicability to higher Reynolds numbers and simple flap deflections. The most significant discrepancies resulted from the conservative transition criterion and the previously-mentioned boundary-layer displacement effect. In addition, this test case saw the introduction of a figure of merit to guide the development of a particular airfoil concept.

To improve the prediction of zero-lift angles and pitching-moment coefficients, a boundary-layer displacement iteration was incorporated into the method (Ref. 22). Initially, the scheme was limited to a single iteration although the latest version of the program allows multiple iterations.

The next step in the development of the method was to extend the range of applicability to even higher Reynolds numbers. For most aircraft applications, higher Reynolds numbers are usually accompanied by higher Mach numbers. Once the free-stream Mach number exceeds 0.4 or so, compressibility effects can no longer be ignored at cruise lift coefficients. Accordingly, a compressibility correction (Ref. 31) to the velocity (and pressure) distributions was incorporated into the method by Richard Eppler and the author. The correction is valid as long as the local flow is not supersonic.

The boundary-layer analysis remains incompressible, which is justifiable because the effect of compressibility on the boundary layer is of second order for subsonic Mach numbers.

## NLF(2)-0415 Airfoil

In 1985, the NLF(2)-0415 airfoil (Ref. 32) was designed by Karl-Heinz Horstmann and the author as part of a cooperative agreement between NASA Langley Research Center and DFVLR, Braunschweig, Federal Republic of Germany. The airfoil is intended for an advanced-technology, 30-passenger, commuter aircraft. This application requires low profile-drag coefficients for the climb lift coefficient ($c_l = 0.38$) at a Reynolds number of $18 \times 10^6$ and a Mach number of 0.38 as well as for the cruise lift coefficients ($c_l = 0.30$ to 0.36) at lower Reynolds numbers ($R = 11$ to $17 \times 10^6$) and higher Mach numbers ($M \leq 0.67$). Detailed objectives and constraints were provided by Louis J. Williams of NASA Langley Research Center. Based on experience, it was felt that the sea-level, climb condition ($R = 18 \times 10^6$ and $M = 0.38$) represented the most difficult condition to satisfy with respect to the achievement of extensive laminar flow.

The design specifications for this third test case can be summarized, relative to the first two test cases, as follows: very extensive laminar flow, much higher Reynolds numbers, and compressible or, in other words, very unconservative.

The NLF(2)-0415 airfoil shape and selected theoretical (inviscid) velocity distributions are shown in Figure 9.

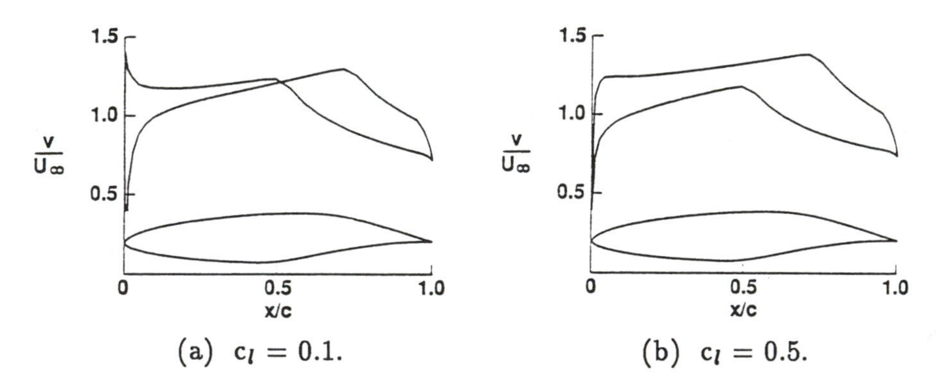


(a) $c_l = 0.1$. (b) $c_l = 0.5$.

Figure 9.- NLF(2)-0415 airfoil shape and inviscid velocity distributions for $M = 0.38$.

This design exhibits the unusual characteristic that transition is imminent over the entire forward portion of the upper and lower surfaces at the upper and lower limits of the low-drag range, respectively. This characteristic is illustrated in Figure 10 which shows the boundary-layer development along the lower surface at the lower limit of the low-drag range for the sea-level, climb condition, which corresponds to the velocity distribution in Figure 9(a) and the inset. In Figure 10, the Reynolds number based on momentum thickness is plotted against the shape factor based on energy and momentum thicknesses. Note that, although $H_{32}$ is a function only of the more conventional $H_{12}$ which contains displacement instead of energy thickness, it has the opposite tendency. Thus, $H_{32}$ decreases toward separation. Certain $H_{32}$ values correspond to specific, laminar boundary-layer phenomena. An $H_{32}$ of 1.620 corresponds to stagnation; 1.573, to the Blasius (flat-plate) boundary layer; and 1.515, to laminar separation. While these values are relatively precise, the value for turbulent separation, 1.46, is less so. The transition-criterion curve should also be considered a band because it is merely a fairing through a 'cloud' of wind-tunnel and flight data. Thus, the shape factor for laminar separation, 1.515, and the transition-criterion curve together form a boundary for the laminar boundary-layer development. If this boundary is reached, the method switches from the laminar to the turbulent boundary-layer equations. Also note that, because the scale for the Reynolds number based on momentum thickness is logarithmic, the boundary-layer development is 'expanded' near the leading edge and increasingly 'compressed' downstream. (Additional details are given in Ref. 21.) Thus, the boundary-layer development begins in the lower right portion of Figure 10 with a stagnation point (point A). The steep adverse gradient immediately downstream of the velocity peak near the leading edge (point B) nearly causes laminar separation. The boundary-layer development then essentially follows the transition-criterion curve. The Blasius solution occurs at point C. The beginning of the pressure recovery (point D) causes the transition criterion to be satisfied which, in turn, invokes the turbulent equations. This then represents an extreme application of the transition concept previously exploited on the lower surface of the NLF(1)-0215F airfoil. At the high Reynolds numbers of this design, this concept allows a wide low-drag range to be achieved.

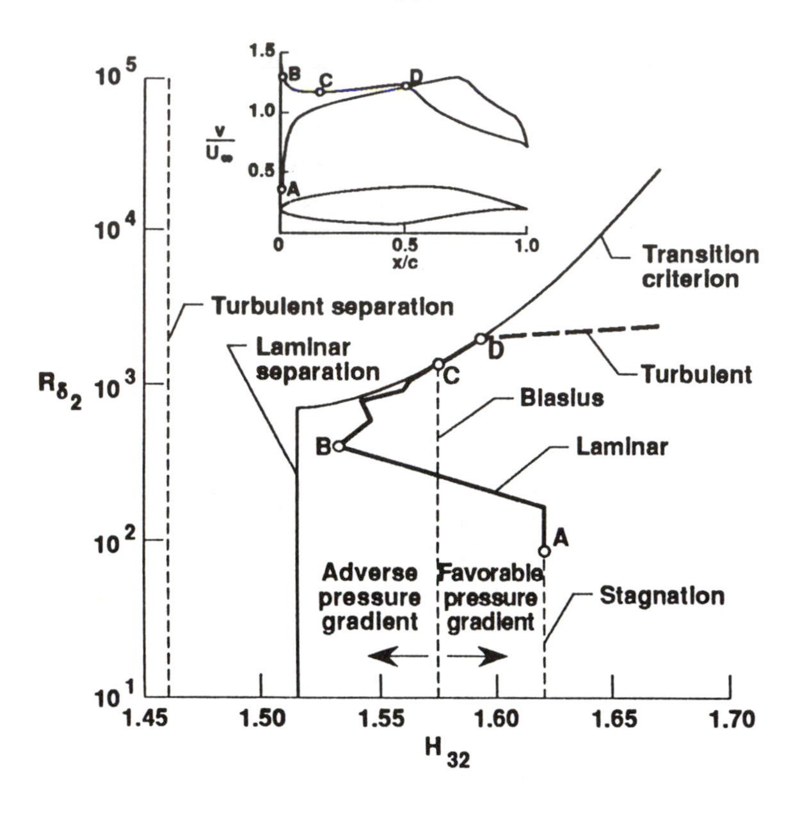


Figure 10.- Theoretical boundary-layer development along lower surface of NLF(2)-0415 airfoil for $c_l = 0.1$, $R = 18 \times 10^6$, and $M = 0.38$.

For this airfoil, laminar flow should extend to 70-percent chord on the upper surface and to 50-percent chord on the lower surface. Having more extensive laminar flow on the upper surface than on the lower is atypical of most, practical laminar-flow airfoils but desirable for this design for two reasons. First, a longer region of favorable pressure gradient on the upper surface produces a higher drag-divergence Mach number than does a shorter one. Second, because the drag due to the upper surface is normally higher than that due to the lower surface because of locally higher velocities, a greater extent of laminar flow on the upper surface than on the lower results in a lower total profile-drag coefficient. Normally, such a concept cannot be employed because of a requirement for a high maximum lift coefficient which is in conflict with extensive laminar flow on the upper surface. For this application, however, the achievement of the required maximum lift coefficient, which is relatively low ($c_{l,max} = 1.65$), is aided by the relatively high Reynolds number ($R = 9 \times 10^6$) of the minimum-speed condition. This concept also leads to a relatively large pitching-moment coefficient, which is permissible, for this design, because the pitching-moment coefficient was unconstrained.

The theoretical and experimental pressure distributions for $c_l = 0.34$ and $M = 0.38$ are compared in Figure 11. The theoretical pressure distribution is inviscid whereas the experimental pressure distribution was obtained at $R = 6 \times 10^6$ in the DFVLR Transsonischer Windkanal Braunschweig (TWB) (Ref. 33). The magnitude of the pressure coefficients and the pressure gradients agree remarkably well.

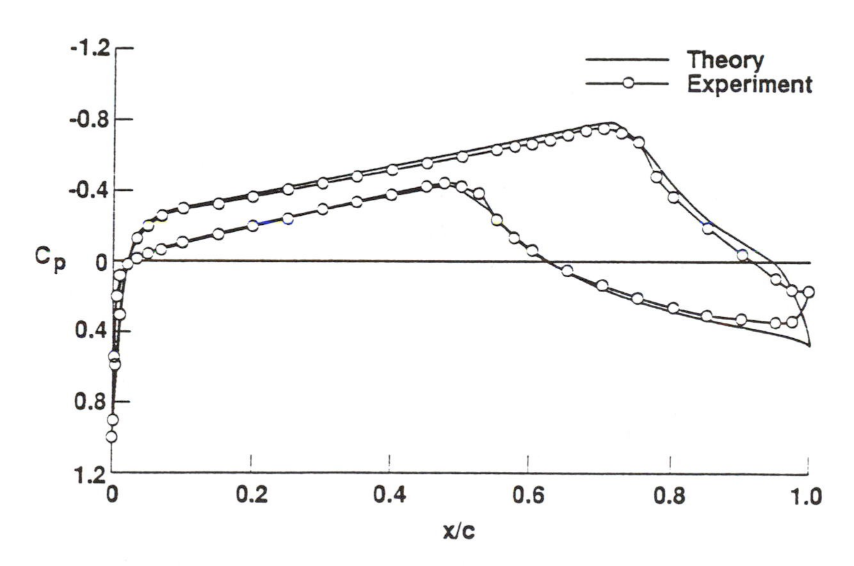


Figure 11.- Comparison of theoretical and experimental pressure distributions of NLF(2)-0415 airfoil for $c_l = 0.34$ and $M = 0.38$.

In summary, the third test case extended the range of applicability of the method to higher, but still subcritical, Mach numbers. For most aircraft applications, even higher Mach numbers (supercritical) imply swept wings which generate crossflow instabilities which the boundary-layer method cannot analyze. Thus, the third test case represents a certain practical, upper limit for the method. Unfortunately, to date, the predicted boundary-layer developments have not been experimentally verified at the design Reynolds number. Such verifications are essential for the calibration of transition criteria.

Thus, the frontier represented by the third test case—high Reynolds number and Mach number—has been reached. Accordingly, we have shifted our attention to lower Reynolds numbers, below $1 \times 10^6$.

## NLF(1)-1015 Airfoil

In 1986, the NASA NLF(1)-1015 airfoil (Ref. 34) was designed by Mark D. Maughmer of The Pennsylvania State University and the author for a high-altitude, long-endurance vehicle. This application requires low profile-drag coefficients for the endurance lift coefficient ($c_l = 1.5$) at a Reynolds number of $0.7 \times 10^6$ as well as for the cruise and dash lift coefficients ($c_l = 1.0$ and 0.4, respectively) at higher Reynolds numbers ($R = 0.9$ and $2.0 \times 10^6$, respectively).

Two primary objectives were identified for this airfoil. First, the airfoil should produce a maximum lift coefficient of 1.8 at $R = 2.0 \times 10^6$. The requirement that $c_{l,max}$ not decrease with transition fixed near the leading edge also applies to this design because all-weather takeoff and landing capability was deemed desirable. The second objective for this airfoil was to obtain low profile-drag coefficients from the dash lift coefficient of 0.4 to the endurance lift coefficient of 1.5.

Two constraints were placed on this design. First, the thickness must be 15-percent chord in order to be compatible with the vehicle design. Second, the zero-lift pitching-moment coefficient should be no more negative than $-0.20$. This constraint grew out of concerns based on previous experience with large pitching-moment airfoils.

The above objectives and constraints are not sufficient to design the airfoil, however, because of the unconstrained extents of laminar flow on the upper and lower surfaces. To evaluate the effect of these variables, a figure of merit was developed to guide the airfoil design process. Later analysis (Ref. 35) indicated that the figure of merit used, $c_{l,max}/c_{d,endurance}$, was actually a limiting case of a more general expression. The use of a figure of merit is desirable for the efficient integration of the airfoil/aircraft design process.

Because this design combines high lift coefficients and low Reynolds numbers, the velocity distributions incorporate two features to achieve the high lift and inhibit the formation of significant (drag-producing) laminar separation bubbles. The first feature is the 'separation ramp,' originally proposed by F. X. Wortmann, illustrated in Figure 12. This feature confines turbulent separation to a small region near the trailing edge. By controlling the movement of the separation point at high angles of attack, high lift coefficients can be achieved with little drag penalty. This feature has the added benefit that it promotes docile stall characteristics. The second feature is embodied in the curved transition ramps (Ref. 36) which alleviate the laminar separation bubbles that would otherwise contribute significantly to the drag of both the upper and lower surfaces. This feature restricts the height of the bubbles and, therefore, their severity by limiting the velocity gradients across the bubbles.

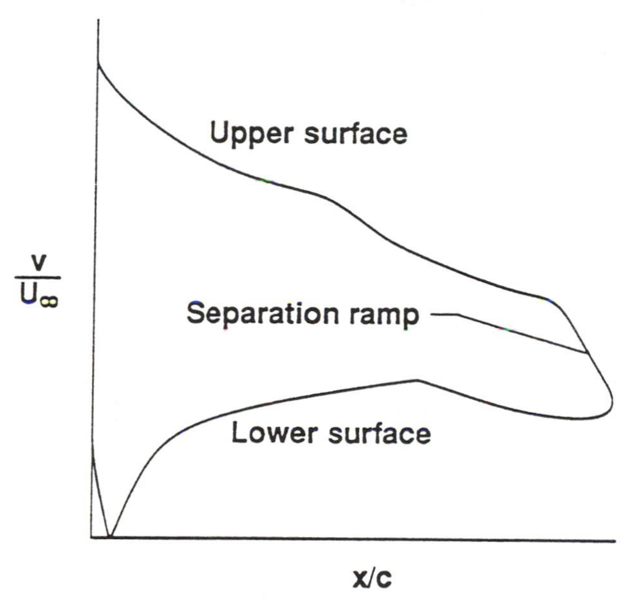


Figure 12.- Separation ramp.

The design specifications for this fourth test case can be summarized, relative to the first three test cases, as follows: lower Reynolds numbers, higher lift coefficients, and incompressible.

The NLF(1)-1015 airfoil shape and selected theoretical (inviscid and incompressible) velocity distributions are shown in Figure 13. The curved transition ramps in the upper- and lower-surface velocity distributions and the separation ramp in the upper-surface velocity distributions are all evident.

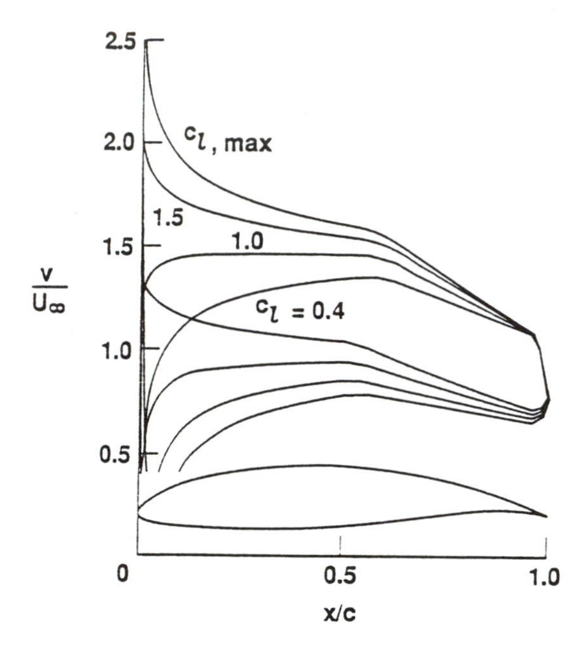


Figure 13.- NLF(1)-1015 airfoil shape and inviscid, incompressible velocity distributions.

The comparison of theoretical and experimental section characteristics for the endurance condition ($R = 0.7 \times 10^6$) is shown in Figure 14. Of primary interest are the predicted and measured drag polars. The bowed shape of the measured polar (higher drag between the limits of the low-drag range than at them) indicates that the upper- and lower-surface laminar separation bubbles have adversely affected the drag. Note, however, that this effect is indeed quite small at the upper limit which corresponds roughly to the maximum endurance point ($c_l = 1.5$ *and* $R = 0.7 \times 10^6$). The fact that significant laminar separation bubbles exist at lower lift coefficients is inconsequential *for this application* because the corresponding flight Reynolds numbers are higher. The measured section characteristics for those Reynolds numbers (not shown) indicate that the drag is not influenced by bubbles. These results illustrate the degree to which an airfoil can be tailored with respect to a specific aircraft. The small differences between the predicted and measured drag coefficients at the limits of the low-drag range do show, however, that some additional drag can be attributed to the laminar separation bubbles.

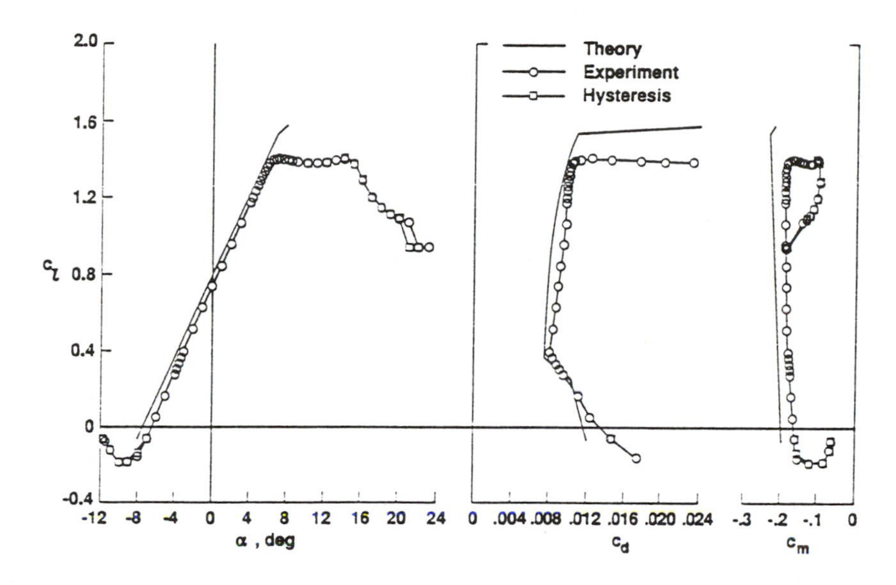


Figure 14.- Comparison of theoretical and experimental section characteristics of NLF(1)-1015 airfoil for $R = 0.7 \times 10^6$.

In summary, the fourth test case extended the range of applicability of the method to lower Reynolds numbers. Elimination of the discrepancies between the theoretical and experimental drag coefficients requires more accurate prediction of the occurrence of significant laminar separation bubbles. This is currently being pursued (Refs. 37 and 38).

## 5. Concluding Remarks

Airfoil design has progressed considerably over the past century. The first airfoils were mere copies of birds's wings. These airfoils were followed by cut-and-try shapes, some of which were tested in simple, low-Reynolds-number wind tunnels. The NACA systematized this approach by perturbing successful airfoil geometries in order to generate series of related airfoils. These airfoils were carefully tested in a more sophisticated wind tunnel which could replicate flight Reynolds numbers. Eastman Jacobs then recognized the need for a theoretical method which would determine the airfoil shape that would produce a specified pressure distribution which would exhibit the desired boundary-layer characteristics. This idea represents the basis of modern airfoil design: the desired boundary-layer characteristics result from the pressure distribution which results from the airfoil shape.

The inversion of an airfoil analysis method provided the means of transforming the pressure distribution into an airfoil shape. The transformation of the desired boundary-layer characteristics into a pressure distribution was left to the imagination of the airfoil designer. Since that time, 50 years ago, Richard Eppler, through his computer program, has developed a much more direct connection between the boundary-layer development and the pressure distribution.

We have adopted the philosophy of Eppler that a reliable theoretical airfoil design method should be developed instead of catalogs of experimental section characteristics. The method can then be used to explore many concepts with respect to each specific application. The success of this philosophy hinges on the verification of the method.

We have designed several airfoils to test Eppler's method. By investigating the airfoils in a low-turbulence wind tunnel, we have established the range of applicability of the method. Initially, we investigated the classical, low-speed Reynolds-number range of 3 to $9 \times 10^6$. From there, we explored higher Reynolds numbers ($\sim 20 \times 10^6$) and Mach numbers ($\sim 0.7$). Most recently, we have investigated lower Reynolds numbers ($\sim 0.5 \times 10^6$). The latest indications are that the method is also applicable at even lower Reynolds numbers ($\sim 0.1 \times 10^6$). The method has been steadily improved in response to inadequacies revealed during these experimental investigations.

In summary, an experimentally-verified, theoretical method has been developed which allows airfoils to be designed for almost all subcritical applications.

On that note, NASA once again fades from the airfoil scene. Further improvements in the science of airfoil design await more accurate theoretical methods. These improvements require fundamental experiments aimed at improving the prediction of the boundary-layer phenomena of transition and separation. But these improvements are not likely to lead to large increases in airfoil performance. Major improvements in airfoil characteristics await advances in the art of airfoil design, which is still the domain of the designer's creativity.

## 6. Symbols

| | |
|---|---|
| $C_p$ | pressure coefficient |
| $c$ | airfoil chord, m |
| $c_d$ | section profile-drag coefficient |
| $c_l$ | section lift coefficient |
| $c_m$ | section pitching-moment coefficient about quarter-chord point |
| $H_{12}$ | boundary-layer shape factor, $\delta_1/\delta_2$ |
| $H_{32}$ | boundary-layer shape factor, $\delta_3/\delta_2$ |
| $M$ | free-stream Mach number |
| $R$ | Reynolds number based on free-stream conditions and airfoil chord |
| $R_{\delta_2}$ | Reynolds number based on local conditions and boundary-layer momentum thickness |
| $U_\infty$ | free-stream velocity, m/s |
| $v$ | local velocity on airfoil, m/s |

| | |
|---|---|
| x | airfoil abscissa, m |
| $\alpha$ | angle of attack relative to chord line, deg |
| $\delta_f$ | flap deflection, positive downward, deg |
| $\delta_1$ | boundary-layer displacement thickness, m |
| $\delta_2$ | boundary-layer momentum thickness, m |
| $\delta_3$ | boundary-layer energy thickness, m |

Subscripts:

| | |
|---|---|
| cruise | cruise condition |
| endurance | maximum endurance condition |
| max | maximum |
| T | transition |
| 0 | zero lift |

Abbreviations:

| | |
|---|---|
| DFVLR | Deutsche Forschungs- und Versuchsanstalt für Luft- und Raumfahrt e.V. |
| LTPT | Low-Turbulence Pressure Tunnel |
| NACA | National Advisory Committee for Aeronautics |
| NASA | National Aeronautics and Space Administration |
| NLF | natural laminar flow |
| TWB | Transsonischer Windkanal Braunschweig |

# References


[1] National Advisory Committee for Aeronautics: First Annual Report of the National Advisory Committee for Aeronautics. 1915.

[2] National Advisory Committee for Aeronautics: Aerodynamic Characteristics of Aerofoils. NACA Rep. 93, 1920.

[3] Hansen, James R.: Engineer in Charge. NASA SP-4305, 1987.

[4] Munk, Max M.; and Miller, Elton W.: Model Tests with a Systematic Series of 27 Wing Sections at Full Reynolds Number. NACA Rep. 221, 1925.

[5] Jacobs, Eastman N.; Ward, Kenneth E.; and Pinkerton, Robert M.: The Characteristics of 78 Related Airfoil Sections from Tests in the Variable-Density Wind Tunnel. NACA Rep. 460, 1933.

[6] Jacobs, Eastman N.; and Pinkerton, Robert M.: Tests in the Variable-Density Wind Tunnel of Related Airfoils Having the Maximum Camber Unusually far Forward. NACA Rep. 537, 1935.

[7] Jacobs, Eastman N.; Pinkerton, Robert M.; and Greenberg, Harry: Tests of Related Forward-Camber Airfoils in the Variable-Density Wind Tunnel. NACA Rep. 610, 1937.

[8] Theodorsen, Theodore: Theory of Wing Sections of Arbitrary Shape. NACA Rep. 411, 1932.

[9] Jacobs, Eastman N.: Preliminary Report on Laminar-Flow Airfoils and New Methods Adopted for Airfoil and Boundary-Layer Investigations. NACA WR L-345, 1939 (formerly, NACA ACR).

[10] Abbott, Ira H.; Von Doenhoff, Albert E.; and Stivers, Louis S., Jr.: Summary of Airfoil Data. NACA Rep. 824, 1945. (Supersedes NACA WR L-560.)

[11] Abbott, Ira H.; and Von Doenhoff, Albert E.: Theory of Wing Sections. Dover Publ., Inc., c. 1959.

[12] Truckenbrodt, E.: Die Berechnung der Profilform bei vorgegebener Geschwindigkeitsverteilung (The Calculation of the Profile Shape from Specified Velocity Distribution). Ingenieur-Archiv, Bd. 19, Heft 6, 1951, pp. 365–377.

[13] Truckenbrodt, E.: A Method of Quadrature for Calculation of the Laminar and Turbulent Boundary Layer in Case of Plane and Rotationally Symmetrical Flow. NACA TM 1379, 1955. (Translated from Ingenieur-Archiv, Bd. 20, Heft 4, 1952, pp. 211–228.)

[14] Althaus, Dieter; and Wortmann, Franz Xaver: Stuttgarter Profilkatalog I. (Stuttgart Profile Catalog I.) Friedr. Vieweg & Sohn (Braunschweig), 1981.

[15] Eppler, R.: Direct Calculation of Airfoils from Pressure Distribution. NASA TT F-15,417, 1974. (Translated from Ingenieur-Archiv, Bd. 25, Heft 1, 1957, pp. 32–57.)

[16] Eppler, R.: Practical Calculation of Laminar and Turbulent Bled-Off Boundary Layers. NASA TM-75328, 1978. (Translated from Ingenieur-Archiv, Bd. 32, Heft 4, 1963, pp. 221–245.)

[17] Whitcomb, Richard T.; and Clark, Larry R.: An Airfoil Shape for Efficient Flight at Supercritical Mach Numbers. NASA TM X-1109, 1965.

[18] McGhee, Robert J.; Beasley, William D.; and Whitcomb, Richard T.: NASA Low- and Medium-Speed Airfoil Development. NASA TM-78709, 1979.

[19] Naegele, H.; and Eppler, R.: Plastic Sailplane FS-24 Phoenix. Soaring, Vol. 22, No. 4, July-Aug. 1958, pp. 2–5. (Translated from Aero Revue, 33, Mar. 1958, pp. 140–143.)

[20] Eppler, Richard: Some New Airfoils. Science and Technology of Low Speed and Motorless Flight, NASA CP-2085, Part I, 1979, pp. 131–153.

[21] Eppler, Richard; and Somers, Dan M.: A Computer Program for the Design and Analysis of Low-Speed Airfoils. NASA TM-80210, 1980.

[22] Eppler, Richard; and Somers, Dan M.: Supplement To: A Computer Program for the Design and Analysis of Low-Speed Airfoils. NASA TM-81862, 1980.

[23] Eppler, Richard: Airfoil Design and Data. Springer-Verlag (Berlin), 1990.

[24] Von Doenhoff, Albert E.; and Abbott, Frank T., Jr.: The Langley Two-Dimensional Low-Turbulence Pressure Tunnel. NACA TN 1283, 1947.

[25] McGhee, Robert J.; Beasley, William D.; and Foster, Jean M.: Recent Modifications and Calibration of the Langley Low-Turbulence Pressure Tunnel. NASA TP-2328, 1984.

[26] Somers, Dan M.: Design and Experimental Results for a Natural-Laminar-Flow Airfoil for General Aviation Applications. NASA TP-1861, 1981.

[27] Somers, Dan M.: Design and Experimental Results for a Flapped Natural-Laminar-Flow Airfoil for General Aviation Applications. NASA TP-1865, 1981.

[28] Payne, Henry E.: Laminar Flow Rethink—Using Composite Structure. [Preprint] 760473, Soc. Automot. Eng., Apr. 1976.

[29] Pfenninger, Werner: Investigations on Reductions of Friction on Wings, in Particular by Means of Boundary Layer Suction. NACA TM 1181, 1947. (Translated from Mitteilungen aus dem Institut für Aerodynamik an der Eidgenössischen Technischen Hochschule Zürich, Nr. 13, 1946.)

[30] Eppler, R.: Laminar Airfoils for Reynolds Numbers Greater Than $4 \times 10^6$. B-819-35, Apr. 1969. (Available from NTIS as N69-28178; translated from Ingenieur-Archiv, Bd. 38, Heft 4/5, 1969, pp. 232–240.)

[31] Labrujere, Th. E.; Loeve, W.; and Sloof, J. W.: An Approximate Method for the Determination of the Pressure Distribution on Wings in the Lower Critical Speed Range. Transonic Aerodynamics. AGARD CP No. 35, Sept. 1968, pp. 17-1–17-10.

[32] Somers, Dan M.; and Horstmann, Karl-Heinz: Design of a Medium-Speed, Natural-Laminar-Flow Airfoil for Commuter Aircraft Application. DFVLR IB 129-85/26, 1985.

[33] Stanewsky, Egon; Puffert-Meissner, Wolfgang; Müller, Rudolf; and Hoheisel, Heinz: Der Transsonische Windkanal Braunschweig der DFVLR (The Transonic Wind Tunnel Braunschweig of DFVLR). Z. Flugwissenschaften und Weltraumforschung, Bd. 6, Heft 6, Nov./Dec. 1982, pp. 398–408.

[34] Maughmer, Mark D.; and Somers, Dan M.: Design and Experimental Results for a High-Altitude, Long-Endurance Airfoil. J. Aircr., Vol. 26, No. 2, Feb. 1989, pp. 148–153.

[35] Maughmer, Mark D.; and Somers, Dan M.: Figures of Merit for Airfoil/Aircraft Design Integration. AIAA Paper 88-4416, Sept. 1988.

[36] Eppler, Richard; and Somers, Dan M.: Airfoil Design for Reynolds Numbers Between 50,000 and 500,000. Conference on Low Reynolds Number Airfoil Aerodynamics. UNDAS-CP-77B123, Univ. of Notre Dame, June 1985, pp. 1–14.

[37] Eppler, R.: Recent Developments in Boundary Layer Computation. Conference on Aerodynamics at Low Reynolds Numbers $10^4 < Re < 10^6$, Vol. II, Roy. Aeronaut. Soc. (London), Oct. 1986, pp. 12.1–12.18.

[38] Dini, Paolo; and Maughmer, Mark D.: A Computationally Efficient Modelling of Laminar Separation Bubbles. Conference on Low Reynolds Number Aerodynamics. Univ. of Notre Dame, June 1989, pp. 368–382.

# DESIGN PHILOSOPHY OF LONG RANGE LFC TRANSPORTS WITH ADVANCED SUPERCRITICAL LFC AIRFOILS

*Werner Pfenninger*[1] *and Chandra S. Vemuru*[2]

Analytical Services & Materials Inc.
107 Research Drive
Hampton, Virginia 23666

## 1. Introduction, Formulation of Goal, Overall Design Approaches

With the interest in long range air transportation, the question arises concerning the design philosophy of long range LFC transports, which can reach any point on earth at high subsonic speeds with relatively large payloads without refuelling. Figure 1 shows as an example a 180,000 kg take-off gross weight LFC transport airplane with 50,000 kg payload (250 passengers plus cargo) cruising at $M_{cruise}= 0.83$. Cruise lift to drag ratios of 39.4 appear feasible with 70% laminar flow on the wing-, tail-, nacelle- and strut surfaces by means of suitable geometry and boundary layer suction in the upstream parts of these surfaces, while accepting a turbulent fuselage. At the same time the induced drag to lift ratio $D_i/L = W/\pi q b^2$ is minimized by raising the wing span b and aspect ratio $b^2/S$. Advanced overall and detailed design concepts were used to minimize the wing structural weight [Ref. 1].

The design philosophy leading to this airplane layout will now be discussed in more detail. The airplane range $R = \eta_{overall} \times L/D \times H \times ln(W_A/W_E)$ is proportional to the airplane lift to drag ratio L/D, the overall efficiency $\eta_{overall}$ of the airplane propulsion system and the natural log of the airplane weight ratio $W_A/W_E$ between start and end of cruise (H = specific heat content of fuel). Assuming for simplicity constant airplane zero lift drag $C_{D,o}= C_{D,\infty}+ C_{D,parasite}$ for different $Re_c$'s and $C_L$'s, the maximum L/D and corresponding optimum cruise lift coefficient $C_{L,opt}$ are: $(L/D)_{max} = \sqrt{\frac{\frac{\pi}{4}\times b^2/S}{C_{D,o}}}$, $C_{L,opt} = \sqrt{\frac{\pi \times b^2}{S} \times C_{D,o}}$.

[1] Senior Research Scientist.
[2] Senior Research Engineer.

The variation of $(L/D)_{max}$ and $C_{L,opt}$ versus the laminar airplane wetted area ratio $A_{laminar}/A_{total}$ is shown in the Figures 2 and 3 [Ref. 2], assuming a linear decrease of $C_{D,o}$=0.012 for the fully turbulent airplane to 0.002 for the all laminar airplane with a relative fuselage size of a B52. $(L/D)_{max}$ grows progressively rapidly as $A_{laminar}/A_{total}$ increases, reaching phenomenally high values with fully laminar flow over the airplane exposed surfaces, especially with higher wing aspect ratios. $C_{L,opt}$ decreases with increasing $A_{laminar}/A_{total}$ ratios, raising accordingly the airplane design Mach number or enabling less wing sweep. Sweep-induced boundary layer crossflow instability is thereby alleviated. Thus, to maximize L/D, $C_{D,o}$ should be minimized by extensive airplane laminarization; furthermore, to obtain the full benefit from the lower laminar airplane friction drag, the airplane induced drag to lift ratio $D_i/L = W/\pi q b^2$ should be reduced simultaneously by lowering the span loading $W/b^2$, i.e., raising the wing span and with it the wing aspect ratio $b^2/S$, compatible with a low wing structural weight.

Besides pure performance considerations the design of large LFC airplanes is critically influenced by suction laminarization problems [Ref. 1], which are easier to handle at lower length Reynolds numbers. Also, the permissible surface roughness and tolerances for laminar flow are inversely proportional to the airplane unit length Reynolds number $U_\infty/\nu$. Therefore, larger LFC airplanes should preferably be designed such that performance optimization is compatible with the desire to alleviate the laminarization problems involved by reducing $Re_c$ and $U_\infty/\nu$ . Since $Re_c = \frac{U_\infty c}{\nu} = \frac{2W^{0.5}}{a\mu C_L M} \times \sqrt{\frac{W/S}{b^2/S}}$ and $\frac{U_\infty}{\nu} = \frac{2(W/S)}{a\mu C_L M}$ it follows that $Re_c$ decreases by lowering the wing loading W/S, raising the wing aspect ratio $b^2/S$ as well as $C_L$. Thus, the reduction of the induced drag to lift ratio $D_i/L = W/\pi q b^2$ by lowering the span loading $W/b^2$ for superior performance is well compatible with the desire to reduce $Re_c \approx \sqrt{W/b^2}$.

## Design Approaches Towards Larger Span Wings With Reasonably Low Structural Weights [Refs. 1,2]

The desire to increase the wing span, while keeping the wing structural weight within bounds, profoundly influences the design of high performance LFC airplanes. Advanced overall and detail design approaches are needed for this purpose, borrowing if necessary design approaches of the past, which become attractive again in the light of new aerodynamic refinements. For example, the wing span can be substantially increased by taking bending- and torsional moments and deformations out of the wing by suction laminarized wide chord external struts (Fig. 1) of low parasite drag. A rigid strut-braced inboard wing is thus added to a cantilever outer wing. Wing loads decrease further by placing an external laminarized fuel nacelle in the outer part of the wing and bracing it against the wing by laminarized struts. Active deflection of a small chord trailing edge cruise flap decreases wing gust-, maneuver-, and dynamic loads and aeroelastic deformations, leading to a further decrease in structural weight. Such a cruise flap allows a variation of $C_L$ at constant $\alpha_{wing}$. When no suction is being considered in the rear pressure rise area of the upper surface a slotted cruise flap could enable a steeper pressure rise in this area than a plain cruise flap, resulting in a more extensive supersonic flat rooftop pressure distribution with laminar flow to raise accordingly $M_{\infty,Design}$ and lower $C_{D,\infty}$. Slotted supercritical wings were first proposed by Whitcomb [Ref. 3].

Aeroelastic wing torsional deformations can be alleviated by an active horizontal control surface located on a boom at the downstream end of the external fuel nacelle, which equalizes the wing angles of attack at the location of these nacelles and the wing root, using for example inertial platforms as sensors. Aeroelastic wing angle of attack changes induced by cruise flap deflection and bending- and torsional deformation of the swept wing can be largely compensated by the active control surfaces of the fuel nacelles. Such an active control of the wing angle of attack along the span is particularly important for supercritical LFC wings, for which the flow in the embedded supersonic zone of the upper surface is especially sensitive to angle of attack changes.

To minimize weight advanced composites of hight strength and stiffness will be used extensively for the main- and secondary structure of the airplane. Advanced supercritical airfoils with particularly high design Mach numbers minimize wing sweep for high subsonic speed airplanes to enable increased wing span and reduce induced drag accordingly.

## Drag Reduction of Wings and Bodies by Laminarization

The variation of the equivalent wing profile drag $C_{D,\infty}$with transition location $(x/c)_{Tr}$ and $Re_c$ is shown in Figs. 4-5 for a $23^0$ swept supercritical LFC wing. At $Re_c = 30.0 \times 10^6$ $C_{D,\infty}$ decreases from 0.0067 with fully turbulent flow to 0.0024 with 70% laminar flow and 0.0010 with 100% laminar flow (including $C_{D,suction}$). As compared to an unswept subcritical LFC wing, $C_{D,\infty}$ of the $23^0$ swept supercritical LFC wing is somewhat larger especially at higher $Re_c$'s (Fig. 5). This is primarily a result of higher suction rates in the front and rear part of the wing for control of sweep-induced boundary layer crossflow instability.

Since the turbulent fuselage drag represents a large percentage drag contribution of an otherwise laminar LFC airplane the question arises concerning the possible suction laminarization of the fuselage at high length Reynolds numbers $Re_L$. Figure 6 shows NASA Ames 12 ft tunnel drag results of the Northrop Reichardt LFC body of revolution (8:1 fineness ratio, 12 ft long) with $C_{D,min} = 0.00026$ (based on body wetted area, including $C_{D,suction}$) [Refs. 4,5]. The resulting body drag reduction with all laminar flow is percentage-wise larger than for all laminar flow wings and therefore rather tempting. The question then arises concerning the possible laminar flow $Re_L$-values of an LFC fuselage in flight at high subsonic speeds. In view of the practically non- existent atmospheric microscale turbulence, $Re_{L,laminar}$ of the Reichardt LFC body of revolution may be safely doubled to $120.0 \times 10^6$ for incompressible flow. The stabilizing influence of compressibility on the growth of amplified TS-waves may again double this value to $200.0 \times 10^6$ to $240.0 \times 10^6$ in flight at high subsonic cruising speeds.

For the present study a fully turbulent fuselage was assumed, applying a 7% and 10% equivalent fuselage drag reduction by riblets and fuselage boundary layer air propulsion in the rear part of the fuselage, respectively.

## 2. Laminarization Problems of Swept LFC Wings

Full chord laminar flow and low profile drags ($C_{D,\infty} = 0.0010$, including $C_{D,suction}$) have been obtained on a $30^0$ swept LFC airfoil [Ref. 6]. Boundary layer crossflow instability and spanwise turbulent contamination along the front attachment line critically affect the design of more strongly swept LFC wings at higher $Re_c$'s [Refs. 5-7]. The question, therefore, arises as to the alleviation of these sweep-induced boundary layer stability problems. In addition, 3-

dimensional leading edge roughness (flyspecks, atmospheric ice crystals) is inherently more sensitive on swept laminar flow wings than without sweep: Besides higher local flow velocities in the leading edge region of the swept wing, the streamwise disturbance vorticity induced by 3-dimensional roughness superimposes adversely with the sweep-induced boundary layer crossflow and its disturbance vortices to precipitate transition. Indeed, leading edge flyspecks often caused extensive loss of laminar flow on the X-21 LFC wing with its $33^0$ swept leading edge at $M_\infty = 0.75$ and 12000 meters altitude [Ref. 8], while full chord laminar flow on the F94 LFC wing glove with $10^0$ swept leading edge has been consistently observed at $M_\infty = 0.65$ and altitudes above 6000 meters to 7000 meters [Ref. 9]. Similarly, atmospheric ice crystal contamination had not been noticed on the F94 LFC glove, while often causing extensive loss of laminar flow on the X-21 wing [Ref. 8].

To alleviate these boundary layer crossflow stability problems wing sweep should be reduced by maximizing $M_{\infty,Design}$ of the airfoil, while maintaining a satisfactory off-design behavior and hopefully easing suction laminarization. This is possible for SC LFC airfoils with upper surface $C_p$-distributions with an extensive low supersonic flat rooftop, preceded by a far upstream supersonic pressure minimum and followed by a steep subsonic rear pressure rise to the trailing edge with low drag suction [Ref. 10]. $M_{\infty,Design}$ increases further by front- and aft loading the airfoil, accomplished by undercutting the front and rear lower airfoil surface, resulting in a relatively sharp leading edge [Refs. 10,11]. To simplify the wing design and minimize the wing weight penalty due to LFC, suction should be limited or preferably avoided in the area of the main wing structure [Ref. 10].

The question arises concerning the choice of the optimum thickness ratio of swept SC LFC airfoils of different chords from the standpoint of $(\frac{C_L}{C_{D,\infty}})$, $M_{\infty,Design}$ and the suction laminarization problems involved, for a given wing span and -structural weight, considering wing bending strength- and stiffness- as well as wing torsional stiffness requirements. According to a boundary layer stability analysis of all laminar flow unswept wings, with TS-instability dominating, $\frac{(t/c)_{opt}}{\sqrt{1-M_\infty^2}} = 0.16$ for $(\frac{C_L}{C_{D,\infty}})_{max}$ in the typical cruise $C_L$- range of LFC transports [Ref. 12], i.e., $(t/c)_{opt} \approx 0.09$ to $0.10$ for $M_\infty$=0.80 airplane. Somewhat lower $(t/c)_{opt}$-values might result when sweep- induced boundary layer crossflow instability dominates over TS-type boundary layer instability. These relatively low $(t/c)_{opt}$- values enable at the same time higher $M_{\infty,Design}$'s or enable less wing sweep. On the other hand, these larger chord and relatively thin wings raise the gust load factors as well as wing torsional deformations to increase the weight of the wing torsion box unless alleviated by active control. These structural disadvantages are partially compensated by the simpler high lift system of the larger chord wings. Based on these considerations (t/c)- ratios of the order 10% were tentatively chosen for the cantilevered areas of the wing.

Next, the suction requirements were studied for full chord suction laminarization at the design point of the supercritical airfoil X66 for $Re_c = 30.0 \times 10^6$ and $23^0$ and $37^0$ wing sweep, corresponding to $M_{flight} = 0.83$ and 0.97, respectively. This airfoil is a conservative derivative of airfoil X63T18S [Ref. 10] with substantially lower flow Mach numbers in the bulge area of the lower surface. Its 2-dimensional design $C_p$-distribution and supersonic bubble are shown in Fig. 7 ($M_{\infty,Design} = 0.781$, $C_{L,Design} = 0.624$).

The Figs. 8,9 show as an example the $C_p$- and suction distribution and the corresponding growth of amplified boundary layer crossflow- and TS-disturbance vortices for the upper and lower surface at $M_\infty =$ 0.97 and $\varphi = 37^0$ sweep, using COSAL code [Ref. 13].

No suction is needed for boundary layer crossflow control in the leading edge area for $\varphi = 23^0$ and $Re_c = 30.0 \times 10^6$. Increasing $\varphi$ to $37^0$ requires suction in a narrow spanwise strip in the front acceleration zone of the upper surface, such that the crossflow remains about neutrally stable in the suction zone, where the flow Mach number varies from low to moderately high subsonic values.

Since the crossflow in the flow deceleration area downstream of the upper surface pressure minimum cancels the crossflow in the preceding acceleration zone, crossflow disturbance vortices and their interaction with amplified TS-waves are practically eliminated in the flat rooftop area, at least as long as the flow in the rooftop area of the upper surface is only slightly accelerated. Even a modest flow acceleration in the rooftop area of the upper surface does not necessarily produce yet a critical boundary layer crossflow disturbance vortex growth, as long as boundary layer crossflow is absent at the beginning of the nonsuction region of the rooftop. Weak suction ($c_Q = -1.2 \times 10^{-4}$) from x/c = 0.05 to 0.30 is then adequate at the design point of airfoil X66 for boundary layer stabilization against amplified TS-disturbances in the rooftop area of the upper surface at $Re_c = 30.0 \times 10^6$ (Fig. 8a). $n_{TS,max}$=6.5 at M=0.97 and $\varphi = 37^0$ appears conservatively high in the practical absence of sweep- induced boundary layer crossflow, at least at the airfoil design point. For comparison, transition experiments in the NASA Ames 12 ft tunnel have given $n_{TS}$-values at transition of 10 (COSAL analysis, Ref. 13) on an ellipsoid of revolution at high subsonic speeds, when the tunnel test section was choked [Ref. 14]. If full chord laminar flow on the wing is desired, i.e., maintaining laminar flow through the rear pressure rise zone by suction, it appears advisable not to exceed the critical threshold level of amplified TS-disturbances in the rooftop area of the upper surface, i.e., providing a margin in $n_{TS}$ of about 4 below $n_{TS,Tr}$. The strongly stabilizing influence of compressibility on the TS- disturbance growth [Ref. 13] enables surprisingly large suction interruptions in the wing structural area, at least at the airfoil design condition. At off-design, small changes in $M_\infty$ and $\alpha_{wing}$ add flow decelerations in the rooftop area of the upper surface (see for example Fig. 10) to adversely affect laminarization in this area. Additional suction strips may then be needed in the main wing box area for further TS-boundary layer stabilization.

Similar to the upper surface, the growth of amplified boundary layer crossflow disturbance vortices in the front part of the lower surface may be alleviated by partially cancelling the boundary layer crossflow, generated in the steep second acceleration zone, by boundary layer crossflows of opposite sign generated in local pressure rise areas immediately up- and downstream of the second acceleration zone between x/c $\approx$ 0.10-0.14 and 0.22-0.30. No suction is needed for boundary layer crossflow control in the accelerated flow region of the lower surface up to x/c $\approx$ 0.10. Relatively strong suction is then needed to minimize the growth of amplified boundary layer crossflow disturbance vortices in and downstream of the second acceleration zone, especially for $\varphi = 37^0$ sweep. Crossflow disturbance vortices for $\varphi = 37^0$ are damped between x/c $\approx$ 0.10-0.20 on the lower surface, i.e., $n_{CF}$ should drop in this area and grow again beyond x/c = 0.20 (the COSAL method in its present form has been unable to analyze such damping, hence this area is shown by dotted horizontal lines). The $n_{CF}$-values on the lower surface up to x/c $\approx$ 0.40 are probably conservatively low, i.e., lower suction rates appear adequate for boundary layer crossflow control between x/c = 0.10-0.30.

The growth of amplified Taylor-Görtler (TG) type disturbance vortices in the concave curvature zones of the front and rear lower surface can be substantially reduced by turning the flow in a series of "corners", instead of using a continuous gradual turn [Ref. 11]. Even though the local Goertler number $Re_\theta\sqrt{\frac{\theta}{r}}$ and the corresponding local TG- growth rates are thus higher the distance, over which TG-vortices travel, is much shorter to minimize accordingly TG-vortex growth. Maintaining laminar flow through the pressure rise areas preceeding these "corners" requires local suction.

The steep flow deceleration and relatively thick boundary layer in the rear pressure rise areas of the upper and lower surface generate severe local boundary layer crossflow. To maintain full chord laminar flow relatively strong local suction is needed for boundary layer crossflow stabilization in the rear pressure rise zones. The total suction mass flow involved decreases by reducing the crossflow disturbance vortex growth distance or time, accomplished by decelerating the flow over a particularly short chordwise distance. At the same time the corresponding crossflow Stokes layer is thinner, shifting the maximum crossflow velocity closer to the surface to raise accordingly $\chi_{min}$.

Crossflow stabilization in the rear pressure rise areas of the upper and lower surface is optimized by minimizing or avoiding amplified crossflow disturbance vortices in the upstream part of the pressure rise. This is possible by sufficiently strong local suction in this region, such that the local $Re_n$ is kept close to the corresponding $\chi_{min}$. In the rear pressure rise area of the upper surface the crossflow disturbance vortices can grow then to its transition value at the termination of the rear pressure rise at the trailing edge (Fig. 8b). This is not as easily possible on the rear lower surface, where the crossflow generated in its rear pressure rise area continues into the downstream high pressure zone to contribute a further crossflow disturbance vortex growth in spite of the absence of a pressure rise in this area. The crossflow on the rear lower surface of airfoil X66 must then be stabilized by sufficiently strong local suction towards the termination of its rear pressure rise at x/c = 0.86 (Fig. 9b), such that the local $Re_n$ is not appreciably larger than the corresponding $\chi_{min}$, thereby avoiding an excessive growth of $n_{CF}$ between x/c = 0.86 and 1.0. Higher overall suction rates are then needed in the rear pressure rise area of the lower surface to raise accordingly its equivalent suction- and profile drag contribution.

Alternately, a local flow acceleration may be added in a "corner" area on the lower surface downstream of 0.86c for partial crossflow cancellation and reduction of $n_{CF}$ between x/c = 0.86 to 1.0. Additional weak suction may be needed in the high pressure zone of the rear lower surface to minimize or avoid amplified TS-disturbance growth and the resulting adverse interaction with amplified crossflow-disturbances in this area.

As a result, the termination of the rear pressure rise on the lower surface at x/c = 0.86 (instead of the trailing edge on the upper surface) requires increased suction for full chord laminarization of the lower surface. The resulting increase in equivalent suction- and profile drag is thus a consequence of the airfoil aft loading, which in turn is due to the design towards high $M_{\infty,Design}$ (without such aft loading either $C_{L,Design}$ or $M_{\infty,Design}$ would be lower).

Similarly, the front loading on the lower surface, desirable to maximize $M_{\infty,Design}$ and at the same time avoid excessive wing pitching moments, leads to a strongly accelerated flow on the lower surface, whether accomplished in a 2-step acceleration (X66 airfoil) or a long gradual one, to further aggravate the boundary layer crossflow problem of the lower surface. Relatively strong local suction is therefore needed to control crossflow around the second acceleration zone of airfoil X66. Likewise, a long gradual acceleration on the front lower surface would require relatively strong suction in this area over a considerable distance for crossflow control. Thus, the design towards high $M_{\infty,Design}$'s inevitably raises the equivalent suction- and profile drag of the lower surface, and vice versa.

The equivalent wing profile drag values $C_{D,\infty}$ with full chord laminar flow (including the equivalent suction drag $C_{D,suction}$, adding 10% for suction duct- and mixing losses) at $Re_c = 30.0 \times 10^6$ are as follows:

$$\varphi = 23^0, M_{flight} = 0.83,$$

$$C_{D,\infty,upper} = 0.488 \times 10^{-3},$$

$$C_{D,\infty,lower} = 0.478 \times 10^{-3},$$

$$C_{D,\infty,total} = 0.966 \times 10^{-3}.$$

$$\varphi = 37^0, M_{flight} = 0.97,$$

$$C_{D,\infty,upper} = 0.505 \times 10^{-3},$$

$$C_{D,\infty,lower} = 0.515 \times 10^{-3},$$

$$C_{D,\infty,total} = 1.02 \times 10^{-3}.$$

As compared to supercritical wings with partially and especially fully turbulent flow, these profile drags are extremely low. As compared to equivalent unswept LFC wings $C_{D,\infty,total}$ is about 20% higher, due primarily to the relatively large percentage drag contribution of the lower surface, being equal to that of the upper surface (with the lower flow velocities on the lower surface, its $C_{D,\infty}$-contribution should be about 25% lower than for the upper surface of an unswept all laminar LFC wing). The relatively large $C_{D,\infty}$-contribution of the lower surface is thus a consequence of the design compromises towards high $M_{\infty,Design}$'s, while avoiding excessive negative wing pitching moments.

For X66 type supercritical LFC airfoils without suction in the rear pressure rise areas (but with a slotted small chord trailing edge cruise flap to improve the rear pressure recovery), extensive laminar flow back to 70% and $C_{D,\infty}$= 0.0025 to 0.0030 appear possible at $Re_c$= 30.0 × $10^6$. $C_{D,\infty}$ could decrease 20% by extending suction laminarization downstream of the main wing structural box for a short distance into the rear pressure rise zone at relatively modest additional suction rates.

## 3. Three-Dimensional Airplane Design Integration

### Optimum upper surface pressure distributions with high design Mach numbers on tapered swept supercritical wings

With wing sweep needed to raise the design Mach number of high subsonic speed LFC transports, the question arises concerning the pros and cons of tapered swept-forward or swept-back wings from the standpoint of design Mach number, laminarization, and structural and aeroelastic characteristics.

A decisive aerodynamic advantage of tapered swept-forward LFC wings is a smaller leading edge sweep angle. Crossflow- and attachment line boundary layer instability as well as leading edge contamination by flyspecks and atmospheric ice crystals are thus alleviated.

Wing taper modifies the optimum chordwise $C_p$- distribution of swept wings laid out for high design $M_{\infty,Design}$'s: On tapered swept-back or swept-forward supercritical wings the isobar sweep decreases or increases from the wing leading- to the trailing edge to superimpose an additional streamwise flow deceleration or acceleration, respectively, when the flow Mach number component $M_{\perp}$ in the direction normal to the isobars is kept constant, (see Fig. 11 for $M_{flight}= 0.83$ and 0.97). Assuming upper surface isobars along constant percentage lines, a tapered swept-back supercritical LFC wing, optimized for a high cruise Mach number, will have a sloping down upper surface rooftop pressure distribution, preceded by a more pronounced far upstream supersonic pressure minimum (Fig. 12), as compared to an equivalent yawing LFC wing. As a result, TS- type disturbances on such tapered sweptback LFC wings are more strongly amplified to require additional suction in the flow deceleration area of the upper surface rooftop for boundary layer stabilization against amplified TS-disturbances. The additional suction rates required for this purpose are modest and can be applied in spanwise suction strips. (Of course, a tapered swept-back supercritical LFC wing does not necessarily have to be designed with such a decelerated flow on the upper surface at the cruise design point, at a corresponding penalty, though, in $M_{\infty,Design}$). In contrast, assuming upper surface isobars along constant percentage lines, tapered swept-forward supercritical LFC wings will have an optimum upper surface design pressure distributions for high $M_{\infty,Design}$ with slightly accelerated flow in their rooftop zones (Fig. 12) to reduce accordingly the TS-disturbance growth.

$M_{\infty,Design}$ or $C_{L,Design}$ of tapered swept-back and swept-forward supercritical LFC wings may be raised further 1) by increasing the upper surface isobar sweep beyond the constant percent chord line sweep, 2) by a favorable 3-dimensional aerodynamic design integration of the airplane configuration and its components at high subsonic speeds.

1) Since the isobar sweep at the start of the rear pressure rise on the upper surface is equal to the sweep angle of a possible shock front during cruise, $M_{\infty,Design}$ would increase by raising this isobar sweep beyond the sweep angle of the local constant percent chord line. On a tapered swept-back wing this is, in principle, possible by shifting the start of the rear pressure rise further forward in the inboard and further aft in the outer wing, and vice versa on a tapered swept-forward wing, by suitably tailoring the airfoil sections along the span. When the rear pressure rise of the upper surface is thus shifted further aft in the inboard area of a tapered swept-forward wing, where wing bending- and torsional moments are largest, the

local $M_{\infty,Design}$ increases further, and vice versa for a tapered swept-back wing. In the outer part of a swept-forward wing, where the structural loads are much less critical, somewhat thinner supercritical airfoils are easily feasible with but minor wing weight penalties to compensate for the loss in $M_{\infty,Design}$ due to the earlier pressure rise on the upper surface. It thus appears well worth to raise $M_{\infty,Design}$ of tapered swept-forward supercritical wings by increasing the isobar sweep at the start of the rear pressure rise on the upper surface over the corresponding sweep angle of the local constant percent chord line. In contrast, much less can thus be gained in raising $M_{\infty,Design}$ for tapered swept-back supercritical wings.

Tapered swept forward wings allow increased isobar sweep at the start of the rear pressure rise of the upper surface and as a result a higher critical Mach number by shifting the rear pressure rise further aft on the upper surface. A satisfactory rear pressure recovery without flow separation must be ensured when laminar flow is lost, such as in ice clouds. Low drag boundary layer suction in the rear pressure rise area of the upper surface, applied for full chord laminarization, is highly effective in preventing premature flow separation when laminar flow is lost. Without suction, a satisfactory steep rear pressure rise appears possible with a slotted trailing edge cruise flap [Ref. 3], optimally subdividing the rear pressure rise on the wing and flap. Alternately, vortex generators, located at the beginning of the rear pressure rise and retracted during laminar flow cruise, may be extended to ensure a satisfactory rear pressure recovery on the upper surface when laminar flow is lost.

To further increase the design Mach number of tapered swept-forward wings it would be ideal if the same isobar sweep could be maintained over most of the upper surface not only at the start of the rear pressure rise but also further upstream into a zone where off-design shocks start far upstream at lower $M_\infty$'s, when the upper shock-free low drag $C_L$-limit starts dropping. This is possible by tailoring the supercritical airfoil sections in the outer wing close to the external fuel nacelle such that the upper surface pressure minimum in this area is located especially far upstream, followed by a particularly rapid flow deceleration, combined with slightly lower local (t/c)'s and $C_L$'s. The upper surface isobars are thus pulled further upstream in the front part of the outer wing to increase accordingly the local isobar sweep angle and maintain a larger isobar sweep over a substantial percentage of the upper surface in the area upstream of the rear pressure rise. The resulting upper surface static pressures decrease thus further over most of the upper surface (Fig. 12) to raise either $C_{L,Design}$ (at given $M_\infty$ and $\varphi$ ) or $M_\infty$(at given $C_{L,Design}$and $\varphi$ ). The resulting upper surface rooftop $C_p$-distribution of such tapered swept-forward wing is more flat; its boundary layer is accordingly more sensitive to TS-disturbances to possibly require several spanwise suction strips in the rooftop area. This is easier possible for swept-forward LFC wings in composite structure, for which suction can be installed in the area of the diagonal composite plies of the wing torsion box, where the shear stresses are relatively low (suction would not be applied in the area of the heavily stressed spanwise wing bending plies).

Figure 13a shows the design KG-plot of such a modified and relatively conservatively laid out supercritical LFC airfoil with a particularly sharp front pressure minimum. The corresponding upper surface hodograph streamline (Fig. 13b) slopes down especially slowly and continuously over a surprisingly wide range of flow inclination angles. As a result, the upper shock-free low drag $C_L$-limit extends to relatively high $C_L$'s over a wide $M_\infty$-range (Fig. 14). This airfoil is superior with respect to the upper shock-free low drag $C_L$-limit over supercritical LFC airfoils with an extensive flat rooftop without a front pressure minimum (Fig. 15).

With this approach, the resulting isobar sweep over most of the upper surface of tapered swept-forward supercritical LFC wings is considerably larger than the structural sweep. In contrast, this is not as easily possible with swept-back supercritical LFC wings. Thus, swept-forward wings may be designed with larger aerodynamic spans (for a given structural span) and correspondingly lower induced drag. Wing aeroelastic divergence must be delayed far beyond any flight dynamic pressures to minimize or preferably eliminate additional divergence induced wing gust loads. For composite wings this is possible at a relatively minor structural weight penalty by sweeping the spanwise bending fibers ahead of the elastic axis, strut-bracing the wing against bending and torsion, sweeping the wing outboard of the external fuel nacelle backwards and actively controlling wing bending- and torsional deformations in gusts by the cruise flap and the control surfaces of the external fuel nacelle.

2) $M_{\infty,Design}$ or $C_{L,Design}$ of the airplane increases further if it should prove possible to reduce the flow Mach number component

($M_\perp$) normal to the upper surface isobars and thereby enabling lower static pressure especially on the front upper surface of high subsonic speed LFC airplanes with swept-forward wings (especially when cruising close to $M_\infty = 1$). This appears possible with a favorable 3-dimensional aerodynamic integration of the airplane configuration and its components. Negative perturbation velocities can be induced on the upper wing surface (especially in its front part) by suitably located laminarized superfan- and fuel- or outrigger gear nacelles, mounted upstream of the wing, and a 3-dimensional integration of the wing with these nacelles and the fuselage. Alternately one may look at the problem as follows: Area ruling will be applied individually for each nacelle and the corresponding local wing as well as for the entire airplane configuration, combined with additional local streamline contouring. At $M_\infty$ close to 1 the combination of wing plus nacelles behaves then to a first approximation as if the wing thickness were distributed over the entire length from the nacelle nose to the wing trailing edge. The wing thickness induced perturbation velocities thus decrease substantially to further increase $M_{\infty,Design}$ of such a 3-dimensionally integrated airplane configuration (Fig. 16). The above described 3-dimensional integration of wing and nacelles is worthwhile only if such nacelles are needed anyhow, or if they alleviate wing structural loads and deformations, and especially if their drag were minimized by suction laminarization.

With such an approach a careful study of the flow interference between the wing and nacelles and their supporting struts is needed to ensure satisfactory pressure distributions on the wing, nacelles and struts.

## 4. Propulsion Considerations

The question arises concerning the choice of the cruise propulsion system from the standpoint of a high overall efficiency $\eta_{overall}$ and reduced acoustic disturbances, particularly in the range of strongly amplified TS- and other types of boundary layer disturbances on the laminarized airplane surfaces, which might adversely affect airplane laminarization[3]. This is particularly important for LFC airplanes

[3] With the atmospheric microscale turbulence being usually too weak to affect transition, the initial disturbances introduced into the boundary layer are created primarily by the airplane itself, especially its propulsion system, as well as by suction induced disturbances.

with extensive natural laminar flow in the flat rooftop area of the upper surface without the stabilizing influence of distributed suction. (Distributed suction along the entire chord strongly stabilizes the boundary layer to allow correspondingly increased initial disturbances.) Of course, propulsion induced disturbances are inherently less critical for aerodynamically efficient airplanes with high (L/D)'s and correspondingly lower propulsion power.

Up to $M_{cruise} = 0.80$ nonregenerative high pressure ratio turboprops, driving high speed counter-rotating propellers, are superior in $\eta_{overall}$ over turbofans. Propeller tone noise frequency is usually below the frequency range of strongly amplified TS-waves on the wing, i.e., propeller tone noise does not appear critical in this respect. Its frequencies, though, are in the range of amplified travelling crossflow- or highly oblique TS-disturbance vortices in the rear pressure rise area of all laminar swept LFC wings to possibly induce amplified travelling boundary layer crossflow disturbance vortices.

Beyond $M_{cruise}= 0.80$ the compressibility problems of the propeller blades can be alleviated by installing the propeller in the decelerated internal flow field of a duct (or possibly in the decelerated rear flow field of a fuselage), especially if extensive laminar flow by means of suitable geometry and suction can be maintained on the external duct surfaces and even in the fan inlet up to the fan rotor. Such a ducted propeller or superfan (bypass ratio 15 to 20) was first proposed by Ackeret in 1938. The fan blades rotate then at relatively low tip speeds to allow accordingly a substantial axial decay of the rotating fan rotor pressure field and of many rotating fan rotor-stator acoustic interference modes in the fan duct. Most of the tone noise of the superfan is generated at frequencies considerably below those of amplified TS-oscillations but above those of amplified travelling boundary layer crossflow disturbance modes in the rear pressure rise area of all laminar swept LFC wings. In contrast, the present high bypass ratio turbofans of bypass ratio 5 to 6 contribute a substantial percentage of their tone- and shock noise in the frequency range of strongly amplified TS-oscillations on the wing.

The influence of engine noise on transition in the presence of amplified TS- and crossflow boundary layer disturbances has been investigated to some extent by Horstmann et al. on the Attas F614 natural laminar flow wing glove [Ref. 15]. Variation of engine r.p.m. and -noise did not seem to affect transition induced by boundary layer crossflow, in contrast to transition due to TS-disturbance growth, when increased engine r.p.m. and -noise moved transition upstream. Based on previous flight transition experiments the TS-disturbance growth rate at transition with but insignificant crossflow disturbances seemed to be somewhat lower than expected, explainable partially by engine noise.

The B757 acoustic flight tests, which could have shed further light on the effect of engine noise, seemed to be masked by self-induced transitional flow over the noise measuring gages and appear, therefore, inconclusive.

It is not clear whether to favor directly driven aft-mounted counter-rotating ducted propellers (or superfans), as proposed by Rolls-Royce [Ref. 16], or geared front superfans. The Rolls-Royce approach could allow a 3-shaft gas generator to enable increased engine pressure ratios with correspondingly higher thermodynamic efficiencies. Furthermore, avoiding the large diameter fan shaft in the gas generator with a rear-mounted superfan permits larger blade dimensions and correspondingly higher stage efficiencies in the high pressure compressor spool [Ref. 16].

With the tone noise frequency of the superfan essentially below the frequency range of amplified TS-disturbances on the wing, such a superfan may be mounted in front of the wing without seriously affecting wing laminarization. To alleviate compressibility problems particularly on the external fan nacelle surfaces and at the same time minimize fan air exit shock noise, especially at near sonic cruising speeds ($M_\infty = 0.97$ LFC transport), the fan nacelle inlet and exit may have to be sufficiently swept parallel to the wing (as seen in planview). On both sides of such swept nacelle inlets and exits the flow is similar as at the wing-fuselage juncture of a swept-back and swept-forward wing, respectively. To maintain shock-free flow in these areas, local area-ruling and streamline contouring is needed as in the wing-root area of a high speed airplane with a waisted fuselage, using body fairings which extend on both sides of the fan nacelle for a considerable distance fore and aft of the fan inlet and exit (see example of $M_{cruise} = 0.97$ airplane, Fig. 16).

In addition, the superfan nacelle, located upstream of the wing in the area of the wing-strut juncture, combined with the fuel nacelles (Figs. 1 and 16), induces a negative perturbation velocity in the area of the downstream wing. Either a thicker less heavier wing is thus possible especially in the area of the wing-strut juncture, where the wing bending moments are largest, or a larger wing span with lower induced drag could be used for the same wing weight, thereby partially compensating for the nacelle parasite drag.

Figure 17 shows the variation of the engine thermodynamic and overall efficiency of the propeller and superfan versus flight Mach number for nonregenerative high pressure ratio gas-turbine engines, driving high speed counter-rotating propellers and superfans (high component efficiencies were assumed). Up to $M_{cruise} = 0.83$ the turboprop is superior over the superfan, while at higher speeds the superfan engine has higher $\eta_{overall}$. Other aspects, though, (blade containment, engine location, weight, etc.) affect the choice of the propulsion system.

For the $M_{cruise}= 0.83$ and 0.97 LFC airplanes of $W_o = 180{,}000$ kg take-off gross weight wing mounted superfans of the Rolls-Royce type with about 18,000 kg take-off thrust were selected, combined with a fuselage boundary layer propulsion engine in the rear fuselage (assuming a turbulent fuselage), whose gas generator is fed with undisturbed ram air. The equivalent turbulent fuselage drag may thus decrease by 10%.

A relatively heavily loaded and correspondingly small counterrotating propeller or unducted fan of the G.E.-type, mounted in the decelerated flow field of the rear fuselage and operating accordingly at rather low resultant flow Mach numbers, could reaccelerate the boundary layer of the rear fuselage close to flight speed with a particularly high propulsive efficiency. Such a fuselage boundary layer propeller appears attractive up to an airplane cruise Mach number of 0.87. In addition to fuselage boundary layer air, this propeller would also use undisturbed ram air for propulsion. To maintain a uniform inflow into the propeller in circumferential direction boundary layer suction could be applied in the rear fuselage such that the fuselage boundary layer upstream of the propeller is reasonably uniform in circumferential direction.

Since the thrust produced by the suction compressors contributes particularly efficiently to the total cruise thrust of an LFC airplane they should be driven by thermodynamically highly efficient and not by thermodynamically inferior engines, i.e., either directly from the main propulsion engine via a mechanical or perhaps a bleed and burn cycle drive, possibly with regeneration. At lower flight speeds the suction compressors should preferably be geared down to reduce their power input. If separate smaller suction drive engines are considered either high pressure ratio nonregenerative or better yet particularly efficient low pressure ratio regenerative engines with a highly efficient regeneration should be chosen. Accepting 70% laminar flow on wing, empennage, struts and tail surfaces (turbulent fuselage), relatively limited suction is needed in the front part of these surfaces, i.e., no suction is required in their rear pressure rise areas. A direct mechanical suction compressor drive from the main engines appears then feasible.

## 5. Airplane Design Examples

The Figs. 1 and 16 show examples of strut-braced $M_\infty = 0.83$ and 0.97 LFC transport airplanes[4] of $W_o = 180{,}000$ kg take-off gross weight and 50,000 kg payload (250 passengers plus cargo). 70% laminar flow was assumed over wing-, empennage-, strut- and nacelles surfaces with fully turbulent flow on the fuselage (60 meters long, $\approx$ 6 meters in diameter). Riblets and boundary layer propulsion reduce the equivalent fuselage drag by 7% and 10%, respectively. A swept-forward wing was chosen between the fuselage and the external fuel nacelle, while the outer wing is swept-back, primarily to alleviate wing divergence. At the same time the structural wing span of such an M-type wing is smaller than that of a fully swept-forward wing (for the same aerodynamic span), thereby reducing the bending moments in the inboard wing and allowing accordingly a somewhat larger aerodynamic span. An additional wing torsional moment is introduced, though, in the inboard part of such an M-type wing by the lift of the sweptback outer wing. It can be reduced by shortening the span of the sweptback outer wing as well as compensating its aerodynamic lift by the inertial load of a small (laminarized) reserve fuel nacelle, located at the juncture of the wing with its split tip. Furthermore, this additional wing torsional moment may be partially compensated by a negative lift on the control surface of the external fuel nacelle. To avoid additional induced drag from this negative control surface lift, the sum of the wing- and control surface circulation should be chosen such as to minimize the induced drag of the total lift carrying system; for the same reason the vertical wake displacement of the wing and control surface should be kept small. The outboard wing is particularly thin to minimize local wing sweep and improve high

[4]Near sonic transport airplanes have been proposed and successfully tested in the 8-foot Langley Transonic Wind Tunnel by Whitcomb and his co-workers [Ref. 17].

speed buffeting towards the wing tip, where wing deflections are the largest.

Excessive pressure minima in the front part of the swept-forward wing towards the wing-fuselage juncture are alleviated by a suitable area-ruling of the fuselage or streamline contouring of the upper fuselage section in the area of the wing-fuselage juncture and possibly reducing the wing sweep towards the root. This is feasible by reducing the wing thickness towards the fuselage, which in turn is structurally easily possible in view of the rapidly decreasing wing bending moments in the strut-braced inboard wing area. At the same time $C_{D,\infty}$ of this thinner inboard wing is somewhat lower to partially compensate for the strut parasite drag.

The rapid reduction of the inboard wing thickness decreases at the same time the local Mach number of the upper wing surface in the area of the wing-strut intersection (the local flow is essentially 3-dimensional) to allow a correspondingly thicker wing in this structurally most critical area, where wing bending moments are largest. Flow choking between the lower wing surface and the strut is avoided by suitable cutouts on the lower surface, using local area-ruling and streamline contouring [Ref. 1]. This was demonstrated in the Wright Field transonic wind tunnel on a strut-braced $35^0$ swept wing airplane configuration of aspect ratio 9 with empennage and simulated nacelles (designed and built by the Northrop LFC research group). The measured lift to drag ratio (L/D) of the entire airplane configuration at $M_\infty = 0.89$ and $Re_c = 1.2 \times 10^6$ was 21.2 [Ref. 18].

Wing mounted superfans, mounted in front and below the wing, combined with the external fuel nacelles, induce a negative perturbation velocity at the location of the wing to reduce essentially the local "free-stream" Mach number on the wing. The interaction of a high aspect ratio swept-forward and particularly thin wing towards the fuselage with the front-mounted superfan- and fuel nacelles enables a favorable cross-sectional area distribution of the airplane configuration with relatively minor area cutouts on the fuselage (see area distribution and fuselage contour of M = 0.97 LFC transports, Fig. 18). A Reichardt type body of revolution was chosen for the equivalent body of revolution. Its calculated pressure distributions and supersonic bubbles are shown in Fig. 19 for M = 0.97 and 0.98. A careful area distribution of the outer fuel nacelle with the thin outer wing enables a high local design Mach number in the juncture region of the wing and external fuel nacelle. The $M_\infty = 0.83$ and 0.97 LFC airplanes require $23^0$ and $37^0$ aerodynamic wing sweep, respectively, at the start of the rear pressure rise. To maximize $M_{\infty,Design}$ the wing was laid out such that the isobar sweep of the swept-forward wing is larger than the sweep angle of the constant percent chord lines. This can be accomplished by suitably tailoring the supercritical airfoils along the span (as discussed above).

Figure 20 shows a plot of L/D versus $C_L$ for the M = 0.83 LFC airplane with wing aspect ratio of 19 and 70% laminar flow on wing, empennage, struts and nacelles (fuselage turbulent); with $(L/D)_{max}$ = 39.4 at $C_L \approx$ 0.6. One might reduce $C_L$ somewhat to raise the cruise Mach number at a slight penalty in L/D. For comparison, Fig. 20 shows L/D of the same airplane both with fully turbulent flow (L/D = 27.5) as well as with fully laminar flow on the wing, empennage, struts and nacelles by means of suction and various degrees of suction laminarization on the fuselage. With increasing fuselage laminarization L/D could increase substantially to exceptionally high values ($\approx$ 80) in the ideal case of 100% laminarization. Figure 21 shows a comparison of L/D and $Re_c$ of the strut-braced $M_\infty$= 0.83 airplane and a cantilever comparison LFC airplane (aspect ratio 12) versus $C_L$, showing superior (L/D)'s and substantially lower $Re_c$'s at optimum cruise for the strut- braced design. Assumptions for the $M_\infty$= 0.83 airplane range estimate:

$W_o$ = take-off gross weight = 180,000 kg,
Payload = 50,000 kg = 0.278 $\times W_o$,
Gross weight empty = 0.38 $\times W_o$,
Fuel reserves for take-off, climb, loitering, etc. = 0.06 $\times W_o$,
$(L/D)_{average}$ = 37,
Specific fuel consumption b = 0.48 kg/kg thrust.

The resulting unrefuelled range is 21,564 kilometers or 11,606 nautical miles.

With the larger wing sweep of the M = 0.97 LFC airplane (as compared to the M = 0.83 airplane) its span and aspect ratio decrease to raise $C_{D,i}$ and reduce $(L/D)_{cruise}$ to 35 (39.4 for the $M_\infty$=0.83 airplane). This reduction in L/D is partially compensated by a 3% higher powerplant overall efficiency of the M=0.97 airplane.

Figure 22 shows a plot of L/D versus $C_L$ for the M = 0.97 near sonic LFC transport airplane with wing aspect ratio of 14.3 and 70%

laminar flow on wing, empennage, struts and nacelles (fuselage turbulent); with $(L/D)_{max} = 35$ at $C_L \approx 0.5$. Figure 22 includes (L/D) versus $C_L$ with fully laminar flow by means of suction on wing, empennage, struts and nacelles and various degrees of fuselage suction laminarization. Assumptions for the $M_\infty = 0.97$ airplane range estimate:

$W_o = 180{,}000$ kg take-off gross weight,

Payload = 50,000 kg,

Gross weight empty = $0.38 \times W_o$,

Fuel reserves = $0.06 \times W_o$,

$(L/D)_{average} = 33$,

$\eta_{overall}$ is 3 % higher than for the $M_\infty = 0.83$ airplane.

The resulting unrefuelled range of the near sonic M = 0.97 LFC transport is 19,810 kilometers (10,708 nautical miles), which is about 8.5% smaller than for the $M_\infty = 0.83$ LFC transport.

## 6. Conclusions

70% laminar flow by means of modest boundary layer suction on wing, empennage, nacelles and struts of long range LFC transports, combined with lower span loadings and correspondingly larger wing spans, could enable an unrefuelled range halfway around the world up to near sonic cruise speeds with large payloads even with a turbulent fuselage. Increasing $M_{cruise}$ from 0.83 to 0.97 reduces the range by 8.5%.

The structural weight of such large span wings is minimized by bracing the wing externally by laminarized struts, alleviating wing bending- and torsional loads and deformations by placing laminarized fuel nacelles in the outer wing with active horizontal control surfaces.

The airplane performance optimization with large span LFC wings is compatible with the reduction of $Re_c$ and $\frac{U_\infty}{\nu}$, desirable to alleviate the laminarization problems involved.

Supercritical LFC airfoils with undercut front and rear lower surfaces and upper surface $C_p$-distributions with an extensive low supersonic flat rooftop, preceded by a far upstream supersonic overexpansion and followed by a steep subsonic rear pressure rise (either with suction or a slotted small chord trailing edge cruise flap) raise

$M_{\infty,Design}$ and ease at the same time sweep-induced boundary layer crossflow instability on the upper surface.

Assuming equal $M_{\perp}$ (normal to the isobars) as on a constant chord wing, taper on sweptback wings adds a streamwise flow deceleration to complicate their laminarization and possible require additional suction strips in their upper surface rooftop zone for TS-disturbance control. In contrast, taper is favorable for laminarization of tapered swept-forward wings. Their isobar sweep at the rear shock location is larger than their structural sweep to allow increased wing span and lower induced drag than for equivalent sweptback wings. Swept-forward wing aeroelastic divergence must be eliminated by sweeping the spanwise composite fibers ahead of the elastic axis, sweeping the outboard wing back and applying active control, using for this purpose the cruise flap and fuel nacelle control surface.

$M_{\infty,Design}$ of tapered swept-forward LFC wings can be raised further by suitably tailoring the airfoils along the span, such that the upper surface isobar sweep is larger than the constant percent chord line sweep. A favorable 3-dimensional aerodynamic design integration of the airplane configuration and its components at high subsonic speeds maximizes $M_{\infty,Design}$ of the airplane (for a given wing sweep).

In the high subsonic speed range wing mounted superfans, possibly combined with a fuselage boundary layer propulsion engine, reduce fuel consumption and engine tone noise in the range of amplified boundary layer disturbances. The fan nacelles may be used to reduce the local flow velocities in the area of the swept-forward wing and thus appear particularly attractive especially if their surfaces can be suction laminarized.

## 7. Appendix

Performance estimates were conducted for a M=0.83 LFC transport(slightly smaller than that of Fig. 1).

Take-off gross weight $W_o$ = 180,000 kg

Take-off wing loading $\frac{W_o}{S}$ = 500 kg/$m^2$

Wing projected area S = 360 $m^2$

Aspect ratio $b^2/S = 18$

Wing span b = 80.5 meters

$M_{cruise} = 0.83$

Assumptions in the drag analysis:

An induced drag factor $\kappa = 1.02$ was assumed for the plain wing without tip feathers. Tip feathers reduce $C_{D,i}$ by 8%, leading to $\kappa = 0.94$. For the fully turbulent fuselage with $Re_L = 270.0 \times 10^6$ at the start of cruise ($C_L = 0.6$) a fuselage drag coefficient $C_{D,fuselage} = 2.0 \times 10^{-3}$ (based on fuselage wetted area) was assumed (At $Re_L = 270.0 \times 10^6$ the incompressible turbulent flat plate drag coefficient is $1.8 \times 10^{-3}$. At high subsonic speeds this value is somewhat lower). Riblets decrease this value by 7%. Using boundary layer propulsion in the rear fuselage decreases the equivalent fuselage drag coefficient by another 10%. Hence, with a fuselage wetted area A = 848 $m^2$ the fuselage drag is $C_{D,fuselage} = 0.00394$(based on wing projected area S = $360 m^2$). Thus, the airplane parasite drag coefficients at $C_L = 0.6$, $S = 360 m^2$ are as follows:

Fuselage (turbulent) = 0.00394

Empennage (70% laminar) = 0.00080

Struts (75% laminar) = 0.00050

Nacelles ($\sim$ 2/3 laminar) = 0.00100

Total parasite drag = 0.00624

Some turbulent wedges have been included in this parasite drag estimate.

Start of Cruise:

| $C_L$ | .6 | .55 | .5 | .65 | .7 |
|---|---|---|---|---|---|
| $Re_{\bar{c}} \times 10^{-6}$ | 19.8 | 21.6 | 23.8 | 18.3 | 17.0 |
| $C_{D,i} \times 10^3$ | 5.984 | 5.028 | 4.155 | 7.022 | 8.144 |
| $C_{D,\infty,wing} \times 10^3$ | 3.000 | 2.910 | 2.814 | 3.085 | 3.167 |
| $C_{D,parasite} \times 10^3$ | 6.240 | 6.100 | 5.950 | 6.370 | 6.490 |
| $C_{D,total} \times 10^3$ | 15.224 | 14.038 | 12.919 | 16.477 | 17.801 |
| L/D | 39.41 | 39.18 | 38.70 | 39.45 | 39.32 |
| $(\frac{U_\infty}{\nu}) \times 10^{-6}/m$ | 4.43 | 4.83 | 5.32 | 4.09 | 3.80 |
| H in ft | 43500 | 41600 | 39600 | 45100 | 46700 |
| in m | 13246 | 12670 | 12060 | 13730 | 14220 |

Practically all turbulent comparison airplane (b=80.5 meters, $S = 360\ m^2$, $W_o = 180{,}000$ kg).

Parasite drag at $C_L = 0.6$: Fuselage without riblets and boundary layer propulsion

$$C_{D,fuselage} = 0.0045$$
$$C_{D,empennages} = 0.0020$$
$$C_{D,struts} = 0.0005 \text{ (75\% laminar)}$$
$$C_{D,nacelle} = 0.0020$$
$$C_{D,parasite} = 0.0090 \text{ (based on } S = 360\ m^2)$$

| $C_L$ | 0.5 | 0.6 | 0.7 | 0.8 |
|---|---|---|---|---|
| $C_{D,i}$ | 0.004155 | 0.00598 | 0.00815 | 0.01064 |
| $C_{D,\infty}$ | 0.007230 | 0.00750 | 0.00775 | 0.00797 |
| $C_{D,parasite}$ | 0.008670 | 0.00900 | 0.00928 | 0.00952 |
| $C_{D,total}$ | 0.021450 | 0.02248 | 0.02518 | 0.02813 |
| $L/D$ | 23.3 | 26.7 | 27.8 | 28.4 |

Since $(L/D)_{max} \sim \sqrt{\frac{b^2/S}{C_{D,\infty+parasite}}}$, $(L/D)_{cruise}$ of the strut-braced turbulent airplane of aspect ratio 18 is considerably higher than that of present high subsonic speed turbulent flow transports. Using the Boeing 707 transport($b^2/S = 7$) and assuming the same $C_{D,\infty+parasite}$, $(L/D)_{max}$ would extrapolate to 28.9 for $b^2/S = 18$. Similarly, using Airbus A340 ($b^2/S = 9.4$ and L/D = 21) L/D at $b^2/S = 18$ would extrapolate to 29. These extrapolated L/D- values are close to the values of the strut-braced turbulent airplane of aspect ratio 18, taking into account its additional strut parasite drag in the performance estimate.

Similar airplane performance estimates were conducted with full length laminar flow by means of boundary layer suction on wing, empennage surfaces, struts, nacelles and various percentage amounts of laminar flow on the fuselage by means of distributed boundary layer suction.

## 8. Acknowledgements

The first author is indebted to R. T. Jones, H. Lomax, J. Spreiter, R. Whitcomb, and P. Rubbert for many valuable discussions on 3-dimensional integration concepts of high subsonic and near sonic speed airplanes. The authors would like to thank Jerry Hefner, Assistant Chief of Fluid Mechanics Division, and Richard Wagner and Dal Maddalon of the Laminar Flow Control Project Office for their support. The research reported herein was sponsored by NASA Langley Research Center under Contract NAS1-18235 and NAS1-18599.

### Notation

| | |
|---|---|
| A | Boundary layer disturbance amplitude |
| $A_O$ | Initial disturbance amplitude |
| a | Sound velocity |
| c | Airfoil or wing chord |
| $\bar{c}$ | Average wing chord |
| $C_{D,i}$ | Lift induced vortex drag coefficient |
| $C_{D,o}$ | Zero lift drag coefficient |
| $C_{D,\infty}$ | Equivalent profile drag coefficient of low drag suction wing |
| $C_L$ | Lift coefficient |
| $C_{m,c/4}$ | Airfoil pitching moment coefficient (with respect to c/4) |
| $C_p = \frac{2p}{\rho U_\infty^2}$ | Surface static pressure coefficient |
| D | Drag |
| $D_i$ | Induced vortex drag |
| $\Delta C_p = \frac{2\Delta p}{\rho U_\infty^2}$ | Nondimensional surface static pressure jump across off-design flow discontinuities and shocks |
| $c_Q = \frac{\rho_{wall} v_o}{\rho_{amb.} U_{flight}}$ | Nondimensional local suction mass-flow coefficient |
| $C_Q = \frac{\rho_{wall} v_o}{\rho_{amb.} U_{flight}}$ | Nondimensional total suction mass-flow coefficient |
| h | Height of supersonic bubble |
| $l$ | Length of supersonic bubble |
| $M_\infty$ | Free-stream Mach number |
| $n = ln(A/A_O)$ | Logarithmic growth factor of amplified boundary layer disturbances |

| | |
|---|---|
| $Q_a$ | Total suction volume in incompressible flow |
| r | Radius of curvature |
| $Re_c$ | Reynolds number based on chord |
| $Re_n$ | Boundary layer cross flow Reynolds number based on $w_{n,max}$ and $\delta_{0.1}$ |
| $Re_\theta$ | Reynolds number based on boundary layer momentum thickness |
| S | Wing projected area |
| $t/c$ | Airfoil thickness ratio |
| $U_\infty$ | Free-stream velocity (normal to leading edge) |
| $U_{flight}$ | Flight speed |
| $v_o$ | Equivalent area suction velocity |
| $w_n$ | Boundary layer crossflow velocity |
| $w_{n,max}$ | Maximum boundary layer crossflow velocity |
| x | Chordwise distance |
| y | Vertical coordinate |
| $\alpha$ | Effective wing angle of attack |
| $\delta$ | Total boundary layer thickness |
| $\delta_{0.1}$ | Boundary layer thickness where $w_n = 0.1 w_{n,max}$ |
| $\varphi$ | Wing sweep angle |
| $\theta$ | Boundary layer momentum thickness |
| $\eta_{overall}$ | Overall efficiency of propulsion system |
| $\mu$ | Absolute viscosity |
| $\nu = \frac{\mu}{\rho}$ | Kinematic viscosity |
| $\rho$ | Density |
| $\chi_{min}$ | Minimum boundary layer crossflow stability limit Reynolds number based on $w_{n,max}$ and $\delta_{0.1}$ |

Subscripts

| | |
|---|---|
| amb | Ambient |
| A | Start of cruise |
| CF | Crossflow |
| max | Maximum |
| min | Minimum |
| opt | Optimum |
| Tr | Transition |
| TS | Tollmien-Schlichting |

# References

[1] Pfenninger, W., "Design considerations of large global range high subsonic speed LFC transport airplanes," AGARD-654, Agard/VKI Special Course on Concepts for Drag Reduction, Rhode-St. Genèse, Belgium, 1977.

[2] Pfenninger, W., "Design considerations of long range and endurance LFC airplanes with practically all laminar flow," George Washington University Report, August 1982.

[3] Whitcomb, R. and Clark, L., "An airfoil shape for efficient flight at supercritical Mach numbers," NASA TM X-1109, July 1965.

[4] Gross, L. W. and Pfenninger, W., "Experimental and theoretical investigation of a Reichardt body of revolution with low drag suction in the NASA Ames 12-foot pressure tunnel," Northrop Report NOR63-46, BLC148, 1963.

[5] Pfenninger, W., "USAF and NAVY sponsored Northrop LFC research between 1949 and 1967," AGARD-654, Agard/VKI Special Course on Concepts for Drag Reduction, Rhode-St. Genèse, Belgium, 1977.

[6] Pfenninger, W. and Bacon, J. W., "About the development of swept laminar suction wings with full chord laminar flow," Boundary Layer and Flow Control, G. V. Lachmann, editor, Vol. 2, pp. 1007-1032, 1961.

[7] Pfenninger, W., "Some results from the X-21 program. Part I: Flow phenomena at the leading edge of swept wings," Recent Developments in Boundary Layer Research, Agardograph 97, Vol. IV, Naples, May 1965.

[8] Fowell, L. R. and Antonatos, P. P., "X-21 laminar flow control flight test results," Recent Developments in Boundary Layer Research, Part IV, Agardograph 97, Naples, May 1965.

[9] Pfenninger, W. and Groth, E., "Low drag boundary layer suction experiments in flight on a wing glove of an F94A airplane with suction through a large number of fine slots," Boundary Layer and Flow Control, G. V. Lachmann, editor, Vol. 2, 1961.

[10] Pfenninger, W., Viken, J. K., Vemuru, C. S., and Volpe, G., "All laminar supercritical LFC airfoils with laminar flow in the main wing structure," AIAA Design and Technology Meeting, Dayton, Ohio, AIAA 86-2625, October 20-22, 1986. Turbulence Management and Reattachment, H. W. Liepmann and R. Narasimha (Eds.), Springer-Verlag, IUTAM Symposium, Bangalore/India, 1987.

[11] Pfenninger, W., Reed, H. L., and Dagenhart, J. R., "Design considerations of advanced supercritical low drag suction airfoils," Viscous Flow Drag Reduction, edited by G. R. Hough, Vol. 72 of Progress in Astronautics and Aeronautics, 1980. Presented at the Symposium for Viscous Drag Reduction, Dallas, Texas, November 1979.

[12] Pfenninger, W., "All laminar sailplanes with low drag boundary layer suction," 18th Ostiv conference, Hobbs, New Mexico, 1983.

[13] Malik, M., "COSAL - A black box compressible stability analysis code for transition prediction in 3-dimensional boundary layers," NASA Contractor Report 165925, Contract NAS1-16919, May 1982.

[14] Vijgen, P., Dodbele, S., Pfenninger, W., and Holmes, B., "Analysis of wind tunnel boundary layer transition experiments on axisymmetric bodies at transonic speeds using compressible boundary layer theory," AIAA Paper No. 88-0008, Reno, Nevada, January 1988.

[15] Horstmann, K. H., Redecker, G., Quast, A., Dressler, U., and Bieler, H., "Flight tests with a natural laminar flow glove on a transport aircraft," AIAA Paper No. 90-3044, Portland, Oregon, August 1990.

[16] Lecture presented by the Rolls-Royce Chief Engineer at the Delft Technical University, September 1985.

[17] Langhans, R. A. and Flechner, S. G., "Wind-tunnel investigation at Mach numbers from 0.25 to 1.01 of a transport configuration designed to cruise at near-sonic speeds," NASA TM-X-2622, August 1972.

[18] Bacon, J. W., Fiul, A., and Pfenninger, W., "WADC 10-foot transonic wind tunnel tests on strut-braced boundary layer airplane," Northrop Report NAI-57-826, BLC-99, 1957.

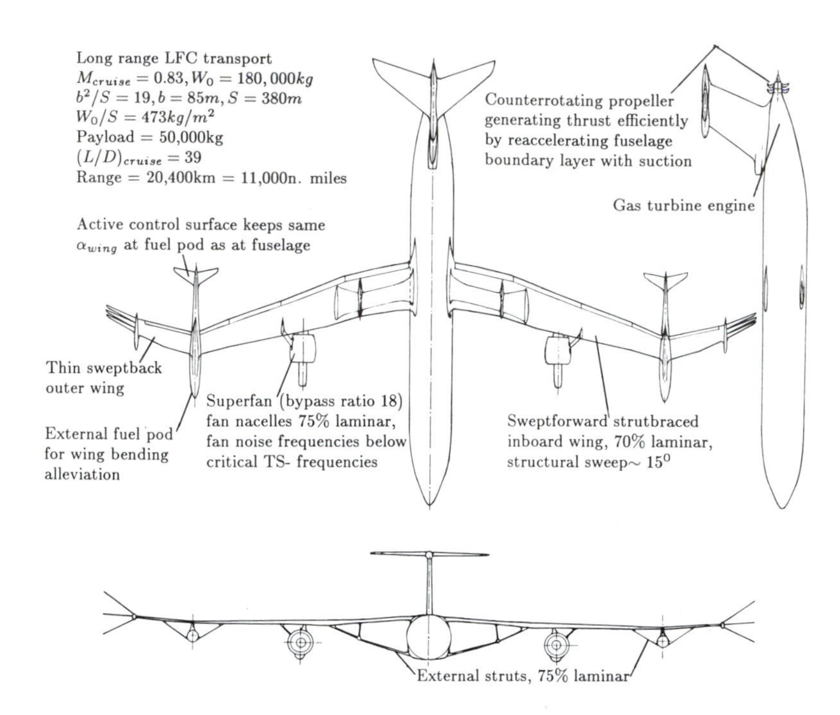


**Fig. 1.** $M_\infty$ **= 0.83 long range LFC transport airplane.**

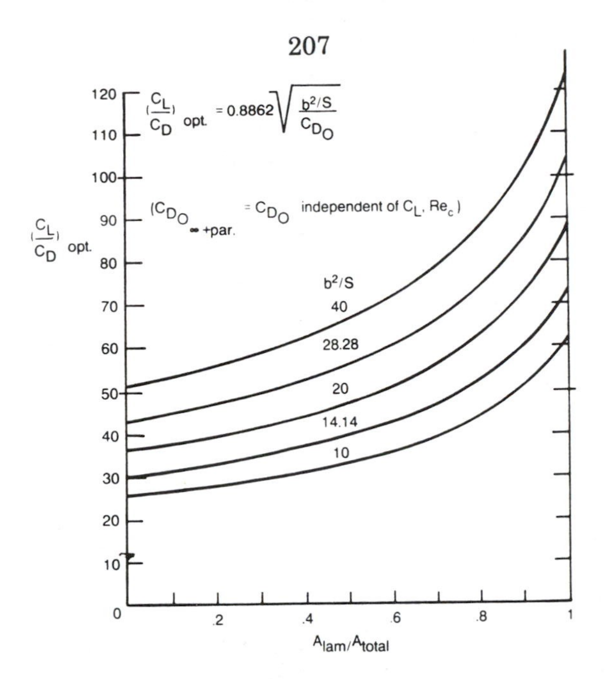


Fig. 2. $(C_L/C_D)_{opt}$ versus $A_{Laminar}/A_{total}$ for various wing aspect ratios $b^2/S$, with $C_{D,o} = C_{D,\infty} + C_{D,parasite}$ varying linearly from 0.012 to 0.02 for $A_{Laminar}/A_{total} = 0$ to 1.0.

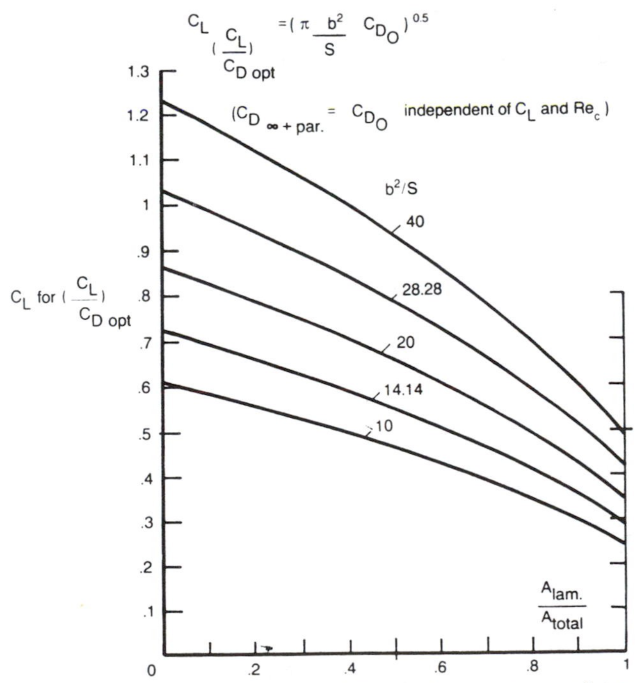


Fig. 3. $C_{L,opt}$ versus $A_{Laminar}/A_{total}$ for various $b^2/S$ values.

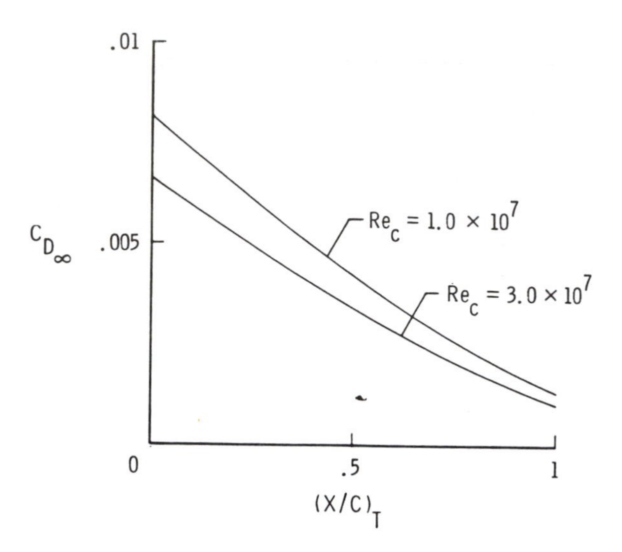


Fig. 4. Variation of $C_{D,\infty}$ with $(x/c)_{T_r}$ of $23^0$ swept SC LFC airfoil for $Re_c = 10.0 \times 10^6$ and $30.0 \times 10^6$.

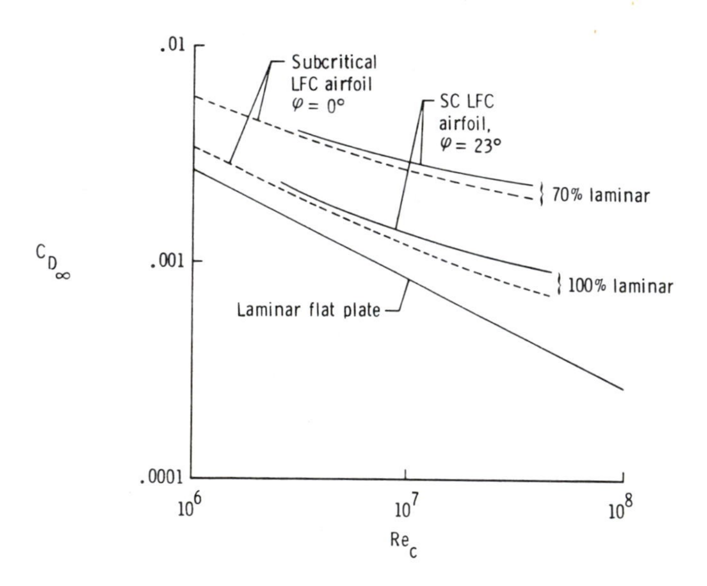


Fig. 5. Variation of $C_{D,\infty}$ with $Re_c$ for $(x/c)_{T_r} = 0.7$ and 1.0.

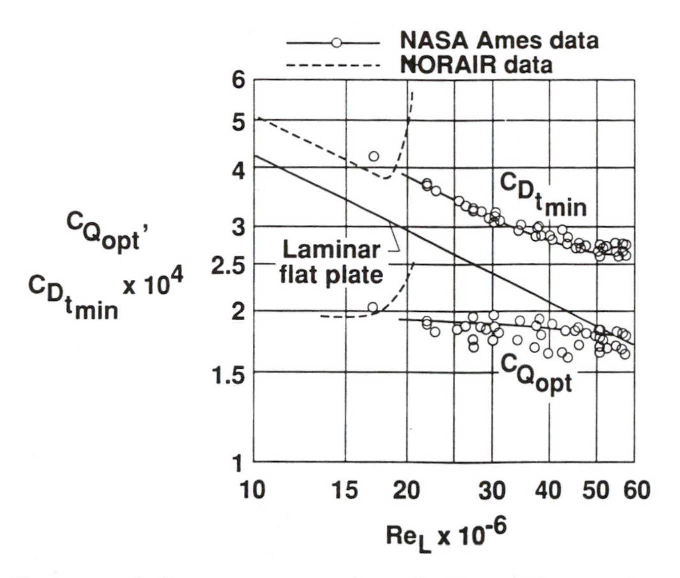


Fig. 6. $C_{Q,opt}$ and $C_{D,t,min}$ versus length Reynolds number $Re_L$ for Reichardt LFC body of revolution. $C_Q = \frac{Q_a}{u_\infty S}$.

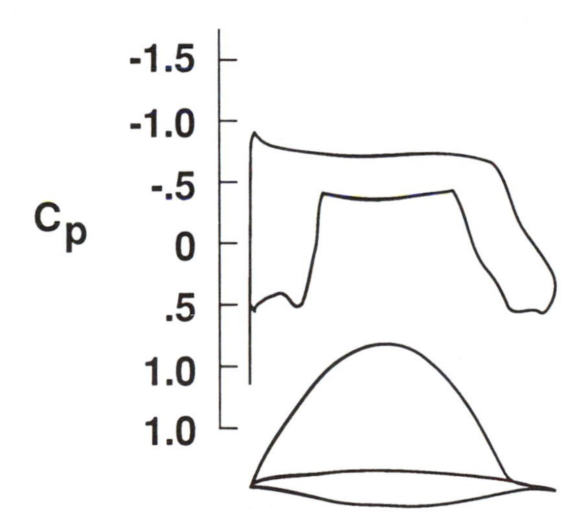


Fig. 7. Pressure distribution and supersonic bubble on X66 SC LFC airfoil at the design condition. $M_\infty = 0.780$, $\alpha = -0.65^0$ and $C_L = 0.608$.

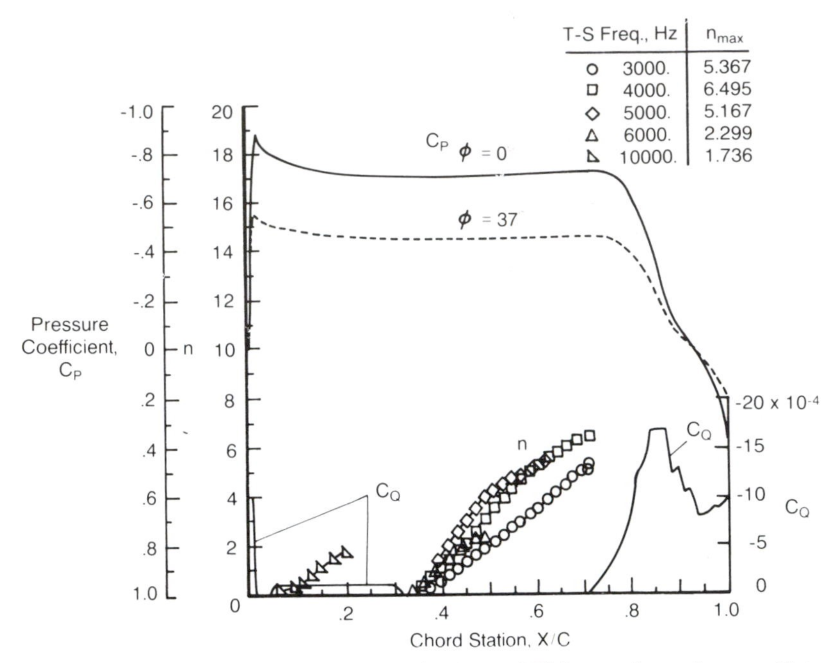


Fig. 8a. X66 SC LFC airfoil at design: TS-boundary layer disturbance vortex growth $n_{TS}$ and suction massflow distribution $c_Q$ on upper surface for $\varphi = 37^0$ sweep, $M_\infty = 0.98$.

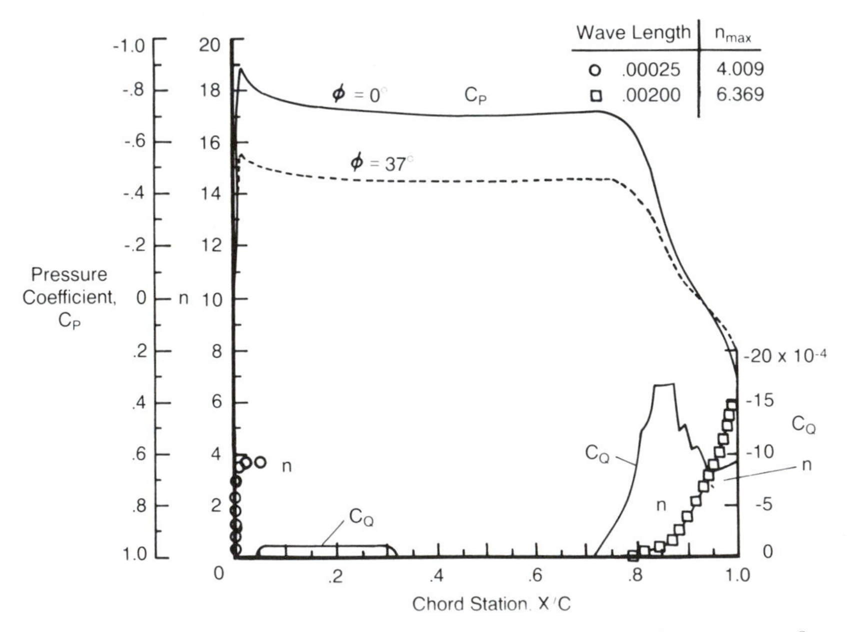


Fig. 8b. X66 SC LFC airfoil at design: Boundary layer crossflow disturbance vortex growth $n_{CF}$ and suction massflow distribution $c_Q$ on upper surface for $\varphi = 37^0$ sweep, $M_\infty = 0.98$.

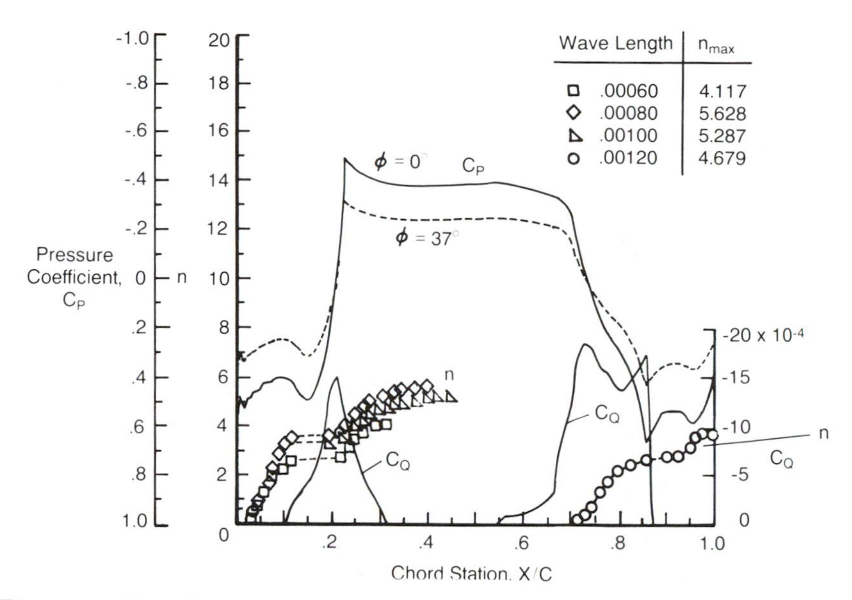


Fig. 9a. X66 SC LFC airfoil at design: TS-boundary layer disturbance vortex growth $n_{TS}$ and suction massflow distribution $c_Q$ on lower surface for $\varphi = 37^0$ sweep, $M_\infty = 0.98$.

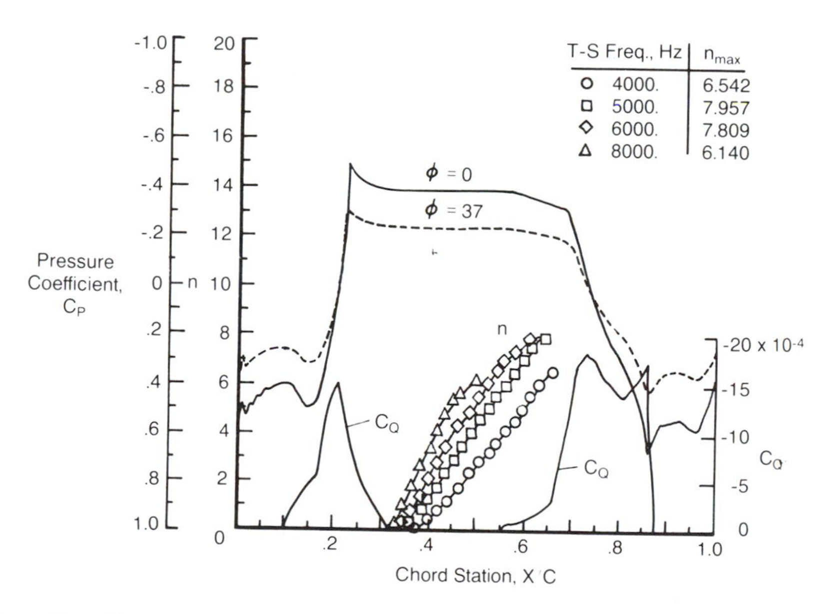


Fig. 9b. X66 SC LFC airfoil at design: Boundary layer crossflow disturbance vortex growth $n_{CF}$ and suction massflow distribution $c_Q$ on lower surface for $\varphi = 37^0$ sweep, $M_\infty = 0.98$.

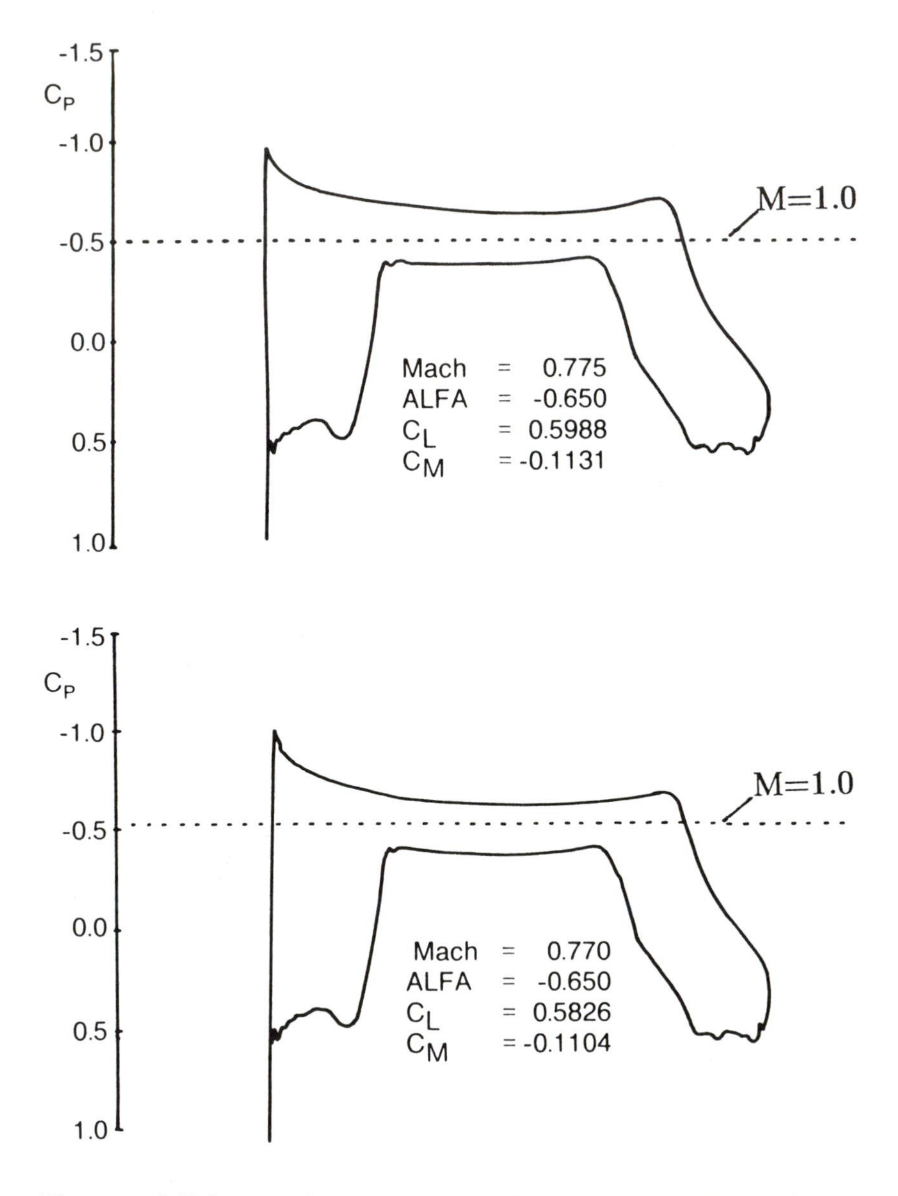


Fig. 10. Off-design $C_p$-distributions for X66 SC LFC airfoil.

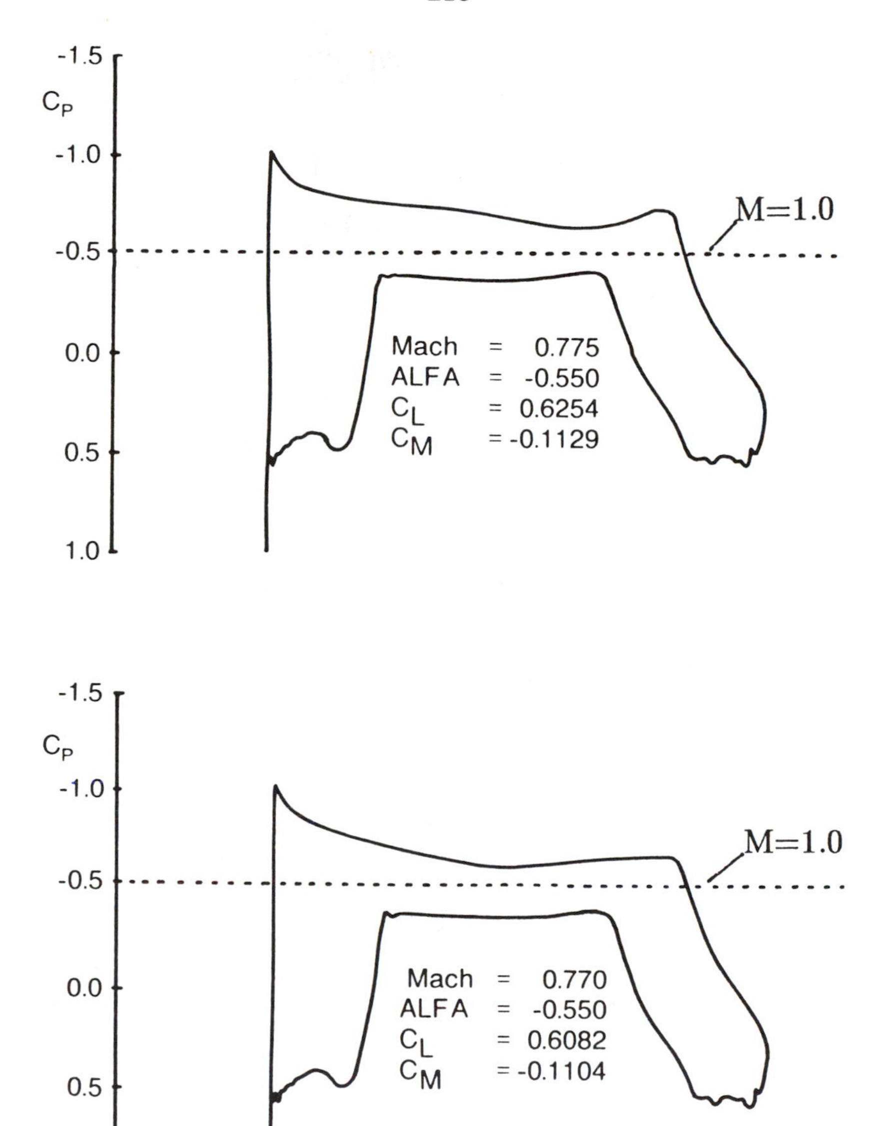


Fig. 10. Off-design $C_p$-distributions for X66 SC LFC airfoil.

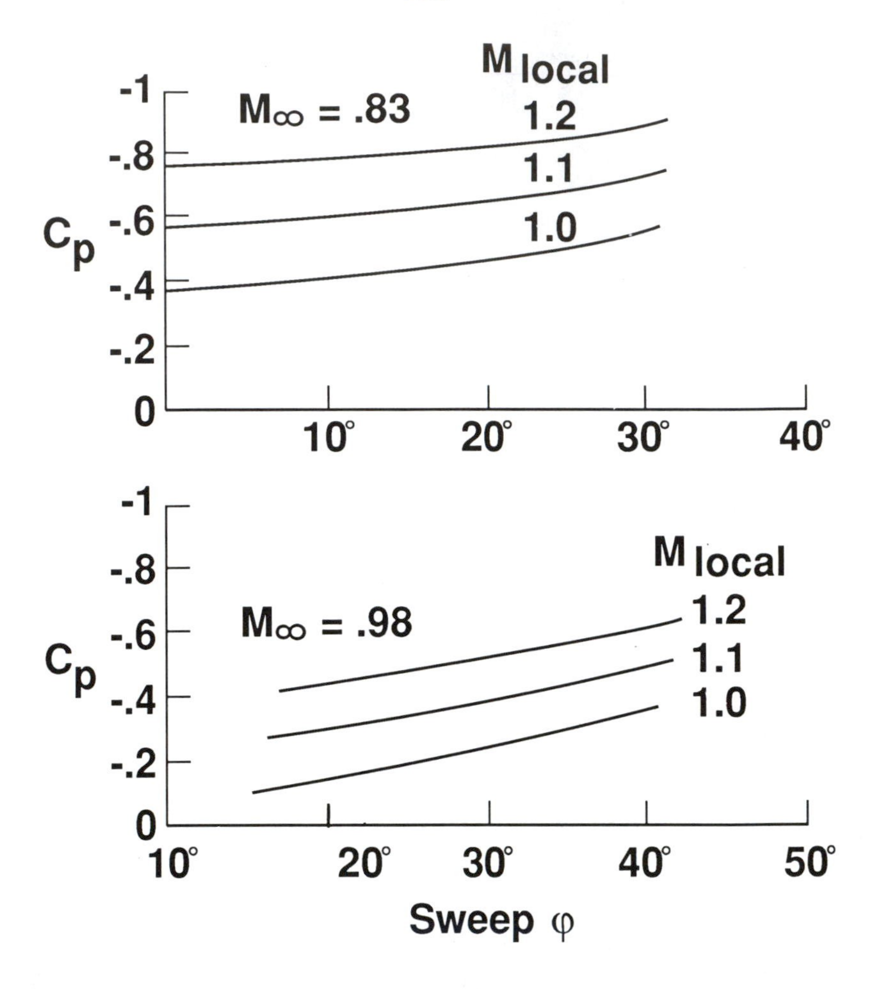


Fig. 11. Variation of $C_p$(based on $\frac{\varrho}{2}U_\infty^2$) with isobar sweep $\varphi$ for $M_\perp$ = constant, $M_{flight}$ = 0.83 and 0.97.

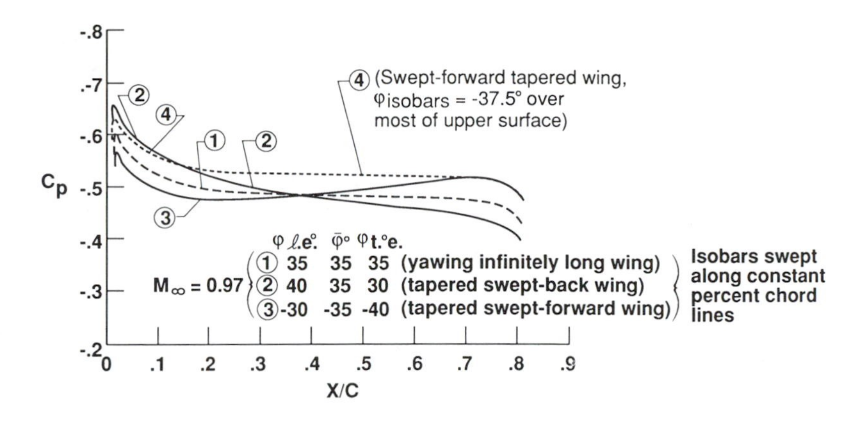


Fig. 12. $C_p$(based on $\frac{\varrho}{2}U_\infty^2$) versus (X/C) for tapered sweptback and swept-forward wing with high $M_{\infty,Design}$.

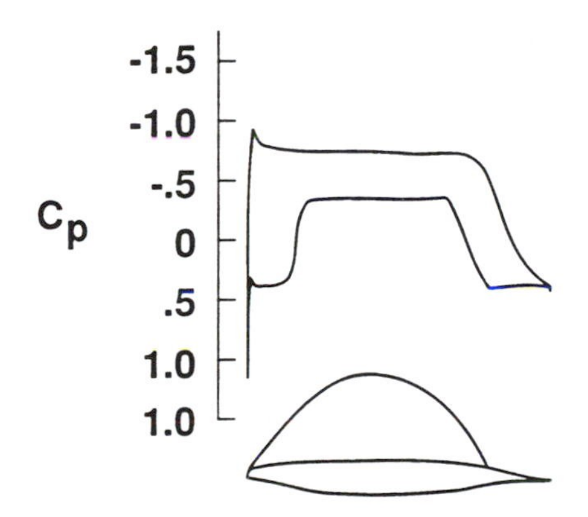


Fig. 13a. Design pressure distribution of PFNIR2 airfoil. $M_\infty =$ 0.764, $\alpha = 0^0$ and $C_L = 0.530$.

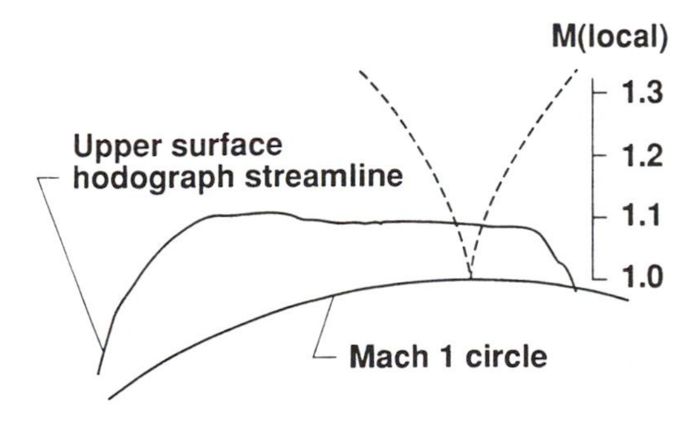


Fig. 13b. Upper surface hodograph plot at design for PFNIR2 airfoil.

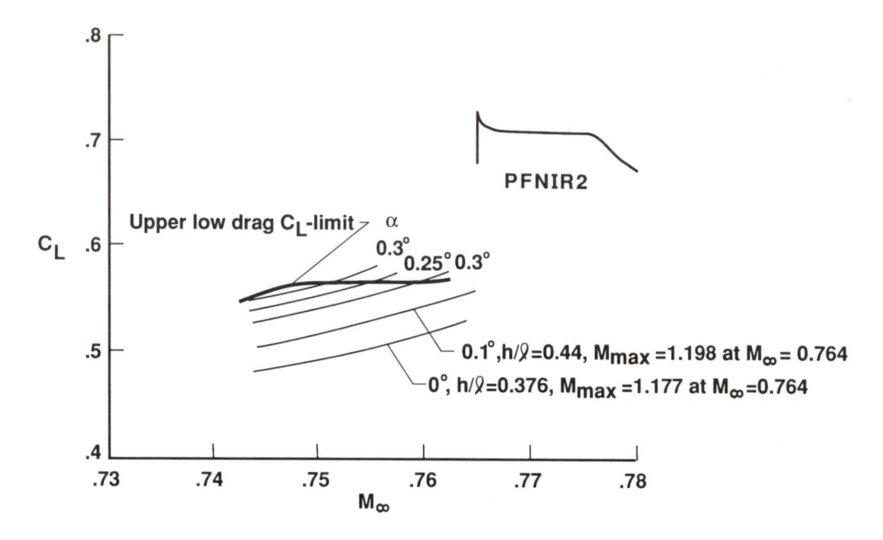


Fig. 14. Upper shockfree low drag $C_L$-limit for PFNIR2 SC LFC airfoil.

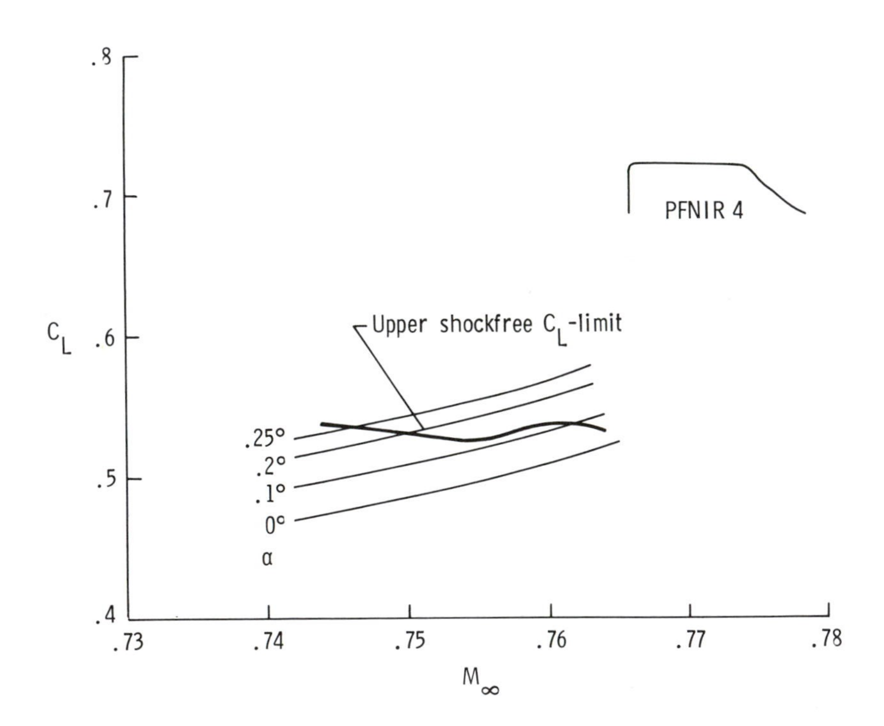


Fig. 15. Upper shockfree low drag $C_L$-limit of SC LFC airfoil with a flat upper surface rooftop without a front pressure minimum.

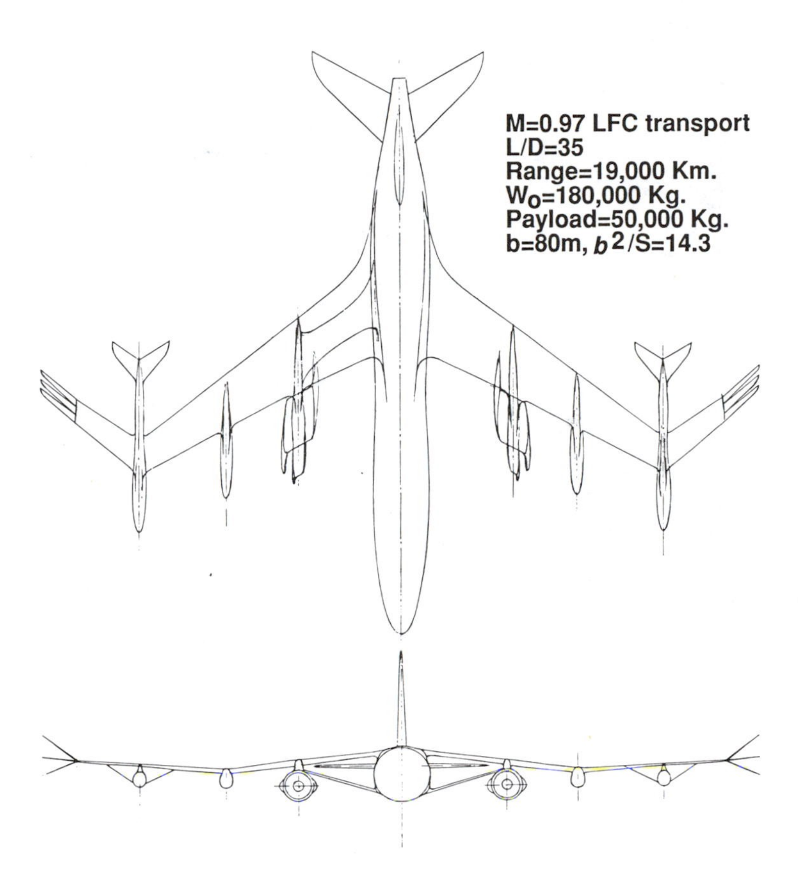


Fig. 16. $M_\infty = 0.97$ near sonic long range LFC transport airplane.

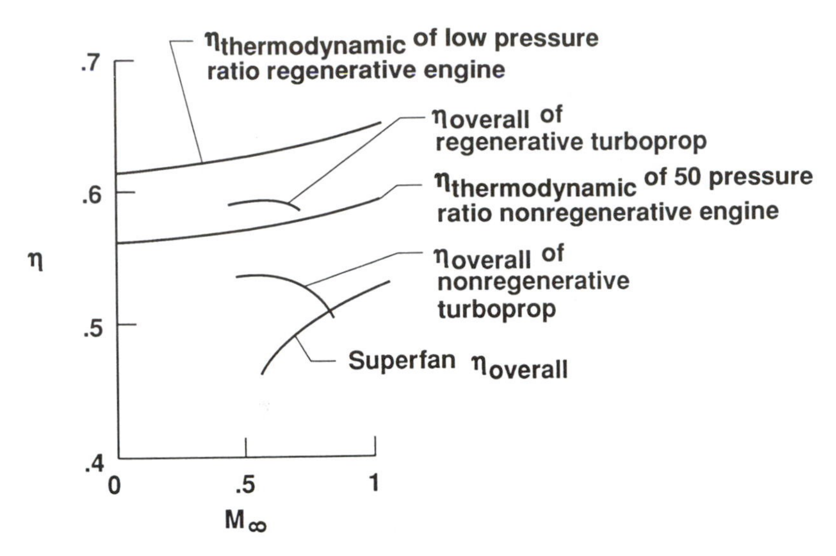


Fig. 17. Thermodynamic and overall efficiency versus $M_\infty$ of superfans (BPR = 18) and high speed turboprops.

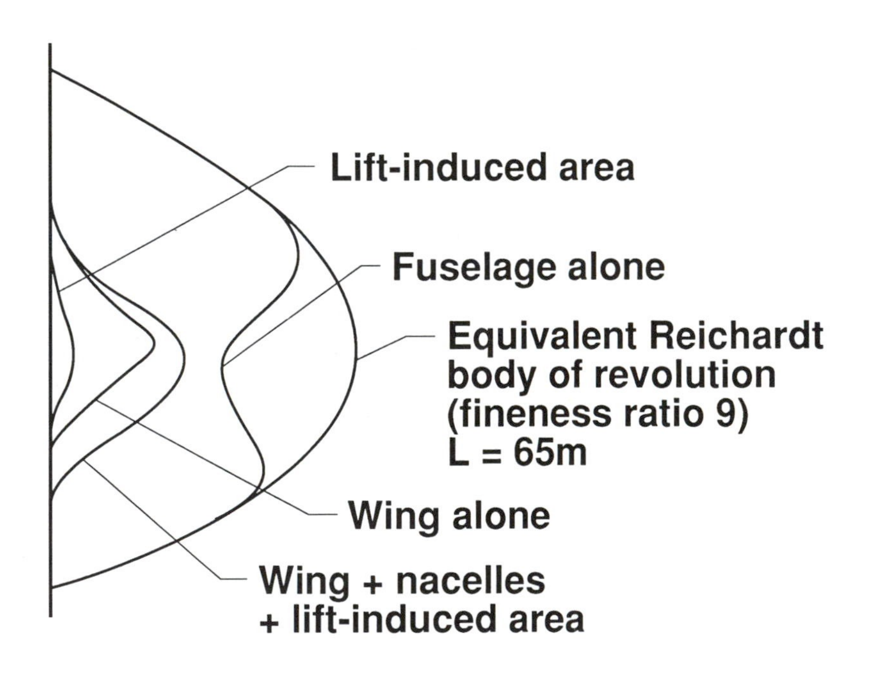


Fig. 18. Crossectional area distribution of $M_\infty = 0.97$ LFC transport (transonic area ruling).

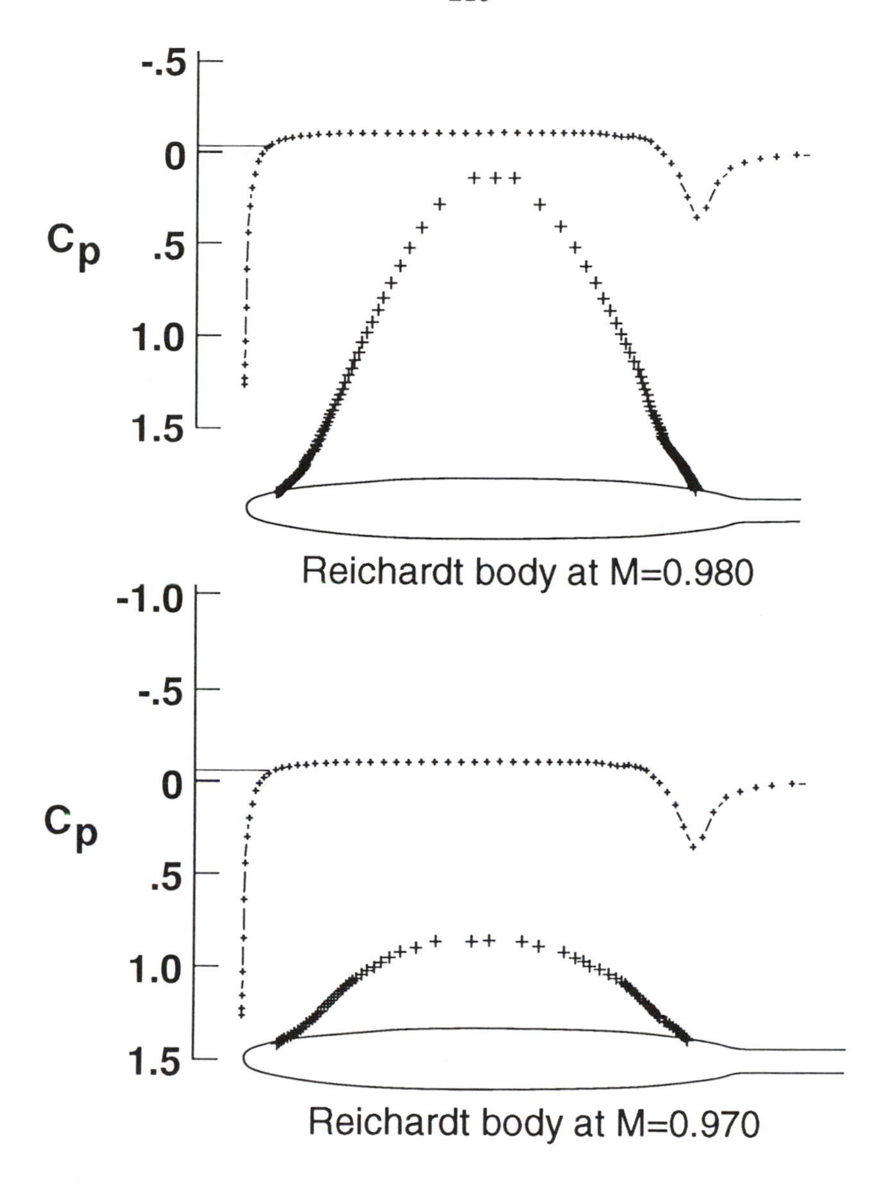


Fig. 19. Pressure distribution and supersonic bubble of equivalent Reichardt body of revolution at $M_\infty = 0.97$ and 0.98.

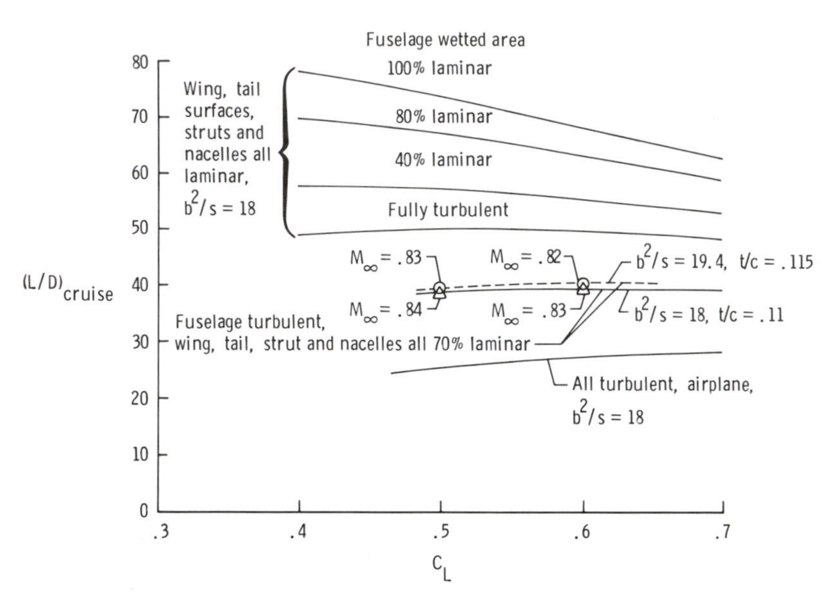


Fig. 20. $M_\infty = 0.83$ LFC transport airplane: $(L/D)_{cruise}$ versus $C_L$ for different extents of laminar flow.

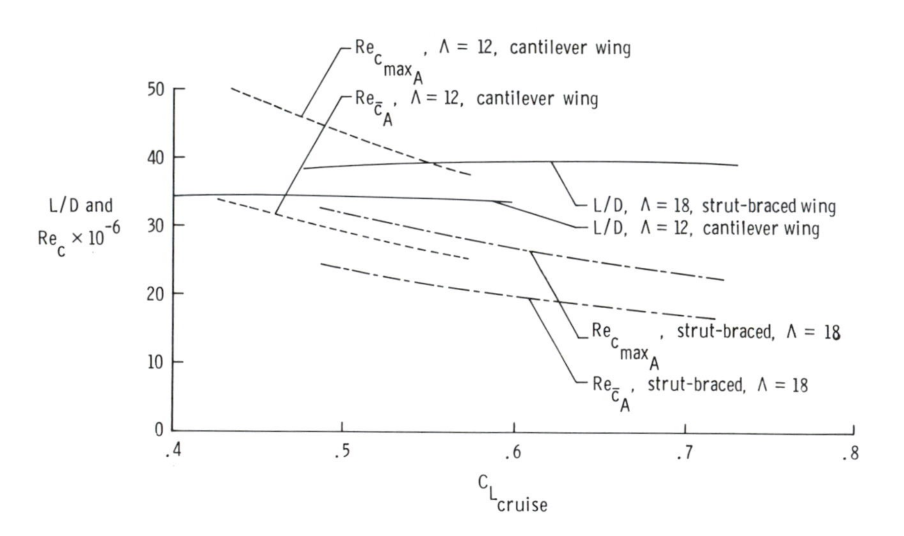


Fig. 22. $M_\infty = 0.97$ near sonic LFC transport airplane: $(L/D)_{cruise}$ versus $C_L$ for different extents of laminar flow.

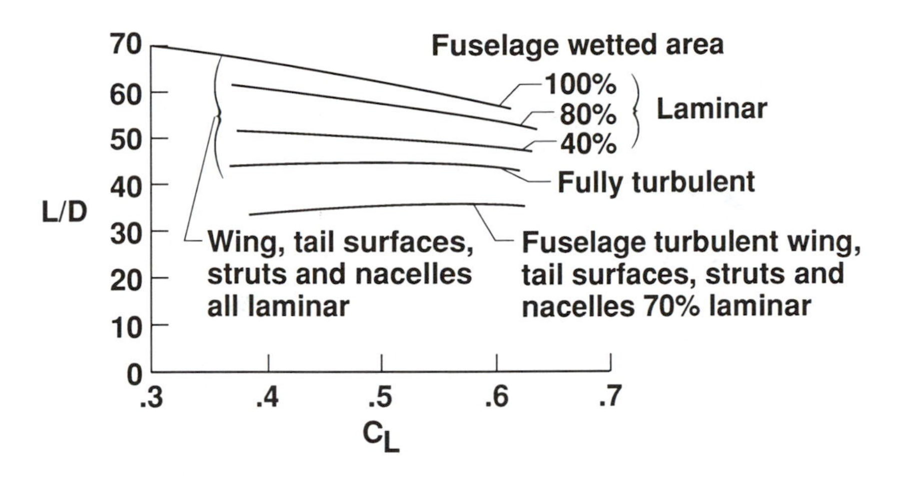


Fig. 21. $M_\infty = 0.83$ LFC transport airplane: $(L/D)_{cruise}$ and $Re_{\bar{c}}$ for strut-braced airplane ($b^2/S = 18$) and cantilever airplane ($b^2/S = 12$).

# WAVE INTERACTION THEORY AND LFC

*Philip Hall*

Department of Mathematics
North Park Road
Exeter, Devon, EX4 4QE, UK
and
Institute for Computer Applications in Science and Engineering
NASA Langley Research Center
Hampton, VA 23665

## 1. Introduction

The complex high Reynolds number flows over modern swept wings can support at least three types of hydrodynamic instability; namely Görtler, Crossflow, and Tollmien-Schlichting instability mechanisms which can all cause transition to turbulence in many flows of practical importance. In the LFC context, it is important for us to know whether the simultaneous presence of these mechanisms can cause the premature onset of transition through their mutual interaction. However, since the mechanisms can only interact in the nonlinear regime, and transition to turbulence quickly follows the onset of nonlinearity, it is clear from the outset that the massive destabilizations of flows attributed to wave interactions cannot be explained by any nonlinear interaction theory. Nevertheless, it is important to understand whether the nonlinear interactions which occur when an instability mechanism has produced a disturbance of finite size can lead to the catastrophic growth of other instabilities which would otherwise have remained small. Given that our present tools for transition prediction are so crude that they have no input from nonlinear theories, the work we discuss here is perhaps most relevant to a situation where the aim is to control a particular type of disturbance once nonlinear effects have been induced. Indeed, we will see below that nonlinearly interacting disturbances can produce the explosive growth of wave systems which would have equilibrated in the absence of any interaction. Nevertheless, it should be remembered that in the aerodynamic situation our discussion is relevant to perhaps only the last 10% of the laminar flow region on a wing. Thus, given the margin of error associated with available transition prediction methods, it is not surprising that these methods have no input from nonlinear theories.

In the following section, we shall discuss the interaction of crossflow vortices with either other crossflow disturbances or with Tollmien-Schlichting waves. In Section 3, we discuss the interaction of Görtler vortices and Tollmien-Schlichting waves and show that the theories formulated for such interactions lead naturally to a self-consistent account of the early stages of transition in a flat plate boundary layer. In this latter context, the interaction of oblique Tollmien-Schlichting waves plays the role of curvature in the Görtler problem. In Section 4, we discuss some further wave interaction problems and draw some conclusions.

## 2. Interactions in the Presence of Crossflow Vortices

The importance of the crossflow vortex mechanism was shown many years ago by Gregory, Stuart, and Walker (1955). This inviscid disturbance has the property that the effective velocity profile associated with the mode has zero velocity at the inflexion point. This means that in the neutral state this mode is stationary. There is in fact a second stationary mode of instability for three-dimensional boundary layers, Hall (1986). This mode is caused by viscous effects and, since it corresponds to an effective velocity profile with zero shear stress at the wall, is perhaps most relevant to separating boundary layers. Various nonlinear aspects of the Gregory, Stuart, and Walker mode have been made clear by Bassom and Gajjar (1988) and, as one would expect, the nonlinear structure of this mode is governed by nonlinear critical layer theory. Clearly, any nonlinear interaction theory must take account of this structure; unfortunately, this was not the case in early investigations of wave interactions involving crossflow vortices. Finally, before giving a discussion of the various theories, we note that the stationary crossflow vortex is not in general the most amplified inviscid disturbance of a three-dimensional boundary layer. However, most experimental investigations of three-dimensional boundary layers reveal the presence of the stationary mode; the role of the more amplified unsteady modes has yet to be elucidated .

Perhaps the first discussion of a wave interaction problem involving crossflow vortices was that given by Reed (1984). In that calculation, Reed considered the interaction of a crossflow vortex with a subharmonic of the primary mode. Reed found that in the presence of the crossflow vortex the subharmonic was significantly amplified. However, this result has not been reproduced by Navier-Stokes calculations, and Reed's expansion procedure is open to severe criticism since the subharmonic mode is not nearly neutral. Thus, since the type of secondary instability analysis performed in Reed's work can only be justified if the interacting waves are nearly neutral, the validity of her results are not established. The alternatives to the type of secondary instability analysis performed by Reed are (a) full numerical simulations or (b) asymptotic wave interaction theories.

The numerical simulations of Malik (1986a) did not reproduce the strong subharmonic growth predicted by Reed and observed experimentally by Saric and Yeates (1985). In fact, other experiments have not indicated that subharmonic growth is the dominant mechanism for the destabilization of crossflow vortices.

An analysis of the type performed by Reed was also carried out by Fischer and Dallmann (1987) who found that crossflow vortices suffer a secondary instability to oblique travelling waves. As in Reed's work, the amplitude of the primary vortex was specified and a Floquet analysis of this linear vortex structure was performed. This type of approach ignores the influence of the crossflow vortex harmonics and the mean flow correction which in fact determine the amplitude of the primary wave. Thus, the results of Fischer and Dallmann (1987) could well be significantly affected by this rather severe approximation. In fact, the error associated with this so-called "shape assumption" is almost certainly larger for crossflow vortices than it is for Tollmien-Schlichting waves because, as Bassom and Gajjar (1988) show, the nonlinear stages of crossflow vortex development are dominated by nonlinear critical layers with phase differences across the layers quite different from their linear values. This structure is not present in the Reed, Fischer, and Dallmann calculations due to their representation of a crossflow vortex as a linear wave of arbitrarily specified amplitude .

More recently Bassom and Hall (1989a,b) have considered interaction problems using a formal asymptotic procedure which utilizes the largeness of the Reynolds number of the flow. In the first paper, it is shown that unsteady crossflow vortices can interact nonlinearly as a resonant triad. There is apparently no triad interaction involving the stationary crossflow vortex. The fact that triads exist in three-dimensional boundary layer flows of practical importance is perhaps significant because the explosive growth of the disturbance amplitudes typical of triad interactions might compress the nonlinear interaction region where transition occurs. Furthermore, if it is indeed this triad interaction which ultimately causes transition, prediction methods should perhaps be more concerned with the linear growth of the triad modes rather than the most amplified linear disturbance.

In the second paper, Bassom and Hall (1989b) considered the linear instability of a crossflow vortex of finite amplitude. The crossflow vortex structure was obtained using the asymptotic framework described by Bassom and Gajjar (1988). In particular, the linear instability of the vortex structure to lower branch Tollmien-Schlichting waves governed by triple-deck theory was investigated. It was found that crossflow vortices can significantly increase the linear growth rates of Tollmien-Schlichting waves. The analysis of Bassom and Hall (1989a,b) suggests that either unsteady crossflow vortices or Tollmien-Schlichting waves might be the secondary mode of instability which causes the breakdown of stationary crossflow vortices.

Thus, as yet, the interaction theories have concentrated on the possible breakdown of pre-existing crossflow vortices; no attempt has been made to find the influence of, say, a finite amplitude Tollmien-Schlichting wave on the linear growth of crossflow vortices.

Finally, in this section, we mention the work of Collier and Malik (1987) who investigated the instability of three-dimensional boundary layers on concave surfaces. Hall (1985) had previously showed that an extremely small spanwise velocity component can destroy Görtler vortices induced by curvature. Collier and Malik found that vortices could exist where the spanwise velocity component is not small. It is possible that the vortex structures found by Collier and Malik were crossflow vortices. If the structures were induced by the curvature terms in the parallel flow approximation equations solved by Collier and Malik, then their results are open to criticism because nonparallel effects for Görtler vortex growth at the wavelengths considered by Collier and Malik are dominant.

## 3. Interactions involving Görtler Vortices

In the presence of surface curvature, a two-dimensional boundary layer is unstable to longitudinal vortex structures known as Görtler vortices. The parameter which governs the growth of this instability is the Görtler number defined by

$$G = R_e \frac{1}{2} \frac{L}{A}. \tag{3.1}$$

Here, $R_e$ is the (large) Reynolds number, $L$ is a length in the flow direction, and $A$ is a radius of curvature of the surface. Instability occurs first for 0(1) values of $G$ and the wavenumber of the induced vortex is comparable with the boundary layer thickness. If the boundary layer grows downstream, and the vortex wavelength is conserved as it is convected downstream then ultimately the vortex wavelength becomes small compared to the boundary layer thickness and the linear asymptotic theory of Hall (1982a) can be used to describe the development of the vortex. In this regime, the neutral Görtler number can be defined uniquely in terms of the vortex wavenumber and Hall (1982a) obtained results consistent with parallel flow theories such as those of Hammerlin (1956), Smith (1955).

However, at 0(1) vortex wavenumbers the linear and nonlinear growth of Görtler vortices is dominated by nonparallel effects, and the old parallel flow theories (and their more recent reincarnations) have no relevance to Görtler vortex development. In this parameter range, there is no alternative to a numerical integration of the disturbance equations, and the parabolic nature of these equations means that at any downstream location there does not exist a uniquely defined growth rate. Such a result causes a difficulty if it is required to predict the transition using, for example, the *en* method; as yet this difficulty has encouraged researchers to ignore the invalidity of the parallel flow equations and use them to predict a unique growth rate. The fact that most of the unstable regime corresponds to relatively large vortex wavenumbers means that approach does not give totally absurd results since at high wavenumbers the parallel flow equations have some validity. Since the only regime where the parallel flow equations have any relevance can be trivially described by (equally accurate) asymptotics, it seems to the present author that, as in Jallade (1989), the asymptotic theory alone should be used to predict growth factors. Moreover, the nonuniqueness associated with the parabolic nature of the disturbance equation can be alleviated by considering the receptivity problem for Görtler vortices.

The brief discussion above indicates clearly that any wave interaction theory involving Görtler vortices is complicated by the nonparallel development of the vortices. For that reason, much of the earlier work concerning the interaction of streamwise vortices with, say, Tollmien-Schlichting waves was carried out in the context of fully-developed flows in curved channels. However, the first attempt to determine the effect of streamwise vortices on the growth of Tollmien-Schlichting waves is due to Nayfeh (1981). In that calculation, Nayfeh arbitrarily assigned a finite amplitude to the linear Görtler vortex eigensolution of the linear parallel flow Görtler vortex equations. The Görtler wavenumber used by Nayfeh was not large so that the parallel flow equations are not appropriate. Moreover, our recent knowledge of Görtler vortices in the nonlinear region, Hall (1982b), Hall (1988), Hall and Lakin (1988), shows that the type of shape assumption used by Nayfeh is also not appropriate since a crucial feature of nonlinear Görtler vortices is their massive restructuring of the original mean velocity profile. Interestingly, Nayfeh showed that Görtler vortices have a significant destabilization effect on Tollmien-Schlichting waves. Later, Malik (1986b) attempted to repeat Nayfeh's calculations and found an inconsistency in Nayfeh's scaling which caused his results to be in error. Thus, within in the context of a spatially periodic approximation to the Navier-Stokes equations, Malik did not find any dramatic destabilization of Tollmien-Schlichting waves by Görtler vortices. Since Malik's calculation also did not take account of the almost certainly crucial streamwise development of the Görtler vortex, it is possible that his results might also be misleading. This possibility can be investigated by full numerical simulation not dependent on streamwise periodicity or a triple-deck analysis of the Tollmien-Schlichting instability of fully nonlinear Görtler vortices. We shall return to discuss the latter problem shortly; the former approach is perhaps still beyond the scope of CFD tools yet available.

Because of the difficulties associated with accounting for the streamwise development of Görtler vortices, Bennett and Hall (1988) and Daudpota et al. (1988) chose to look at the Görtler-Tollmien-Schlichting interaction problem in the context of curved channel flows. Thus, Bennett and Hall looked at the linear instability of nonlinear vortex flows to lower branch Tollmien-Schlichting waves. The nonlinear vortex states were found by numerical integration of the Navier-Stokes equations. They found that even large amplitude vortices have only a modest effect on the growth rate of Tollmien-Schlichting waves. If this result were to carry over to the external flow situation, it would mean that the presence of finite amplitude Görtler vortices in a boundary layer would not significantly alter the length-scale over which small amplitude Tollmien-Schlichting waves would have to grow before they might induce transition. Unlike Bennett and Hall (1988), Daudpota et al. (1988) considered the Görtler vortex interaction problem at finite Reynolds numbers. Moreover, that calculation assumed that the vortex and Tollmien-Schlichting wave were comparable in the interaction region. Such a calculation is certainly relevant only to the "endgame" of transition and is therefore of limited applicability to LFC theory.

More recent work on vortex-wave interactions in boundary layers and channels, Hall and Smith (1988), Hall and Smith (1989) has again been more concerned with the strongly nonlinear stages of transition than the prediction of how a finite amplitude vortex might influence the growth of linear Tollmien-Schlichting waves. Thus, these calculations are probably not of great relevance to LFC theory but their principal results are certainly worth stating here. Firstly, it has been shown in these papers that curvature is not required to sustain longitudinal vortex structures in boundary layers. In fact, interacting oblique Tollmien-Schlichting waves transfer energy into longitudinal vortex structures and in this way take the role of centrifugal effects in the Görtler problem. Moreover, these interaction theories give a self-consistent asymptotic description of the growth of 2D Tollmien-Schlichting waves, their 3D breakdown and the development of longitudinal vortex structures.

## 4. Conclusion

In the previous discussion, we have indicated the possible influence of pre-existing finite amplitude instabilities on the growth of other disturbances. Necessarily such an interaction is important only when nonlinear effects are important at which stage a boundary layer is usually on the point of undergoing transition. Thus, current design tools for LFC take no account of this kind of interaction since their primary aim is of course to prevent any kind of instability reaching a size where nonlinearity is important.

It is clear that as further progress with the receptivity problem for boundary layers is made, and a rational description of the evolution of incoming disturbances into finite amplitude solutions of the equations of motion is completed, then future transition prediction methods will need to take account of the type of wave interactions we have considered in this paper.

Finally, we close by saying a few words about an interaction problem which suggests that in certain cases nonlinearity could well introduce possible bypass mechanisms which might lead to the premature onset of transition. We refer to the work of Hall and Seddougui (1988) who considered the interaction of lower branch Tollmien-Schlicting waves in the attachment region of a swept wing flows. In particular, the interaction of waves propagating along and at an angle to the attachment line was considered. In the absence of the oblique wave, it is known from Hall and Malik (1987) that the spanwise propagating wave equilibrates to finite amplitude as a solution of a Stuart-Landau equation. However, in the presence of an oblique wave, the mutual interaction of the waves causes their amplitudes to become singular as the waves develop in the chordwise direction. This singular behaviour heralds a rapid change in flow structure with significantly larger waves being set up. Clearly, that type of interaction will likely have a significant effect on the location of transition. Thus, it will ultimately become necessary for transition prediction methods to monitor the growth of less amplified waves whose interaction with the most amplified wave might cause a bypass transition.

# References

[1] Bassom, A. P. and Gajjar, J. S. B., 1988, "Non-stationary crossflow vortices in three-dimensional boundary layer flows."

[2] Bassom, A. P. and Hall P., 1989a, "On the interaction of stationary crossflow vortices and Tollmien-Schlichting waves in the boundary layer on a rotating disc," submitted to *Proc. Roy. Soc.*

[3] Bassom, A. P. and Hall, P., 1989b, "The interaction of non-stationary crossflow vortices in a three-dimensional boundary layer," to appear as an ICASE report.

[4] Bennett, J. and Hall, P., 1988, "On the secondary instability of Taylor-Görtler vortices to Tollmien-Schlichting waves in fully developed flows," *J. Fluid Mech.* **186** pp. 445-469.

[5] Collier F. S., Jr. and Malik, M. R., 1987, "Stationary disturbances in three dimensional boundary layers over concave surfaces," *AIAA Paper No. 87-1412.*

[6] Daudpota, Q. I., Zang, T. A., and Hall, P., 1987, "Interaction of Görtler vortices and Tollmien-Schlichting waves in a curved channel flow," *AIAA Paper No. 87-1205.*

[7] Fischer, T. M. and Dallmann, U., 1987, "Theoretical investigation of secondary instability of three-dimensional boundary layer flows," *AIAA Paper No. 87-1338.*

[8] Gregory, N., Stuart, J. T., and Walker, W. S., 1955, "On the stability of three-dimensional boundary layers with applications to the flow due to a rotating disk," *Philos. Trans. Roy. Soc. London Ser. A* **248**, pp. 155-199.

[9] Hall, P., 1985, "The Görtler vortex instability mechanism in three-dimensional boundary layers," *Proc. Roy. Soc. London Ser. A* **399**, pp. 135-152.

[10] Hall, P., 1986, "An asymptotic investigation of the stationary modes of instability of the boundary layer on a rotating disc," *Proc. Roy. Soc. London Ser. A* **406**, pp. 93-106.

[11] Hall, P. and Malik, M. R., 1986, "On the instability of a three-dimensional attachment-line boundary layer: Weakly nonlinear theory and a numerical approach," *J. Fluid Mech.* **163**, pp. 257-282.

[12] Hall, P. and Seddougui, S., 1988, "Wave interactions in a three-dimensional attachment-line boundary layer," ICASE Report No. 88-?.

[13] Hall, P. and Smith, F. T., 1988, "The nonlinear interaction of Tollmien-Schlichting waves and Taylor-Görtler vortices in curved channel flows," *Proc. Roy. Soc. London Ser. A* **417**, pp. 255-282.

[14] Hall, P. and Smith, F. T., 1989, "Nonlinear Tollmien-Schlichting vortex interaction in boundary layers."

[15] Hammerlin, G., 1956, "Zür theorie der dreidimensionales instabilitat laminar Grenzschishten," *Z. Angew. Math. Phys.* **1**, pp. 156-167.

[16] Malik, M. R., 1986a, "Wave interactions in three-dimensional boundary layers," *AIAA Paper No. 86-1129.*

[17] Malik, M. R., 1986b, "The neutral curve for stationary disturbances in rotating-disk flow," *J. Fluid Mech.* **164**, pp. 275-87.

[18] Nayfeh, A. H., 1981, "Effect of streamwise vortices on Tollmien-Schlichting waves," *J. Fluid Mech.* **107** pp. 441-453.

[19] Saric, W. S. and Yeates, L. G., 1985, "Experiments on the stability of crossflow vortices in swept-wing flows," *AIAA Paper No. 85-0493.*

[20] Smith, A. M. O., 1955, "Görtler vortices in boundary layer flows," *Quart. Appl. Math.* **13**, pp. 233-262.

# SUPERSONIC LAMINAR FLOW CONTROL

*D. M. Bushnell*

NASA Langley Research Center
Hampton, VA 23665

## 1. Introduction

For several decades (1920's to 1960's), the development trend of commercial aviation was higher and faster, culminating in the 707 class of conventional takeoff and landing (CTOL) transport (and subsequent derivatives). This extraordinary marriage of the swept wing and jet engines revolutionized long-haul passenger transport and supplanted steam ships, trains, and more recently, even eroded the lower end of the long-haul transport spectrum, buses. The higher and faster trend was abruptly halted in the early '70s by a combination of economic reality and environmental concerns. The next logical step beyond the 707 class aircraft would have been a supersonic transport (SST) or, as they are termed today, a high-speed civil transport (HSCT). Such an aircraft cruises in the Mach number 2 to 3 range and would represent a revolutionary development in long distance transport. An early version of such an aircraft, the Concorde, while a technological marvel for its time, has not proven to be economically viable, and only a small number were produced and operated. The U.S. SST program was canceled in the early '70s in an era of: (a) general technological antipathy, (b) sharply rising fuel costs, and (c) environmental sensitivity/concern. Today there is a resurgence of interest in civilian supersonic long-haul aircraft. A probable major reason for this is the emergence of the Pacific Rim as a major economic entity. The subsonic CTOL flight times associated with passage between some of the major Pacific economic players is the order of 12 hours or greater, and an HSCT capability is urgently needed to foster improved and continued development of the Pacific area and particularly the U.S. involvement in this economic arena.

The Pacific application of an HSCT is more technologically demanding in that longer ranges are required than for many Atlantic flights. Additional technological problems or boundary conditions imposed upon an HSCT for operation in the 1990's and beyond involve: (a) ozone depletion and upper atmospheric pollution concerns,

(b) sonic boom, and (c) sideline noise. Fortunately, several technologies have developed to a considerable degree since the last large-scale SST studies, including: (a) variable cycle engines with higher turbine inlet temperature, (b) lightweight, high-temperature composite materials, and (c) flow control, including automatic load alleviation and laminar flow control (LFC). However, even with currently projected technology levels, the development of an economically viable HSCT will be a formidable task. We are not starting from a surfeit of performance, and the available studies indicate the need for both a significant "fare premium" and innovative drag reduction concepts. There exists a clear and pressing need for improved aerodynamic performance, even 10–percent improvements in L/D would be extremely significant, given the small payload fractions inherent in HSCT design, while a factor of 2 increase in L/D would literally be revolutionary and perhaps alter the entire economic viability issue. The historical (e.g., Ref. 1) mid-term (Refs. 2 and 3) and still current (Ref. 4) HSCT L/D is on the order of 10 (without the inclusion of many "real vehicle" influences) in excellent agreement with Kucheman's empirical rule $\frac{L}{D}\max = \frac{4(M+3)}{M}$ (Ref. 5). The Concorde value is on the order of 7. Values greater than 20 (Ref. 6) have been recently proffered.

The drag breakdown of a typical HSCT design is indicated, for example, in References 7–9. Crudely, assuming no pressure drag associated with flow separations and disregarding trim drag, the drag is (depending upon detailed design) on the order of 1/3 skin friction, 1/3 wave drag (mostly volume wave drag, but including sizable wave drag due to lift), and 1/3 vortex drag due to lift.

The purpose of the present paper is to review the current status of laminar flow control at supersonic speeds. The benefits of sizable supersonic drag reductions via LFC could be used to help alleviate several of the environmental as well as economic concerns in that reduced fuel requirements associated with drag reduction result in lower weight, which could provide benefits in terms of reduced: (a) sonic boom, (b) sideline noise, and (c) pollution as well as reduced initial and direct operating costs. Additionally, LFC could lower skin temperature through reduced stanton number and recovery factor by the order of $100°F$ at $M \sim 3$ which increases structural material options as well as reduces thermal (and sound) insulation weight requirements for the cabin if LFC is applied to the fuselage (Ref. 10). From the OSTP report on "National Aeronautical R&D

Goals/Technology for America's Future," (Ref. 11) "fault tolerant computers that provide local load alleviation and dynamic damping, coupled with the cooler structure allowed by supersonic laminar flow will significantly reduce the aircraft weight per pound of payload as compared with earlier supersonic transports."

## 2. Surface Roughness

Roughness drag and roughness effects per se are usually not included in conventional systems studies but were certainly present on all of the supersonic cruise aircraft produced thus far. Roughness at supersonic speeds is particularly worrisome due to the increased element drag and disturbance amplification caused by element shock wave formation (Refs. 12 and 13). Concomitant with supersonic flight is aerodynamic heating and elevated temperature levels which produce thermal stresses which can be larger than those induced by aerodynamic or mass loading. Design approaches to alleviate these thermal stresses such as tiles, shingles, corrugations, etc., produce steps, gaps, joints, and waviness, i.e., surface roughness (as an example, tile-induced roughness dominates the windward surface transition behavior on the space shuttle). Non-smooth aircraft surfaces not only increase drag directly, but are also responsible for promoting early transition, therefore, smooth surfaces are an enabling condition for supersonic laminar flow control (Ref. 14). Fortunately, as in the subsonic LFC case, contemporary materials and fabrication techniques such as molded sandwich skins made of thermoplastics or SPFDB Titanium (Refs. 15 and 16) appear to be capable of surfaces of sufficient smoothness to minimize direct roughness drag and delay transition/promote LFC. Also, the lower near-wall density at supersonic speeds usually allows a less stringent physical smoothness criteria than the corresponding subsonic case.

## 3. Transition Physics, Estimation, and Delay

Historically, due to innate surface roughness and the tenuous and incomplete status of transition knowledge and research, the existence of turbulent flow on HSCT designs was taken as a given. Research over the last 12 years in connection with the Langley subsonic LFC program and, more recently, NASP, has greatly improved the ability to estimate transition location and delay transition via detailed vehicle aerodynamic design. The essential features of this significant improvement in transition estimation capability include: (1) development of advanced linear stability theories (along with utilization of the advancements in flow field CFD), (2) invention and development of high-speed "quiet tunnels", (3) high quality flight LFC and transition experiments, and (4) the realization (by inference) that the canonical background disturbance levels in both flight and quiet/low-disturbance wind tunnels are similar. These developments have enabled the extension and application of the eN method into high-speed flows and transition initiated by each of the four major linear instability modes (TS, crossflow, Görtler and Macks 2nd mode) with a precision (for transition estimation) generally on the order of 20 percent or better in transition Reynolds number/location (Ref. 17). This compares to a previous uncertainty of an order of magnitude or more. This transition estimation technique parameterizes transition location as a function of the multitudinous variables which affect the mean flow, and in particular the streamwise variation of these parameters, as well as allowing specification of parameter levels required to delay transition. The capability is limited to cases, such as most flight situations, where the background disturbances are low and much of the disturbance amplification leading to transition is linear.

Such a transition estimation capability can be utilized, along with smooth surfaces, to design for transition delay for drag reduction, and thus provide limited "natural laminar flow" over the vehicle nose and other (low sweep) components. The technology can also be applied to design suction LFC systems to delay transition to reasonably large Reynolds numbers.

## 4. Supersonic Laminar Flow Control

Laminar flow control or LFC is used herein to refer to transition delay via active control as opposed to transition delay by aerodynamic design ("natural" laminar flow). Most HSCT surfaces are highly swept for wave drag reduction, thereby negating "natural" laminar flow except in very specialized circumstances/body areas. Recourse therefore must usually be made to suction LFC to achieve extensive laminar flow on high-speed cruise machines. In general, LFC becomes both easier and more difficult as speed increases. Easier due to reduced roughness sensitivity and disturbance amplification (critical layer moves out and wall region Reynolds number is reduced) and harder because increased control levels are required due to the outward movement of the critical layer and because of increased crossflow. For supersonic aircraft, the usual LFC techniques of choice are suction and wall cooling. The wall cooling approach has been demonstrated by the Russians up to $R_s \sim 34 \times 10^6$ at supersonic speeds (Ref. 18), but the technique is limited to: (a) nonhypersonic cryogenic-fueled aircraft, and (b) regions of small crossflow such as the attachment line or fuselage nose as cooling does not significantly damp the crossflow instability (Ref. 19). Cooling does damp the 1st mode TS waves but actually destabilizes Macks 2nd (hypersonic) modes (Ref. 20).

Supersonic suction LFC is an extremely powerful technique, particularly for crossflow dominated regions (which are endemic on the typically highly swept HSCT configurations) and concave curvature (Görtler instabilities). The net benefits (after allowing for the suction and ducting penalties) are greater than 50 percent of the skin friction drag, and research in the 1960's primarily by W. Pfenninger and his group at Northrop (Refs. 21–25) demonstrated that the technique works (using suction through multiple closely spaced slots) essentially up to the limits of the ground facilities employed, which was on the order of 25 million for swept wings and 50 million for axisymmetric bodies. This Northrop supersonic suction LFC research program existed coincident with and following the Air Force–Northrop X-21 (transonic) LFC flight program. The cogent results from the Northrop work are summarized in Table 1. The test series included initial studies of laminarization through shock impingement regions. The success of these tests is remarkable in that they were conducted under the extremely adverse conditions of high unit Reynolds number (stringent smoothness tolerance) and high free-stream noise (radiated nozzle wall turbulent boundary-layer noise, present in all non-quiet supersonic/hypersonic tunnels). These tests demonstrated that, basically, supersonic suction LFC is aerodynamically feasible, and the validity of the technique is further bolstered by the long record of successes produced by the Langley subsonic LFC and natural laminar flow control program (Ref. 26). A summary of the "lessons learned" from these Northrop experiments include: (1) (as expected) required suction rates (and therefore suction drag) increased with sweep, (2) the slot width should be less than 20 percent of the "sucked height," (3) the sucked height (per slot) should be less than the momentum thickness, (4) for axisymmetric bodies even small incidence (1°) can be highly destabilizing due to induced crossflow, and (5) spanwise contamination locus on the lower surface of swept wings may necessitate laminarization of wing fuselage junction.

**TABLE 1 — LFC SUCTION EXPERIMENTAL DATA BASE**
**Mostly Northrop/Pfenninger/AEDC Tunnel A**

| Model | Mach Number | Reynolds Number With Laminar Flow (Using Suction) |
|---|---|---|
| Flat Plate (with and without reflective shock wave, $P_2/P_1 \sim 1.1$ | 2.5 - 3.5 | $25.7 \times 10^6$ |
| 6° half-angle cone | 5 - 8 | $30 \times 10^6$ |
| Axisymmetric model, cylindrical afterbody (with and without reflective shock, $P_2/P_1 = 1.16$) | 2.5 - 3.5 | $51 \times 10^6$ |
| 36° swept wing, 3.0% T/C, biconvex | 2.5 - 3.5 | $25 \times 10^6$ |
| 50° swept wing, 2.5% T/C, biconvex | 2.5 - 3.5 | $17 \times 10^6$ |
| 72.5° swept wing | 1.99 - 2.25 | $9 \times 10^6$ |

In general, Pfenninger comments (private communication) that supersonic LFC was "easier" than corresponding subsonic cases. This is borne out by the performance achieved as a function of stream disturbance level (Fig. 1) and may be at least partially due to decreased stream disturbance/roughness coupling due to the reduced roughness sensitivity/lower wall Reynolds number.

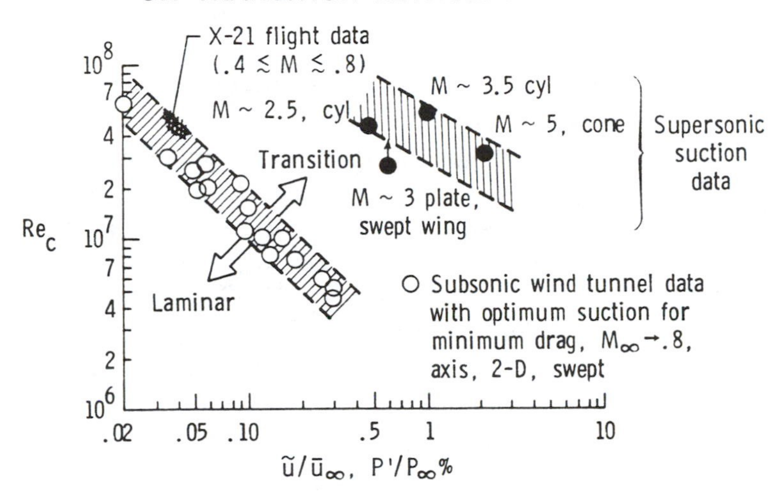


What is particularly intriguing concerning supersonic LFC are the large potential benefits, not only for drag directly but also reduced radiation equilibrium surface temperature and the resizing benefits for sonic boom reduction and possible synergisms with other drag reduction techniques including: (1) reduced skin friction penalty for some favorable interference wave drag reduction approaches (and for strut bracing employed to increase aspect ratio/reduce DDL and increase sweep/reduce wave drag) (Ref. 6), (2) turbulent skin friction reduction and direct wave drag reduction from boundary-layer thickening/wall wake effects of slot injection of LFC suction air, (3) possible use of LFC suction air for vortex DDL reduction via tip blowing, and (4) utilization of LFC suction system for high-lift production/separation control for takeoff and landing. It may be possible to accomplish a portion of the supersonic LFC suction via passive bleed, depending upon the specific design. Various estimates at hypersonic speeds indicate L/D increases of approximately 50 percent for an all-laminar vehicle as compared to an all-turbulent one (Ref. 27). Wave rider configurations, although having higher wave drag would have less crossflow and, therefore, more direct LFC benefits.

Recent estimates for a hybrid LFC system (on a portion of the wings only) indicate (for an HSCT) on the order of 10–percent improvements in L/D. Extensive application of LFC along with use of strut bracing and multiple bodies could yield L/D increases on the order of 100 percent (Ref. 6), see also References 28–30. This does not mean that supersonic suction LFC is in hand, considerable research is required such as attachment region and porous/perforated suction surface physics, including effective 3–D roughness induced by the discrete suction, and further 3–D roughness and waviness (and combination therefor) studies at supersonic speeds in regions of high crossflow, system optimization studies, including minimization of suction drag penalty, the extent to which passive bleed can be used for LFC suction; and supersonic flight suction experiments at appreciable Reynolds number to determine "real world" design, maintenance, and reliability issues including heated air handling/duct seals, and compatibility with high-lift systems, i.e., the known key issues are primarily flight and systems related. Of particular concern are the destabilizing influences of various disturbance fields generated by the (usually turbulent and discretely rough) fuselage and radiated onto the wing LFC surfaces. Even weak waves produced by fuselage joints, etc., would oscillate (and thereby create a dynamic radiated disturbance) due to interaction with the fuselage boundary-layer turbulence.

There are two essential differences between subsonic/transonic and supersonic LFC, much higher sweep and shock waves. Shocks are generated by: (a) vehicle surface features such as juncture regions (wing fuselage, wing-strut, fuselage-strut, nacelle-wing) and (b) the suction process itself. The latter both increases effective suction drag and possibly amplifies boundary-layer disturbance fields (Ref. 31). While the limited subsonic/transonic LFC experience with "modern" perforated surfaces (closely spaced small electron beam/laser produced holes) is favorable, the supersonic LFC perforated surface case, with much higher sweep and additional vorticity production due to curved (suction induced) shock waves (as well as amplification due to imposed/impinging and roughness/waviness induced shocks) must be examined carefully. We currently have no hard information concerning the appropriate physics/performance of such surfaces. Spanwise contamination will also be a non-trivial issue for supersonic LFC due to the finite thickness of the highly swept subsonic leading edges required for wave drag reduction and leading-edge thrust production. Research in Reference 32 indicates that natural bleed attained via chordwise leading-edge slots can relaminarize the leading-edge boundary layer.

Minimization of net suction drag is necessitated by the suction rates required to stabilize highly swept wing flows. Possible approaches include: (a) improved suction surface pressure recovery, (b) use of crossflow — canceling wing design features such as over expansion and subsequent recompression near the leading edge (Ref. 28) and possibly spanwise pressure gradients, (c) utilization of passive bleed rather than active suction (Ref. 14), and (d) utilization of LFC suction air efflux to reduce turbulent skin friction and closure wave drag.

## 5. Concluding Remarks

Cruise drag reduction is an "enabling technology" for an advanced SST/high-speed civil transport, particularly for many transpacific ranges. Supersonic laminar flow can provide on the order of a 10–percent increase in L/D and possibly much more, depending upon how aggressively LFC is developed and applied. The supersonic LFC technology level is considerably behind the subsonic case, with no flight experiments yet conducted and a wind-tunnel data base limited to slot suction downstream of leading edges. What is fundamentally different about supersonic LFC, compared to the subsonic/transonic case, is shock waves (both vehicle and suction induced) and large sweep/crossflow. In general, the associated suction drag is higher than the lower speed cases but the payoff, due to an extraordinary payload sensitivity to zero lift drag, is still tremendous. Supersonic LFC is the direct recipient of over 12 years' concentrated research in such areas as supersonic quiet tunnel development, advanced stability theory and subsonic/transonic LFC flight and wind-tunnel experimentation. This arsenal of tools and experience, applied to the various critical problem areas discussed herein, should establish LFC (after solution of the usual plethora of "surprises" encountered along the way) as a viable and important constituent of an advanced SST/HSCT technology base.

## 6. References

[1] Baals, D. D.; Robins, A. W.; and Harris, R. V., Jr.: Aerodynamic design integration of supersonic aircraft. AIAA Paper 68–1018, AIAA 5th Annual Meeting and Technical Display, Philadelphia, PA, 1968.

[2] Goebel, T. P.; Bonner, E.; and Robinson, D. A.: A study of wing body blending for an advanced supersonic transport. NASA CP-2108, 1980, pp. 149–169.

[3] Roensch, R. L. and Page, G. S.: Analytical development of an improved supersonic cruise aircraft based on wind tunnel data. NASA CP-2108, 1980, pp. 205–227.

[4] Coen, P. G.: The effect of advanced technology on the 2nd-generation SST. AIAA Paper 86–2672, AIAA/AHS/ASEE Aircraft Systems, Design, and Technology Meeting, Dayton, OH, 1986.

[5] Kuchemann, D.: The aerodynamic design of aircraft. Pergamon Press, Oxford, 1978.

[6] Pfenninger, W. and Vemuru, C. S.: Design aspects of long range supersonic LFC airplanes with highly swept wings. SAE Technical Paper 88–1397, SAE Aerospace Technology Conference and Exposition, Anaheim, CA, 1988.

[7] Kuchemann, D.: Aircraft shapes and their aerodynamics for flight at supersonic speeds. Advanced in Aeronautical Science, Vol. 3, 1962, pp. 221–253.

[8] Leyman, C. S. and Markham, T.: Prediction of supersonic aircraft aerodynamic characteristics. AGARD-LS-6, AGARD Lecture Series on Prediction Methods for Aircraft Aerodynamic Characteristics, 1974, pp. 5–1 to 5–52.

[9] Kulfan, R. M.: Application of hypersonic favorable aerodynamic interference concepts to supersonic Aircraft. AIAA Paper 78–1458, AIAA Aircraft Systems and Technology Conference, Los Angeles, CA, 1978.

[10] Gasich, W. E.: Application of laminar flow control to transport aircraft. Aerospace Engineering, Vol. 20, No. 10, 1961, pp. 22, 23, 44, 46–48, 50–52.

[11] National Aeronautics R&D goals/technology for America's future, 1985, Executive Office of the President, Office of Science and Technology Policy.

[12] Gaudet, L. and Winter, K. G.: Measurements of the drag of some characteristic aircraft excrescences immersed in turbulent boundary layers. AGARD Conference Proceedings No. 124, Aerodynamic Drag, 1973, pp. 4–1 to 4–12.

[13] Czarnecki, K. R.: The problem of roughness drag at supersonic speeds. NASA SP-124, NASA Conference on Aircraft Aerodynamics, 1966, pp. 455–468.

[14] Bushnell, D. M. and Malik, M. R.: Supersonic laminar flow control. NASA CP-2487, Pt. 3, Research on Natural Laminar Flow and Laminar Flow Control, pp. 923–946.

[15] McQuilkin, F. T.: Feasibility of SPF/DB titanium sandwich for LFC wings. NASA Contractor Report 165929, 1982.

[16] Ecklund, R. C. and Williams, N. R.: Laminar flow control SPF/DB feasibility demonstration. NASA Contractor Report 165818, 1981.

[17] Bushnell, D. M. and Malik, M. R.: Transition prediction in external flows via linear stability theory. 1988, IUTAM Symposium Transsonicum III, Gottingen, FRG, Springer-Verlag, 1989, J. Zierep and H. Oertel (Eds.). (Also NASP Technical Memorandum 1021, Jun. 1988.)

[18] Kuziminskiy, V. A.: Effect of cooling of the wing surface on the transition of laminar boundary layer into the turbulent at supersonic speeds of flow. Uchenyye Zapiski TSAGI V. 12, No. 1, 1981, pp.1–178.

[19] Lekoudis, S.: The stability of the boundary layer on a swept wing with wall cooling. AIAA Paper 79–1495, AIAA 12th Fluid and Plasma Dynamics Conference, Williamsburg, VA, 1979.

[20] Mack, Leslie M.: Boundary layer stability theory. AGARD R-709, 1984, pp. 3–1 to 3–81.

[21] Bushnell, Dennis M. and Tuttle, Marie, H.: Survey and bibliography on attainment of laminar flow control in air using pressure gradient and suction. V.I. NASA RP-1035, 1979.

[22] Pfenninger, W.: USAF and Navy sponsored Northrop LFC research between 1949 and 1967, Special Course on Concepts for Drag Reduction, Belgium, AGARD Report No. 654, 1977, pp. 3–4 — 3–23.

[23] Burchfield, C. G.: Boundary-layer suction experiments on a 6–deg half-angle cone at Mach numbers 5 through 8. Arnold Engineering Development Company. AEDC-TR-66–230, 1966.

[24] Summary of laminar boundary layer control research, Vol. II, Northrop Corp., Technical Documentary Report ASD-TDR-63–554.

[25] Goldsmith, E. L.; Evans, N. A.; and Smith, G. V. F.: The effect of combined boundary-layer suction and base bleed on the drag of a 10 degree cone at $M = 2.58$. Royal Aircraft Establishment Technical Report 67218, 1967.

[26] Hefner, Jerry N. and Sabo, Frances E.: Research in natural laminar flow and laminar-flow control. NASA CP-2487, Parts 1 to 3, 1987.

[27] Bowcutt, K. G. and Anderson, J. D.: Viscous optimized hypersonic waveriders. AIAA Paper 87–0272, AIAA 25th Aerospace Sciences Meeeting, Reno, NV, 1987.

[28] Pfenninger, Werner and Vermuru, C. S.: Suction laminarization of highly swept supersonic laminar flow control wings. AIAA Paper 88–4471, AIAA/AHS/ASEE Aircraft Design, Systems and Operations Meeting, Atlanta, GA, 1988.

[29] Agrawal, S.; Kinard, T. A.; and Powell, A. G.: Supersonic boundary layer stability analysis with and without suction on aircraft wings. 4th Symposium on Numerical and Physical Aspects of Aerodynamic Flows, Long Beach, CA, 1989.

[30] Parikh, P. G.; Sullivan, P. P.; Bermingham, E.; and Nagel, A. L.: Stability of 3–D wing boundary layer on a SST configuration. AIAA Paper 89–0036, AIAA 27th Aerospace Sciences Meeting, Reno, NV, 1989.

[31] Anyiwo, J. C. and Bushnell, D. M.: Turbulence amplification in shockwave-boundary layer interactions. AIAA Journal, Vol. 20, No. 7, 1982, pp. 893–899.

[32] Bushnell, Dennis M. and Huffman, Jarrett K.: Investigation of heat transfer to leading edge of a 76 swept fin with and without chordwise slots and correlations of swept-leading-edge transition data for Mach 2 to 8, NASA TM X-1475, 1967.

# THE LANGLEY 8-FT TRANSONIC PRESSURE TUNNEL LAMINAR-FLOW-CONTROL EXPERIMENT
## -Background and Accomplishments-

*Percy J. Bobbitt, William D. Harvey*
*Charles D. Harris, and Cuyler W. Brooks, Jr.*

Langley Research Center
Hampton, VA 23665

## 1. Introduction

The NASA's response to the oil shortage and price increases of the 70's was the creation of the Aircraft Energy Efficiency (ACEE) Program. Its objective was to provide aerodynamic and controls technology that would enable the design of commercial transports with substantially better fuel efficiency than those in service at the time. This program had a number of facets, most of which were centered on supercritical airfoil/wing and winglet technology, and included both industry and NASA in-house research components. This paper concerns one of the NASA in-house activities, the Langley Laminar Flow Control (LFC) Project, which was carried out in the Langley 8-foot Transonic Pressure Tunnel (8-ft TPT). The idea for such an undertaking came from Dr. Werner Pfenninger in 1975 and stemmed, primarily, from a desire to know more about the compatability of a high performance supercritical wing with laminar flow control. The possible adverse effect of a large supersonic zone and its associated wave structure on the stability of a suction controlled boundary layer was a particular concern.

Aside from supercritical wings, there were a number of other technology advances which militated for a new look at LFC wings (see Reference 1). Larger (more storage) and faster computers coupled with improved finite difference techniques developed in the early 70's enabled the rapid solution of the nonlinear potential equations governing two- and three-dimensional transonic flow. Newly formulated wing and airfoil codes based on this technology were coupled with boundary layer routines to obtain highly accurate pressure distributions and boundary layers parameters. These, in turn, were employed in improved linear boundary-layer-stability codes to determine transition. On the structures side new composite materials, forming and perforating techniques and coatings led to an array of structural concepts and/or surfaces much more amenable to laminar flow control. It will become apparent in subsequent discussions that all of these new technologies were employed in the Langley 8-ft TPT LFC experiment.

The purpose of this paper will be to describe a number of the considerations that went into the selection of the wind tunnel and the design and test parameters as well as the design of the wind tunnel liner and swept LFC wing. The types and locations of the instrumentation utilized and representative data displays are also discussed. Both slotted and perforated upper surfaces were tested with full- and partial-chord suction and representative results are shown and analyzed for both. The last sections of the paper detail some of the lessons learned, list and discuss a number of suggestions for future work and summarize what was achieved vis-a-vis the initial objectives.

## 2. Background

The use of suction to delay the onset of transition goes back to the work of Prandtl in the early 1900's. Since that time a vast array of analyses and wind tunnel and flight experiments have been carried out (References 2 and 3) to further advance the state of the art and to define the problems and potential of this approach. Prior to the ACEE sponsored research, there was a flurry of activity during the late 40's, throughout the 50's, and into the early 60's that was carried out by industry (sponsored by the Air Force), the NASA Langley Research Center and the Arnold Engineering and Development Center. Particularly relevant to this paper is the research carried out by Dr. W. Pfenninger and his associates in both flight and wind tunnels on large chord wings and airfoils. In the late 50's, a series of LFC tests of $30^\circ$ swept models were conducted in the Ames 12-foot Pressure, the Michigan 5- by 7-ft and the Norair 7- by 10-ft Wind Tunnels (Reference 4). The models had both 7.0 and 10.0-ft chords, thicknesses of 12 and 16 percent and modified NACA 66-012 and NACA 64-016 symmetrical airfoil sections. In the Ames tests, full-chord laminar flow was maintained by suction slots on the upper surface up to a chord Reynolds number of $29 \times 10^6$ and in the Norair and Michigan tests up to $10 \times 10^6$. The superior flow quality in the Ames 12-foot Pressure Tunnel (12-ft PT) at low to moderate subsonic speeds was the most probable cause of the better results in this facility.

Flight tests carried out under Air Force sponsorship included a slotted glove test on the F-94 (Reference 5) which had 10° of wing sweep, and a much more ambitious complete-wing test on the X-21 with 33° of wing sweep (Reference 6). Although both utilized subcritical wing-section designs they were flown at speeds high enough to incur supercritical velocities on the upper surface.

The experiments described in the proceeding paragraphs and the basic research carried out to support them formed a major part of the technology base for transonic LFC at the start of the Langley LFC project. A thorough knowledge of this data and an intuitive understanding of LFC suction-system, and airfoil pressure-distribution, requirements by Dr. W. Pfenninger were also invaluable resources.

Research to design a new LFC supercritical airfoil for a large chord wind tunnel experiment began in the fall of 1975. Dr. W. Pfenninger, Mr. P. J. Bobbitt, and Mr. D. O. Allison went through numerous iterations, starting with a conventional looking Whitcomb/Harris "flat-top" supercritical airfoil, to arrive at an airfoil referred to herein as the "Baseline" Airfoil. Subsequent to this activity, Dr. W. Pfenninger with the support of Dr. H. Reed and Mr. R. Dagenhart produced a second airfoil with higher design performance which was designated as the official LFC airfoil in the fall of 1978.

The LFC project office was established in 1976 under the direction of Mr. Ralph Muraca and it remained under his direction until July of 1978 when Mr. A. L. Nagle assumed the responsibility for about a two month period. Mr. W. D. Harvey became project manager in September of 1978 and he held that position until the project's completion in the fall of 1988. At the same time Mr. W. D. Harvey became project manager the project office was moved to the Transonic Aerodynamic Division of the Langley Aeronautics Directorate. The preliminary and final design work for the airfoil model and its suction system was under the direction of Mr. J. D. Pride of the Systems Engineering and Operations Directorate at the Langley Research Center.

During the years 1976 and 1977, it was intended that the LFC experiment be carried out in the Ames 12-ft PT. Its size and good flow quality made it the logical choice. When the difficulty of the experiment became more apparent, and the NASA Headquarters and the Langley Research Center committed funds for several flow-quality modifications to the Langley 8-ft TPT, the "tunnel of choice" became the Langley 8-ft TPT in June of 1978.

The LFC experiment as originally conceived involved only the use of a slotted surface for boundary layer suction. Sometime during the 1980-81 time frame it appeared (on the basis of ACEE sponsored research) that the technology for the construction of a perforated titanium surface with integral suction ducts could be developed. Subsequently, a perforated upper surface design, fabrication, and test activity was added to the scope of the project.

The scope of the experiment was expanded again around 1985 to include a hybrid laminar-flow-control (HLFC) concept in which suction was applied over the first 20 percent (or less) of the chord and a favorable pressure gradient employed beyond the suction cut off to obtain a run of natural laminar flow. To do this the upper forward perforated section (0-23% chord) of the perforated LFC model was matched to new solid non-suction, middle and aft sections to obtain the desired HLFC pressure distribution. Only the LFC tests will be described in this paper.

Testing of the LFC airfoil with a slotted surface began in the spring of 1981 but a number of substantial hardware problems delayed serious research testing until late in 1983. The slotted LFC tests were completed in the spring of 1985 and the model dismantled and reassembled with a perforated upper surface in the late summer of 1985. These tests were completed in the fall of 1987 and HLFC tests were conducted from the winter of 1987 through September of 1988.

The principal investigator for the LFC project was C. D. Harris, the coinvestigator C. W. Brooks, Jr., the primary data reduction specialist, P. E. Clukey, and the instrumentation and data acquisition specialist was J. P. Stack. Most of the data included in this paper have come from the aforementioned researchers, Mr. W. D. Harvey (Project Manager), and from other researchers involved in various aspects of the experiment including Mr. D. O. Allison, Dr. P. A. Newman, Dr. E. C. Andersen, Mr. J. Peterson, Mr. R. Dagenhart, Mr. P. C. Stainback, Mr. G. S. Jones, and Mr. D. Madalon.

## 3. Technical Considerations

The LFC experiment in the 8-ft TPT was not designed or constructed to replicate or imitate a flight LFC system. It was designed to obtain basic suction-requirements data on a supercritical airfoil with a substantial supersonic zone. As a basic experiment, as much variability and control was built into the model as funds permitted. Near full chord suction was implemented because it offered the opportunity to obtain more understanding of the suction requirements as a function of the extent of laminar flow than any partial chord hybrid concept could provide. Since both the top and bottom were equipped with suction surfaces (top to 96% and the bottom to 85% chord) and each had different pressure distributions, it also provided the opportunity to investigate two radically different suction distributions and an additional stability mechanism (Taylor-Görtler) associated with the concave regions on the bottom surface.

In order to obtain the same flow around the yawed LFC wing in the wind tunnel as in unbounded flight, a a 50-ft long contoured liner for the contraction and test section had to be provided. This liner took into account the displacement effects of the boundary layer and, since the LFC model spanned the tunnel, the complex flow phenomena in the juncture regions where the model intersects the wall liner. Since the airfoil was asymmetric and yawed, the streamlines around the model, and consequently the liner, were different on all four walls.

Another consideration in the liner design was the need to have easy access to the LFC model for maintenance and cleaning. Consequently, the model was mounted vertically in the wind tunnel with the most forward leading edge at the top of the tunnel. The east and west side walls of the tunnel then became the walls "above" and "below" the top and bottom surfaces of the airfoil, respectively.

It was planned from the outset to test at subsonic, as well as transonic, Mach numbers to provide data on suction requirements for both sub and supercritical pressure distributions. At supercritical speeds the supersonic bubble grows with increasing Mach number until it reaches the tunnel-wall liner. At this condition the supersonic bubble is distorted by its interaction with the wall and the results become unreliable. Consequently, the design Mach number and lift coefficient of the airfoil were chosen as high as possible within the constraint that the top of the supersonic bubble above the model upper surface be about a foot away from the wall. It should be noted that the liner was designed for the design-point Mach number and lift coefficient of the airfoil and was non-optimum for any other combination of Mach number and lift coefficient. In addition the liner was designed for the effective shape of the airfoil with full chord laminar flow — a condition not always achieved.

Over the period of time that it was planned to conduct the LFC test in the Ames 12-ft PT the airfoil was designed with a chord of 8 feet. The chord was changed to 7 feet after it was decided to carry out the test in the Langley 8-ft TPT and the responsibility for the test was assumed by the Transonic Aerodynamics Division. This enabled a slightly higher design Mach number, a modest decrease in the distance that the flow had to "set-up" ahead of the model and a small decrease in liner thickness. On the negative side, the smaller chord increased the difficulty of fabricating the airfoil and suction hardware. The space between the slots had to be smaller and hence the ducts were smaller and construction tolerances had to be more rigidly enforced. There was less volume in the airfoil so that suction slots and ducts could only be installed back to 96% of chord (top surface).

Analysis codes for turbulent flow in the juncture of a blunt-leading-edge, swept wing, and a wall did not exist at the time the LFC experiment was designed. It was well known, however, that fillets and suction would be helpful in maintaining attached flow and both were used. Codes that could analyze a two-dimensional boundary layer with pressure gradient were employed to estimate the suction levels required to prevent flow separation on the wall adjacent to the rear portion of the airfoil.

A similar situation existed for the turbulent wedge on the airfoil which originates at the intersection of the leading edge and the tunnel wall boundary layer. The two-dimensional theoretical airfoil pressures were used with a strip boundary layer technique to determine the suction requirements in this region to minimize boundary layer thickness and prevent flow separation.

Tunnel flow quality was always a concern, starting with the 12-ft PT at Ames during the first years of the program and later for the 8-ft TPT at Langley. It was known that noise, vorticity, and thermal spottiness had an adverse affect on transition, and the more turbulent the wind tunnel environment the more degraded the transition results become compared to what one would obtain in free air. As tunnel pressure is increased to increase Reynolds number the noise and vorticity levels also rise and mask some of the Reynolds number effects. A similar phenomena occurs when Mach number is increased to assess compressibility effects. Anti-turbulence screens, honeycombs, sonic chokes and mufflers were available to minimize these problems. In the Langley 8-ft TPT LFC experiment, all of these treatments except mufflers were employed.

## 4. Program Objectives

The primary objective of the Langley 8-ft TPT LFC Program was to conduct a basic experiment on a high performance, swept, supercritical wing to determine the ability of surface suction to maintain laminar flow over a large supercritical-flow region and through weak shocks. Another objective was to evaluate the ability of state-of-the-art stability codes to predict suction requirements and transition locations. After the perforated titanium surface was added to the program, a third objective was added as well, i.e., to determine which type of surface, perforated or slotted, is more conducive (from an aerodynamic standpoint) to obtaining laminar flow. Finally, while we had no stated objective relative to flow quality, there was the expectation that we would increase our understanding of the effect of flow quality on transition.

## 5. Tunnel Selection

There are only a handful of tunnels around the world which have the attributes required to carry out meaningful high chord Reynolds number laminar flow testing at low ($< 4 \times 10^6$) unit Reynolds numbers. There are fewer still that have these attributes and can operate at transonic speeds as well. The desired characteristics that were sought in 1976 were:

- Excellent flow quality, i.e., low vorticity, noise and thermal spottiness
- High test-section aspect ratio ($> 2.5$)
- Large contraction ratio ($> 16$)
- Unit Reynolds number capability to approximately $4 \times 10^6$ at high subsonic speeds
- Reynolds number variability with Mach number fixed

- Good pumping (suction) capability, i.e., high flow rates and high differential pressures
- Ability to maintain tunnel pressure constant for long periods
- Ability to accommodate large models
- Good data system including real time display
- Available for long test programs.

At the start of the Langley LFC Project, the wind tunnel that seemed to satisfy most of these requirements and thus capable of providing useful data for laminar flow wings was the Ames 12-ft PT. It was built with the same technology employed in the Langley Low Turbulence Pressure Tunnel (LTPT) which, at the subsonic speeds it operates, has excellent flow quality. The Ames 12-ft PT also had a large diameter test section and a modest ability to change Reynolds number at a fixed Mach number at high subsonic speeds (Figure 1). However, its unit Reynolds number capability was marginal at the higher speeds. It had one of the highest contraction ratios of any NASA transonic tunnel (25 to 1) but a low-aspect-ratio test section (1.53), typical of subsonic tunnels. Test section access was thought to be a problem for the high levels of suction mass flows required for the LFC experiment.

Hot wire anemometer measurements were carried out in the Ames 12-ft PT in 1978 which indicated that it had excellent flow quality at low subsonic speeds, but at the Mach numbers required for the LFC experiment turbulence levels (mass-flow fluctuations) were in the range from 0.1 to 0.2 percent (see Figure 2). This level of turbulence was expected to yield full-chord-laminar-flow Reynolds numbers in the $4 - 6 \times 10^6$ range based on the Pfenninger correlation curve of Figure 3. Measurements in the Langley 8-ft TPT taken at about the same time showed higher levels of turbulence in the test section at $M_\infty \approx 0.8$ than the Ames 12-ft PT (see Figure 2) but for a tunnel without any anti-turbulence screens the levels were considered quite good. On the plus side it had good unit Reynolds number capability at high speeds, a good contraction ratio (20 to 1), a good test section aspect ratio (3.00), excellent pumping capability and test section access.

Calculations and extrapolations based on earlier screen and honeycomb research (References 7 through 13) indicated that four or five screens in combination with a honeycomb could reduce the turbulence intensity (primarily vorticity) in the 8-ft TPT by roughly a factor of three. In addition, if a downstream choke were employed to prevent diffuser noise from propagating upstream into the test section, RMS $u'/u_\infty$ and $p'/p_\infty$ levels equivalent to those obtained at low subsonic speeds, i.e., $\approx 0.05\%$, might be realized. Referring again to the correlation figure of Pfenninger, Figure 3, it can be seen that the full chord-laminar-flow Reynolds numbers that one might expect from this level of turbulence are on the order of $20 \times 10^6$.

Taking into account the above factors, the difficulty of conducting a very complex experiment in a tunnel three thousand miles from "home," and a decision by NASA to modify the Langley 8-ft TPT to improve its flow quality, NASA Langley management made a decision in June of 1978 to commit the Langley 8-ft TPT to the LFC experiment.

## 6. Flow Quality Improvements in the NASA Langley 8-ft TPT

In the previous section we mentioned the need for good flow quality in carrying out laminar flow research and touched on an activity aimed at improving the flow quality in the 8-ft TPT. The object of this section is to describe further the need for, and results from, this activity including the instruments used. Research aimed at improving the flow quality in the Langley 8-ft TPT had been initiated in 1976 prior to any thought of its being used for the LFC experiment. It was started simply to obtain a high-flow-quality transonic tunnel which would complement the Langley LTPT in conducting general laminar flow research. The fact that the 8-ft TPT had very good flow quality, even without a flow-quality fix (screens and honeycombs), was anticipated due to its large contraction ratio, large test section aspect ratio, and 3-degree half-angle diffuser. In addition, the location and construction of the cooler, with its eight rows of coils and closely spaced fins ($\sim 0.1''$), clearly enabled it to reduce the turbulence level of the flow moving through the settling chamber to the entrance of the contraction.

The goal in 1976 was to quantify the turbulence in the 8-ft TPT using hot-wire anemometry and the AEDC transition cone in a manner similar to that planned for the Ames 12-ft PT. The services of Dr. F. Kevin Owen were secured and he, along with W. D. Harvey and P. C. Stainback of the Langley Research Center made turbulence measurements in both tunnels starting in late 1977 and ending in late 1978. The papers published documenting the results (References 14-15) indicated that the Ames 12-ft PT had lower mass and pressure fluctuations than the 8-ft TPT over the range of Mach numbers and Reynolds numbers for which data was obtained. Figure 2 shows the variation of the mass flow fluctuations as a function of unit Reynolds number for a range of Mach numbers for both the Ames 12-ft PT and the Langley 8-ft TPT. At a Mach number of 0.26 in the Ames 12-ft PT normalized mass flow fluctuations range from $\sim$ 0.04 percent at low unit Reynolds numbers to 0.09 percent at high unit Reynolds numbers. These are impressive numbers. However, as noted earlier, at a Mach number of 0.82 the levels of mass-flow fluctuations are much higher — approaching 0.2 percent. Turbulence levels this high would be expected to limit full chord laminar flow to a chord Reynolds number of approximately $6 \times 10^6$ based on the correlation plot of Figure 3. Since full chord laminar flow Reynolds numbers on the order of 20 to $25 \times 10^6$ were desired for the LFC experiment, this level of turbulence was unsatisfactory. The 8-ft TPT, which had even higher levels of turbulence at a Mach number of 0.8, was also unsuitable.

Pressure fluctuations in the Ames 12-ft PT and the Langley 8-ft TPT are plotted in Figure 4 as a function of unit Reynolds number. The Langley 8-ft TPT data was taken with the test-section slots covered and the Ames 12-ft PT data with the slots on the sides of its strut faired over. At the lowest Mach number the levels of normalized pressure fluctuations are about the same in the two facilities. At a Mach number of approximately 0.8 there is a large difference in the magnitude of $\tilde{p}/p_\infty$ between the Langley and Ames tunnels. Note also the difference in the trends of $\tilde{p}/p_\infty$ with $R_\infty$ at Mach numbers of $\sim$ 0.6 and $\sim$ 0.8.

In both the Ames 12-ft PT and the Langley 8-ft TPT a substantial fraction of the noise in the test section at high speeds comes from the strut at the downstream end of the test section and the boundary layer noise emanating from the high speed diffuser downstream of the test section. Fan noise is also convected upstream through the high speed diffuser and into the test section. A substantial fraction of the noise from these various downstream noise sources can be eliminated by the use of a sonic choke ahead of the strut. Test results for $\tilde{p}/p_\infty$ with and without a sonic choke installed in the Langley 8-ft TPT test section are given in Figure 5. The triangle and diamond symbols represent data for two-wall and four-wall choke tests, respectively. The square symbols are data points for the basic tunnel without chokes while the circle symbols are for data taken in the settling chamber. Results for the various choked-tunnel situations are similar in that once the tunnel is choked the level of pressure fluctuations is reduced to that of the settling chamber — an order of magnitude reduction. This data clearly indicated the need for a sonic choke in the 8-ft TPT to enhance the tunnel's ability to carry out meaningful laminar flow testing. Initially a two-wall permanent choke was designed for the 8-ft TPT to be installed at the same time as the honeycomb and screens (discussed later). However, it was not built since its installation would have slowed up the installation of equipment needed for the LFC experiment including the streamline liner which had its own flexible two-wall choke.

Once the need to reduce vorticity levels and associated mass and velocity fluctuations in the 8-ft TPT was identified an effort was initiated (1978) to determine how such reductions could be achieved. This effort very quickly led to the construction of a pilot facility to simulate the full scale Reynolds number flow through the cooling coil, turning vanes, and various anti-turbulence devices. A sketch of the pilot facility is given in Figure 6 along with a list of parameters that were changed during the tests. Much of the hardware used for the pilot facility was already in existence having been constructed as part of a pilot apparatus for the National Transonic Facility. Once the pilot facility was operative, combinations of honeycombs and screens were tested where the honeycomb and screens were mounted oblique to the flow at the trailing edge of the turning vanes or normal to the flow just downstream of the inside corner. The number of screens and screen-mesh size were varied and two honeycombs with different cell aspect ratios were also tested. Single and cross-wire hot-wire probes were used in combination to determine both the fluctuating longitudinal and lateral velocities. A detailed description of the facility and the data obtained is presented in References 16, 17, and 18.

A general conclusion of the pilot facility research was that the honeycomb was more effective in reducing the lateral velocity fluctuations than the longitudinal. Conversely the screens were more effective in reducing the longitudinal velocity fluctuations than the lateral. The most effective treatment was an array of screens placed downstream of a honeycomb. Details of the configuration finally selected along with the cooler characteristics are listed in Figure 7. A photo of the honeycomb installed in the Langley 8-ft TPT taken from just downstream of the turning vanes is shown in Figure 8. A second photo showing the downstream screen looking upstream from the entrance to the test section is provided in Figure 9. Finally, Figure 10 gives a sketch of the Langley 8-ft TPT circuit with the honeycomb and screens installed.

As a result of the pilot facility research, the earlier flow quality measurements in the 8-ft TPT and analytical expressions that predict the effect of a contraction on mass flow and velocity fluctuations, a forecast could be made of these quantities in the test section with the screens and honeycomb installed. Figure 11 shows the variations of mass flow fluctuations at various locations in the tunnel circuit. Some of the curves are faired wind tunnel data; those labeled three or four screens are predictions. The top two curves on the figure give the measured variations, downstream of the turning vanes and in the settling chamber, of $\tilde{m}/m_\infty$ with unit Reynolds number. Levels of $\tilde{m}/m_\infty$ in the settling chamber at the higher values of $R_\infty$ are slightly above 2%. With the addition of three or four screens and a honeycomb, these levels drop to the 0.7 - 0.9% range. Since five screens were actually installed and most of the testing in the LFC experiment was to be carried out in the unit Reynolds number range of 1.5 to $3 \times 10^6$, the $\tilde{m}/m_\infty$ level projected in the settling chanber was from 0.3 to 0.4%.

The measured variations of $\tilde{m}/m_\infty$ in the test section of the 8-ft TPT in its normal configuration are also plotted on Figure 11. The data are for a Mach number of 0.8 and vary between 0.3 and 0.4%. For comparison, the Ames 12-ft PT test-section measurements at $M_\infty = 0.8$ are also shown (solid symbols). Predictions of the levels in the test section following the installation of the honeycomb and three or four screens are provided by the bottom two curves on Figure 11. For four screens at a unit Reynolds number of $2 \times 10^6$, $\tilde{m}/m_\infty$ has a value of 0.1% which is over a factor of two lower than that of the original untreated tunnel. The five-screen fix actually installed was expected to yield a slightly lower value.

A similar but simpler plot of the fluctuating velocity is given in Figure 12. It is evident from this figure that with the untreated tunnel the values of $\tilde{u}/U_\infty$ range from 0.3 to 0.4%. The level of $\tilde{u}/U_\infty$ with all the flow-quality fixes installed were predicted to be in the range of 0.03 to 0.06% for unit Reynolds numbers of interest. Single wire hot-wire-probe data with the fixes installed are seen to fall within this band. Consequently, our predicted level of $\tilde{u}/U_\infty$ was achieved. Flight data for the AEDC cone tested on an F-15 are also given on this plot and are within the band of $\tilde{u}/U_\infty$ values that we achieved in the tunnel. It should be noted that these data for $\tilde{u}/U_\infty$ were deduced from unsteady pressure measurements. The other flight data points given on the plot were obtained with a hot-wire probe mounted on a Jetstar aircraft at the Ames/Dryden Flight Research Center. They are quite a bit lower than the F-15 cone data and are thought to be more representative of free air turbulence.

## Three-Wire, Hot-Wire Anemometers

At the start of the tunnel flow-quality research in 1976, state-of-the-art anemometry techniques for measuring fluctuating mass flow used single-wire hot-wire probes sometimes augmented by an unsteady pressure sensor to aid in the determination of unsteady velocity. Most diagnosticians agreed, however, that the errors in the calibration of a single wire and the assumptions used in the data reduction at high subsonic speeds could only lead to erroneous results. Nevertheless, these were the types of data conventionally used to quantify flow quality and, consequently, were (and still are) the basis for the empirical equations used to assess the effects of flow quality on transition. It should also be made clear that single wire probes were used to make the $\tilde{m}/m_\infty$ and $\tilde{u}/U_\infty$ measurements shown earlier and these, in turn, were used to forecast the flow quality improvements in the 8-ft TPT.

In 1982, P. Calvin Stainback invented a new hot wire technique based on the use of a 3-wire probe. Enlarged photographs of a three wire probe and probe head are given in Figure 13. The method of calibrating such a probe and of extracting unsteady velocity, density, mass, temperature, and pressure data is documented in References 19 through 23. Figure 14 from Reference 23 gives some example velocity-fluctuation results, obtained in 1988, for the 8-ft TPT along with data obtained using the more familiar two-wire and single wire techniques. The single wire technique labeled E is the one used in the early flow quality measurements in the Ames 12-ft PT and the Langley 8-ft TPT. These data were taken with the flow-quality fixes, streamline liner and LFC airfoil in place. Differences between the three wire technique, labeled A, and the single wire techniques labeled C, D, and E are about an order of magnitude and generally are attributed to the additional assumptions made to extract data from the one- and two-wire probes. These differences are clearly cause for concern.

The variations of mass flow fluctuations with Mach number plotted in Figure 15 are for $p_t = 860 psf$ and $T_t = 540°R$. The level of $\tilde{m}/m_\infty$ in the test section measured by techniques C, D, and E is about the same as that predicted earlier (Figure 11) for a Mach number of approximately 0.75 (no choking). The three-wire data yields, as previously pointed out, much higher levels of $\tilde{m}/m_\infty$ at a Mach number near 0.8 and the difference does not diminish as the Mach number decreases to 0.4.

The variation of $\tilde{u}/U_\infty$ with Mach number is similar to that for $\tilde{m}/m_\infty$ (see Figures 14 and 15). Single wire data (C, D, and E techniques) at the test Mach number show that the levels obtained are about those predicted six years earlier (Figure 11). So the combination of flow fixes provided the conditions in the test section we sought, using single hot-wire technology; however, with the acquisition of the three-wire data, it is not clear that we got what we needed. It may well be that we will have to adopt a whole new scale to quantify good and bad flow quality.

## 7. Airfoil Design

As noted in the Background section, initial airfoil design was carried out by W. Pfenninger, P. J. Bobbitt, and D. O. Allison. The codes employed were those of P. Garabedian, D. Korn, A. Jameson, and F. Bauer and included both direct and inverse calculation procedures based on solutions to the two-dimensional full-potential, nonlinear equation governing transonic flow. The direct calculation codes, where the geometry is input and a pressure distribution determined, contain a turbulent boundary layer routine due to Nash and McDonald which is interacted with the inviscid solution to provide the pressure on the "effective" airfoil shape, i.e., the geometric coordinates plus the boundary layer displacement thickness. The direct codes do not employ a strong interaction solution at the trailing edge but rather a fix developed by P. Garabedian which enables pressures near the trailing edge to be determined with good accuracy. Neither is there a special treatment of the boundary interaction in the region of shocks. While the programs calculate drag, the results are not normally relied on except to establish trends.

The initial design work was carried out using the inverse hodograph method documented in Reference 24. This reference provided a design code for lifting, shockless, supercritical airfoils and was developed under grant to NASA Langley. However, it also required the input of many abstract parameters whose affect on the airfoil geometry was difficult to codify. Several attempts were made to design an airfoil with a concave surface near the leading edge on the lower surface, but they were not successful. Once the analysis code of Reference 25 became available, it became the workhorse for all subsequent airfoil design work.

Based on the supercritical airfoil designs of Whitcomb and Harris (see for example Reference 26) the target design parameters for the LFC airfoil were set as follows:

$$t/c_n > 13\%$$

$$M_{\infty,n} \approx 0.75$$

$$M_{l_{max}} < 1.15$$

$$c_{l,n} \approx 0.6.$$

It was hoped at the outset that the LFC airfoil would be competitive with state-of-the-art turbulent airfoils but it was not a requirement. Supersonic bubble height limitations and the desire for a stable upper surface flow were more of a concern. With respect to the latter it was known that Schlieren images of the flow about flat top supercritical airfoils in the Langley 8-ft TPT showed unsteady wave patterns and shock movement at some conditions.

To indicate some of the "gyrations" that the airfoil designers went through in the early days using the hodograph airfoil program, Figure 16 is provided. A range of Mach numbers, lift coefficients and thickness ratios were explored while at the same time "looking" at the maximum local Mach numbers, acceleration of the flow aft of the leading edge and pressure recovery. The upper surface pressure distribution on the last of the airfoils of Figure 16 (above Figure 17) was finally settled on as having many of the desired characteristics but the lower surface distribution underwent a number of additional modifications through the use of the analysis code of Reference 25.

In order to maintain a smooth upper surface and to install slots or perforations near the leading edge, the use of a high lift device, such as a leading edge slat, is not practical. The high-lift device must be stored on the underside of the airfoil and the attendant cover hinges and gaps make the retention of laminar flow on the lower surface very difficult. It was W. Pfenninger's idea to undercut the lower surface just aft of the leading edge to reduce the velocity and local Reynolds number in this region. The lowering of the Reynolds number makes it easier to maintain laminar flow across steps and gaps and, hopefully, permit laminar flow back to the trailing edge with reasonable suction levels. A number of computer runs made with the design Mach number and lift fixed showed that the upper surface pressures were only minimally affected by undercutting the lower surface.

One of the earlier airfoils designed with the undercut is shown in Figure 17 along with the upper and lower surface pressures. The amount of undercut is modest and there is substantial loading over the whole chord. Nevertheless, the thickness and Mach number of this design are both slightly lower than desired. A later version of this airfoil, often referred to as the "Baseline Airfoil," is shown in Figure 18. The undercut has been increased and the $t/c_n$ has gone from 12.8 to 13.5% yielding higher suction pressures (more negative pressure coefficients) in the mid-chord region of the bottom surface. The design leading-edge-normal Mach number and lift coefficient of the airfoil in Figure 18 is 0.73 and 0.6 respectively and the maximum local Mach number is 1.123. This airfoil was the primary candidate for the Ames 12-ft PT tests and stability codes were applied to determine the suction requirements. In this connection, the stability analysis for the undercut region was of particular concern due to the lack of a validated stability code for Taylor-Görtler instabilities.

A plot often used in the mid 70's to judge the performance of an airfoil is given in Figure 19. The solid symbols on this graph are plotted at the design $t/c_n$ and $M_{\infty,n}$ of the airfoils they represent. The symbol labeled "Baseline Airfoil" is well below the $c_{l,n} = 0.6$ line indicating that it is a conservative design. Some researchers questioned whether a more ambitious design, one that would be competitive with the best turbulent airfoils, would be necessary to prove the "practical value" of LFC. Others wondered whether the lower surface should be more conventional or, perhaps, symmetrical with the upper surface. Foremost among those who thought a higher performance airfoil was required was Dr. W. Pfenninger. Consequently, he continued his airfoil design activities aided by H. L. Reed and J. R. Dagenhart. Their thrust was to increase the design Mach number and, as a consequence, flatten the upper surface pressures. In addition, the turning of the flow in the forward undercut region was changed from a continuous smooth curve to two straight lines with a short transition segment between them. This was done to deal more effectively with Taylor-Görtler instabilities (see Suction Requirements Chapter). The aft cusp region on the lower surface was also modified by flattening the last 20%. Figure 20 shows the modified baseline airfoil with one sharp turn or corner in the undercut region and a second in the aft cusp. Also shown in this figure are the various stability mechanisms which predominate in the leading edge, mid-chord and trailing edge regions. This airfoil was identified as Airfoil A.

Further work by Pfenninger et al. evolved to an airfoil with a more drastic undercut and a lower $t/c_n$ than Airfoil A. It was identified as Airfoil B and is compared to Airfoil A in Figure 21. Also shown in this figure are drag rise curves and design-point pressure distributions. The design conditions are also plotted on the performance plot of Figure 19 for comparison with the Baseline Airfoil.

A selection had to be made in September of 1978 of the final airfoil shape to be used in the LFC experiment. The choice was between the Baseline Airfoil and its conservative design and the more ambitious Airfoil B. It was decided to use Airfoil B but to design the experiment for a Mach number and lift coefficient slightly smaller than those for which it was designed. The conditions chosen are those given in the upper left corner of Figure 22. Other parameters "finalized" at this time were the streamwise chord length, leading-edge sweep and test Reynolds number. The chord selected in September of 1978 was seven feet; some months later a 0.01c extension was added. This yielded a slightly higher Reynolds number for the same free stream conditions.

While it was intended that Airfoil B be the airfoil shape used in the experiment, further changes were made to Airfoil B after September of 1978 but they effected primarily the cusp regions of the lower surface. Specifically, two corners were used in the forward undercut region and six in the aft cusp in addition to the previously mentioned extension of the trailing edge to soften the pressure recovery. It was expected that the latter change would reduce the possibility of flow separation near the trailing edge and on the liner. The geometric details of this airfoil, designated as Airfoil C are shown in Figure 23 along with the lines for Airfoil B. In Figure 24 the ordinates of Airfoil C are magnified by a factor of two to illustrate more clearly the corners and straight-line segments in the forward undercut and aft cusp regions.

Airfoils B and C had a maximum shockless design Mach number normal to the leading edge of 0.758 and a lift coefficient of 0.58. The pressure distributions and two-dimensional supersonic bubbles for the former are plotted in the upper right of Figure 22. For the sweep and free stream Mach number finally selected, 23 degrees and 0.82 respectively, the top of the streamwise supersonic bubble would have impacted the liner opposite to the upper surface and either choked the tunnel or caused a very unsteady flow situation. Consequently the $c_{l,n}$ and $M_{\infty,n}$ values chosen as the wind tunnel "design" conditions were reduced (as previously noted) to 0.55 and 0.755, respectively, and the bubble was reduced in vertical extent to approximately that of Airfoil A. Pressure distributions for Airfoils B and C are compared at these conditions in Figure 25 and show only small differences in the forward and aft cusp regions on the lower surface. Figure 26 gives a computer plot of the streamwise supersonic bubble for Airfoil C with a 7.07 ft. streamwise chord at a free stream Mach number of 0.82 and a sweep of 23 degrees. The bubble comes within about seven inches of the corner fillet of the test section liner wall opposite the upper surface. The supersonic bubble on the lower surface is smaller but slightly closer to the corner fillet of the liner wall opposite the lower surface. Note in Figure 26 that the airfoil is located to one side of the tunnel centerline to account for the larger supersonic bubble on the upper surface of the airfoil relative to that on the lower. Airfoil C was the airfoil design finally constructed and tested; its official designation is SCLFC(1)-0513.

The final free stream test conditions and geometric parameters are:

$$M_\infty = 0.820$$

$$c_l = 0.47$$

$$\Lambda = 23^\circ$$

$$c = 7.07 \text{ ft}$$

$$t/c = 12$$

$$R_c = 20.2 \times 10^6.$$

In the plane normal to the wing leading edge, the final freestream conditions become:

$$M_{\infty,n} = 0.755$$

$$c_n = 6.508 \text{ ft}$$

$$R_{c,n} = 17.1 \times 10^6$$

$$t/c_n = 13.0 \text{ percent}$$

$$c_{l,n} = 0.550.$$

For each of the airfoils shown, plus a large number not shown, extensive off design calculations were made. In addition, the affect of trailing edge flaps of various chords on the pressure distributions at near design conditions were carried out. It was found that small trailing edge flaps could be used to restore the pressure distributions resulting from a loss of laminar flow to near their original shape. Also, it was determined that the onset of shocks could be delayed significantly by the use of small trailing edge flaps and thus increase drag-rise Mach number.

In conclusion, the design of the LFC airfoil was a long and sometimes torturous process. The airfoil selected was indeed competitive with the best turbulent airfoils even when its design conditions were modified to reduce the size of its supersonic bubble. It also came very close to satisfying all of the goals laid down in 1975.

Additional information concerning the development of the "Baseline Airfoil" and Airfoils A, B, and C is contained in References 27, 28, 29, and 30.

## 8. Suction Requirements

Empirical relations, derived from earlier low-speed and low-Reynolds-number experiments, are available and may be used for predicting transition (References 31, 32). However, more sophisticated approaches have been developed to determine the transition location on swept and unswept wings for incompressible (References 33, 34, for example) as well as compressible (References 35-38, for example) flows. These methods are based on temporal or spatial stability analyses for laminar boundary layers where the local amplification rates are obtained as solutions of the governing stability equations as a function of the frequency, wave length, and propagation angle of the disturbance. The amplification ratio $A/A_o$ of the disturbance at any downstream location $x$ can be expressed as an exponential function $e^N$ where $N = \int_{x_0}^{x} \sigma dx$, and $\sigma$ is the spatial growth rate of the most amplified disturbance. When the temporal stability theory is used (References 33 and 36), the spatial growth rate $\sigma$ is computed by using Gaster's group velocity transformation. If $x_0$ and $x_t$ represent the locations of initial instability and onset of transition respectively, then the derived $N$-factor may be correlated with experimental transition data and used, based on experience, to predict transition location. These analysis methods only apply in linear attached flow regions and calculations must be carried out to insure that the $N$-factor is maximized with respect to all frequencies. When using the empirical $N$-factor method for design, one must choose a limiting $N$-value for transition based on experimental correlations.

## Design Cycle

Boundary layer instabilities that may occur on swept wings are leading edge attachment-line contamination, crossflow (CF) vortices, Tollmien-Schlichting (TS) tangential waves, and Görtler vortices that are associated with concave surfaces. These various disturbances are dependent on a variety of factors including geometry and are normally analyzed separately although mutual interactions are possible (References 39-46). For example, in transonic flow, crossflow disturbances may interact with weakly amplified oblique TS waves in the relatively flat, midchord regions on wings causing both the crossflow vortices and TS waves to be distorted. Such nonlinear interaction in the two-disturbance modes will cause the crossflow vortices to grow faster than predicted by linear stability theory. El Hady (Reference 40) has recently shown that, at the leading edge of a swept wing, many possible wave triads may resonantly interact. These triads include combinations of steady CF, unsteady CF, TS, and vortical modes. Because of these interactions the triad components will grow rapidly depending on their initial spectrum of amplitudes and phases, even if their initial amplitudes are small. In the design of the LFC airfoil, CF disturbances were deliberately kept small to prevent CF and TS interactions.

The design and optimization of the LFC airfoil required a number of iteration cycles using advanced computer codes to analyze the aerodynamics, the surface mass transfer and displacement thickness effects, and the stability of the boundary layer. The application of advanced codes was expected to improve the determination of the suction flow requirements and to reduce the uncertainties associated with the design of the suction system. A block diagram of the iterative design cycle is shown in Figure 27 with illustrative outputs from the various codes. The design approach utilized a transonic-airfoil analysis code (Reference 25), a swept-wing boundary layer code (Reference 47), and stability theories (References 33-34) for the prediction and optimization of suction requirements. Successive computations were made to optimize the combination of design Mach number, lift coefficient, and airfoil thickness at $R_c = 20 \times 10^6$. The three-dimensional, boundary layer analysis code (Reference 47) was modified to include the effects of local mass transfer and computation of boundary layer displacement thickness. This code, in combination with the transonic airfoil analysis code, permitted the analysis of transonic flow and three-dimensional boundary layer parameters over an infinite-span, yawed LFC airfoil. Combining results from the boundary layer analysis with the stability codes provided the required suction rates. Solutions of the stability codes were based on the fixed frequency, or envelope, method for various types of local boundary layer instabilities in the different regions of the airfoil upper and lower surfaces.

## Upper-Surface Suction Requirements

Initially, the linearized incompressible, boundary layer crossflow stability codes (References 33, 34) were used in the LFC airfoil design cycle to calculate integrated TS and CF disturbance amplifications based on minimum suction requirements for laminarization over a Reynolds number range. Later, these calculations were checked using linear-compressible spatial stability codes of El Hady (Reference 37) and Mack (Reference 48). For these analyses, an $N$-factor of 7 was used. This value is somewhat lower than the limiting $N$-factors of 9 to 11 obtained from previous calibrations of the linear stability theories (References 49-52). However, the choice of 7 was intended to provide conservative values of the suction requirements for a given transition location. The design results indicated that in the leading edge region of the upper surface, crossflow was minimized by rapidly accelerating the chordwise flow from the stagnation point to the pressure minimum by the use of a relatively "sharp" leading-edge geometry (Figure 23). The choice of leading edge sweep and the low unit Reynolds numbers of the tests also had a favorable effect on CF disturbances. A maximum $N$-factor value of 4.5 in the nose and 6.5 in the aft region were maintained at design conditions based on incompressible stability analyses. These values were found to be 15 percent higher than those calculated by compressible analyses.

Tollmien-Schlichting wave disturbances dominate over the midchord region of the upper surface where there is a slowly decelerating supersonic flow associated with a relatively flat pressure distribution (Figure 25). The growth of TS disturbances is kept in check by the flat top type pressure distribution, and by modest, nearly constant suction level over the surface rearward to about 96 percent chord. Since the CF disturbances in the leading edge region are concentrated over a very short chordwise extent and decays rapidly in slowly decelerating downstream flow, the interaction of TS and CF disturbances in the flat pressure region of the upper surface appeared insignificant.

## Lower-Surface Suction Requirements

The incompressible stability results, based on Tollmien-Schlichting (TS) instabilities, for the lower surface were similar in level and trend to those for the upper surface and indicated that suction requirements in the forward and aft regions might be conservative. However, laminarization of the LFC airfoil lower surface depended primarily on the suction control requirements of centrifugal instabilities in the concave-surface regions referred to as "Taylor-Görtler" (TG) instabilities (Refs. 53 and 54). Results from linearized analyses (Refs. 53, 55, and 56) of the TG instability presented in Reference 56 were used to determine amplification factors for the lower-surface concave regions. Figure 28, taken from these references, shows the variation of the linearized maximum growth factor B with the "Görtler" parameter $G$ for the Blasius boundary layer profile with and without asymptotic area suction. Suction generally tends to pull the TG vortices towards the surface where stronger viscous forces dampen their growth. Thus, by using values of $B$ between the two theoretical curves shown in Figure 28, calculated values of $G$ based on airfoil-surface radius of curvature, and calculated boundary-layer parameters, the TG vortex maximum growth rates in the concave regions were obtained from the following equation:

$$N_{\max} = \int_{(x_N)_1}^{(x_N)_2} \left[ \frac{(\beta\theta R_\theta)_{\max}}{\theta R_\theta} \right] dx.$$

Turning of the flow in the lower-surface concave-curvature regions through several "corners" was believed to minimize growth of the TG vortices. Consequently, the flow stability in corners was analyzed rather than a gradual turn over a large chordwise distance. Figure 29 shows representative values of $N_{\max}$ with small incremental turns of the flow $\Delta x/c_n$ based on the Blasius and asymptotic suction amplification rates shown in Figure 28.

According to Pfenninger, et al. (Reference 30), the wave number of the most amplified TG vortices is relatively large for the large values of $G$ in short chordwise-distance turns or "corners" of small radii. This leads to small vortex wavelengths and heights from the surface (Reference 55) when compared with those of a longer, more gradual turn through the same angle. With potentially smaller vortices, sufficient damping due to dissipation in the corners may tend to minimize overall growth of the disturbances. The insert in Figure 29 illustrates typical variations of normalized curvature $c_n/r$ with $x/c_n$ for several possible turn increments. Even though the Görtler number and local growth rate of the TG vortices with a sharp turn of small radii are large, the integrated growth rate $N_{\max}$ is seen to decrease as the radius of concave curvature decreases.

Since the flow decelerates as it approaches the concave "corner" regions of the model, increased local suction was required to prevent laminar separation. Values of $N_{\max} = 2.5$ to $4.0$ based on the approach described above for TG instabilities were obtained in the front and rear concave regions of the LFC airfoil. The increased suction associated with these corners appears as spikes in the suction distribution. Although disturbance amplitudes may decrease locally in the region of each corner, the overall trend was for the disturbance amplitudes to grow and reach a maximum amplitude at the end of the concave region. Following the concave region was the convex-curvature midchord region and associated favorable pressure gradient that tended to dampen the growth of disturbances.

Thus, for the lower aft concave region, six such corners were incorporated. Two were located in the suction zone, each followed by a slightly convex turn (negative turning angle). There were four additional concave corners in the rear where no suction was provided ($x/c > 0.84$). Spikes occurred in the pressure distribution of Airfoil C for the first two corners (Figure 25) but not for the last four since turbulent flow was assumed to exist in the region $x/c > 0.84$. It should be noted that the same lower-surface slotted suction panels were used for the final test configurations of both the slotted and perforated LFC airfoil models.

## Suction Distributions

Evaluation of the optimum suction rates required to avoid premature transition on the swept LFC airfoil were determined using the design pressure distribution at $R_c = 20 \times 10^6$ and the conservative linearized incompressible stability calculations previously discussed for amplified boundary-layer disturbances. Figure 30a shows the variations in the theoretical chordwise suction-coefficient ($C_Q$) distribution for both the slotted and perforated airfoils upper and lower surfaces in the laminar test regions for a range of chord Reynolds numbers. The increased suction required in the turbulent regions at each end of the airfoil model is discussed in the Liner Design Chapter.

The results shown in Figure 30a are based on boundary-layer stability analysis $N$-factors of 5 to 7 for CF and TS disturbances and 2.5 to 4 for TG disturbances. Suction can be eliminated in the nose region of sharp leading edge airfoils for moderately high Reynolds numbers in contrast to the finite requirements of airfoils with blunt noses. Suction extended rearward to 96 percent chord on the upper and 84 percent chord on the lower surface. Increased suction was required in the upper-surface aft-pressure-rise region and decelerated-flow regions of the lower-surface concave regions. It should be noted that, while the $C_Q$ vs. $x/c$ levels shown in Figure 30a decrease with increasing $R_c$, the value of suction mass flow actually increases.

The $C_Q$ distributions shown in Figure 30a should be considered "idealized" theoretical distributions since they were modified during model design due to practical considerations. The continuous distributions of Figure 30a were integrated over the chordwise extent of each duct to assign discrete suction levels to each duct. These levels were then "smoothed" to arrive at the $C_Q$ distributions used to design the slotted-surface suction-system hardware of the model (Figure 30b).

Figure 30c shows the theoretical chordwise variation of the suction coefficient $C_Q$ over the perforated LFC airfoil upper surface for two chord Reynolds numbers. The results shown in Figure 30c are based on the same practical considerations as for the slotted LFC airfoil (Figure 30b) and the capabilities of the slotted surface itself. The lower surface suction distribution used for the perforated model was essentially the same as for the slotted (Figure 30a).

## 9. Liner Design

Contoured liners have often been used in wind tunnel testing to minimize wall interference effects, but we are not aware that any were as complicated as that designed for the 8-ft TPT LFC tests. Since it represented the streamlines about a lifting, yawed, infinite-span airfoil offset to one side of the tunnel centerline, the contours on all four walls were different. Indeed, the design of the 8-ft TPT LFC liner for the airfoil design Mach number and lift coefficient strained considerably the state-of-the-art of viscous and inviscid flow methodology as it existed in the late 1970's.

The required LFC liner had the general features shown in Figure 31 and consisted of contraction, test section (where airfoil is located), choke and diffuser sections. A preliminary study of the liner was carried out, and indicated that the contraction had to be located much further forward, toward the plenum or "big end," than that of the original tunnel. This was necessary to give the onset flow sufficient distance to achieve the desired uniformity, to accommodate a large chord model, to provide sufficient distance downstream of the model trailing edge to install the sidewall chokes so that they did not interfere with the wing flow and, finally, to fair the liner back to the tunnel walls without increasing the diffuser's angle beyond where attached flow could be achieved. As a point of reference, the origin of the slots in the normal 8-ft TPT test section walls is at the 50 foot station.

Figure 32 shows an overlay of the contraction grid on the existing, circular equivalent 8-ft TPT contraction shape. The term "circular equivalent" is used since the contraction starts out circular in cross-section and ends up square, with filleted corners. The downstream asterisk is the location of the equivalent liner radius at the 36 foot station; the asterisk at the upstream end is on the same streamline and at the maximum radius of the "big end." The use of this streamline in the contraction design insured that mass was conserved over the length of the contraction, a necessary condition to achieving the desired Mach number with the liner contours selected.

It was impractical to design and install a liner that extended all the way forward to the maximum diameter of the contraction. Consequently, it was decided to start the liner in the vicinity of the 24-ft station. The equivalent liner radius (i.e., the streamline with the asterisk on each end) is well off the wall at this station and some modifications of the liner shape were necessary to fair the equivalent radius streamlines to the wall at this location. This will be explained in a little more detail in the next paragraph.

In order to get a "head start" on shaping the onset flow to that of an infinite-span, yawed wing it was decided to adjust the contours of the axisymmetric-contraction liner to account for the upstream flow field of the yawed wing. Perturbation velocities $\delta u, \delta v, \delta w$ to the free stream velocity caused by the presence of the model were added to the axisymmetric-contraction field velocities on the cartesian grid as depicted in Figure 33a. Subsequently, streamlines were determined by integrating upstream, starting with the liner geometry at the 36 foot station, through this perturbed velocity field (see Figure 33b). Generally, these streamlines did not intersect the tunnel wall, as noted previously, and a spline fairing was used to bring them smoothly to the wall in the vicinity of the 24 foot station. These fairings were done in a iterative manner, using the stream-tube curvature (STC) code and equivalent radius concept, to insure that the contraction velocity distribution was smooth, that no boundary layer separation occurred and that the required Mach number at the 36 foot station was achieved. Clearly, with the use of the procedure just described, the contraction liner could not be symmetric. Liner contours in the contraction have been plotted in Figure 34 so that the left to right asymmetry can be easily seen.

It should be noted that the depth of the liner at the 36 foot station was not arbitrarily fixed nor was it dictated entirely by mass-conservation considerations. A number of streamline calculations were made, starting at the 36 foot station, through the test section to insure that there would be enough depth left at the closest approach of a streamline to the test section wall for the supporting substructure and a boundary layer displacement correction. These calculations took place over a period of time as the methodology improved, the location of the airfoil was changed and the nature of the liner substructure became known. So in a very real sense, the test section liner was the "dog" that wagged the contraction liner's "tail."

The LFC airfoil was designed using the code developed at NYU (see Airfoil Design Chapter) which solves the fluid flow problem in a transformed plane. The coordinate system in the physical plane is nearly orthogonal but not rectilinear and somewhat inconvenient for use in liner design. Consequently, an airfoil code developed by Dr. Lee Carlson of Texas A&M, which solves the full-potential equation on a cartesian grid, was employed (Reference 57). This code, which has the acronym TRANDES, does not normally have the grid resolution in the leading edge region equivalent to that of the NYU airfoil analysis code. Calculations made with the TRANDES code typically result in lower suction peaks near the leading edge and slightly lower lift coefficients. In the present application, adjustments were made to the grid until both the pressure distribution and lift coefficient were in good agreement (see Reference 58) with those obtained using the NYU code.

The velocity field used to determine the liner shape in the test section region was provided by the TRANDES code plus a constant-sweep velocity component (see Figure 35a). This velocity field was used to determine the streamline shapes which form the basis of the wall contours. Starting at the 36 foot station and integrating downstream from one hundred and forty-four starting locations around the periphery, the streamlines that form the inviscid shape were defined. On the floor and ceiling the streamlines that intersect the leading edge split to go over the top and bottom airfoil surfaces. As shown in Figure 35b, the streamlines over the top and bottom surface start at the same point on the leading edge but follow different paths and emerge at different spanwise locations at the trailing edge. Of course the liner contours must follow these streamlines over the top and bottom and this resulted in a step in the liner at the trailing edge. Figure 36, which shows cross-sections of the liner contour from just downstream of the leading edge of the model (station 44) to just beyond the start of the sidewall chokes, clearly shows these steps. Starting at station 57 there was a gradual "fairing-out" of the steps until, at station 71, they no longer existed; Figure 37 depicts this phenomena.

The discussion so far has concerned itself with inviscid streamlines. However, the displacement effects of the viscous boundary layer on the liner and on the airfoil, particularly in the turbulent zones at each end adjacent to the wall, had to be accounted for. A two-dimensional boundary layer analysis applied along the inviscid streamlines was thought to be adequate for most of the liner, the juncture regions of the airfoil being the exception. In the turbulent zones at each end of the airfoil, the flow had to be controlled by suction. It reduced the boundary layer thickness and insured that the flow did not separate in the highly adverse pressure-gradient regions near the trailing edge. These calculations were started at the airfoil leading edge with the assumption that the flow remained laminar until the turbulent wedge, emanating from the wing-leading-edge/liner juncture, was reached. The turbulent zone was broken into a number of strips and the procedure applied to each to determine the suction requirements. The required suction levels were higher than those in the laminar zones as depicted in Figure 38a for the lower surface. Consequently, bulkheads were built into the substructure of the airfoil to provide a separate suction system for the turbulent flow regions and to keep the turbulent boundary-layer noise from contaminating the laminar-region suction system. The suction mass flows in the laminar mid-span regions were ramped up to the turbulent levels at the ends of the model as shown in Figure 38a. Some attempt was made to provide the capability for higher than design suction mass flows in order to treat off design conditions but volume constraints did not allow this to be done in a smooth manner. As a result, the final turbulent zone suction distribution was a bit irregular as seen in Figure 38b. A planform view of the wing is given in Figure 39 to show the high suction regions, the turbulent zones and the wall contours on the lower surface. A similar situation existed on the upper surface. Since the model surface near the liner/wing juncture could not be indented to account for the displacement effects of the boundary layer in the turbulent zones, an equivalent area was taken out of the liner in the juncture region. In this way, the required overall streamtube-area distributions above and below the airfoil could be maintained.

Suction requirements on the liner in the wing/liner juncture regions were treated using the same boundary layer methodology but the approach was necessarily different. The streamlines that pass above and below the wing originate far upstream and the boundary layer growth along them has a strong effect on the suction mass flow needed to keep the boundary layer attached as it moves through the pressure field of the airfoil. Consequently, the boundary layer calculations for the juncture regions above and below the airfoil were begun near the start of the contraction and terminated well downstream of the trailing edge. Suction levels required to keep the boundary layer from separating were automatically determined by the code.

Everywhere along each streamline an equivalent boundary layer displacement correction, including suction effects, was calculated. It was determined as a relative effective deficit stream-function correction downstream of a reference plane near the end of the contraction where the boundary layer is thin (see Reference 58 for details). These displacement corrections are, of course, subtracted from the inviscid liner contours to provide the actual liner surface coordinates.

The suction distributions needed on the wall in the juncture regions are given in the contour plots of Figure 40. It is evident from this figure that the suction requirements are much greater in the aft regions on the bottom side than on the top. The suction manifolds and slots required to provide the suction mass flows of Figure 40 are shown in the sketch of Figure 41. Note that in the high suction regions the slots are more closely spaced. The suction slots just ahead of the leading edge were provided to minimize flow separation in this region and enable a smoother flow in the wing/liner junctures downstream.

As will be discussed later there were a number of pressure taps installed in the liner along calculated streamlines. They were used for many purposes including the provision of a check on the liner design itself. Pressures measured along the streamline in the middle of the wall opposite the upper surface, plotted in Figure 42, give a good indication of the success of the liner in providing the desired onset flow. Generally the best airfoil pressure distributions were obtained at Mach numbers slightly higher than design, consequently, Mach numbers near the wall were higher than desired particularly at $R_c = 20 \times 10^6$ (see Figure 42a). At $R_c = 10 \times 10^6$ wall pressures were slightly lower and the agreement was quite good with the design curve as shown in Figure 42b. Clearly the liner performed in an outstanding manner even though the analytical methodology used required many assumptions and approximations.

It was noted earlier that a sonic choke was a very effective device in improving the flow quality in the test section. Consequently, an adjustable sonic choke was specified for the LFC experiment and was designed as an integral part of the liner. It was a two-wall choke (one on each side wall) located between stations 53 and 59. The trailing edge of the model at the floor was 3 feet ahead of the leading edge of the choke; the trailing edge at the ceiling was 5 feet ahead of the choke (see Figure 31). Calculations were carried out to determine the upstream flow field of the choke in order to make an informed judgement on the separation distance required. At 3 feet there was a negligible effect, consequently this became the minimum separation distance between the airfoil trailing edge and choke.

A much more detailed discussion of the liner design is given in Reference 58.

## 10. Model Construction

The LFC experiment consisted of evaluating two different suction-surface concepts on the top side of the airfoil; construction features of the two surfaces were also different. Top and bottom surfaces for both concepts were divided into three separate panels, as shown in Figure 43, and these in turn were bolted to a strongback (also referred to as a wing box). A 10.9% trailing edge flap completed the assembly.

All six panels of the slotted LFC model were slotted (see Figure 44) while the perforated surface was only implemented for the three top-surface panels. The middle and aft panels on the bottom of the perforated model were the same slotted panels used with the slotted-panel model, while the bottom forward panel was a solid surface with slightly different and smoother lines than the slotted one it replaced. A solid lower forward surface was used since it enabled the matching of the perforated top-side forward panel with the lower surface at the same time the slotted-surface tests were being run, a considerable time saving. Midway through the perforated model tests the solid lower forward panel was replaced by the slotted panel from the slotted model. The HLFC test (not discussed) used the combination of panels shown in the sketch at the bottom of Figure 44.

### Slotted Model

Construction features of the slotted surface can be seen in Figures

45 and 46. Each slotted panel started out as a solid piece of aluminum so that a considerable amount of machine work had to be done just to "rough-out" the general shape. Numerically controlled milling machines were used to machine the ducts on the bottom side, as well as the plenum slots on the top side (see Figures 47 and 48), of each panel. The metering holes, and slots, were drilled or cut by machines attached to a carriage which in turn rode on rails fastened on either side of a large flat set-up table. The carriage was computer controlled to enable a faster and more accurate operation. Figure 49 shows a photograph of the lower forward panel mounted on this device during a drilling operation.

Once the panel substructures were completed, aluminum skins $\frac{1}{32}$ inches thick were bonded to the plenum-duct sides. The slots, which ranged in width from 0.002 to 0.0063 inches, were then cut with jewelers circular saw blades. Finally, the surfaces of the upper panels and the lower forward panel were hand polished and coated with a polymeric material known as TUFRAM using a patented deposition process. The TUFRAM coating provided a much harder and erosion resistant surface than the bare aluminum. No coating was applied to the two lower aft panels since they were less exposed and less critical than the other four.

The criteria used to determine the slot sizes and spacing, the plenum duct heights and widths, and surface waviness tolerances were based on earlier work (References 4, 59, 60, 61) and some analytical studies carried out by Pfenninger, Reed, and Dagenhart (Reference 30) using the NYU code. The latter contains results from a high resolution calculation with simulated surface waviness of various heights. These calculations led to the dashed line on Figure 50 and provide a waviness goal for supercritical airfoils. The solid curves on the figure were derived from the X-21 project (References 59 and 60) where the airfoil used was of the subcritical variety. Measured height and wave length data from the LFC model are plotted as the solid triangles. It is evident from Figure 50 that the quality of the LFC slotted panels, in terms of waviness, is quite good. Additional polishing of the surface was carried out after these measurements were made, consequently the data is conservative. The perforated surface was slightly more wavy than the slotted one but still of good quality by X-21 criteria.

The assembly of the LFC airfoil started with the attachment of the upper three panels by bolts from the underside directly to the wing box. Following this the lower forward panel was pinned to the upper forward panel near the leading edge (see Figure 51a). The former was also bolted directly to the strongback. Penetrations were required to bolt the lower middle and aft sections to the strongback. To increase the rigidity of the aft part of the model, the top and bottom sections were bolted together from the bottom side near the 77-percent chord station (Reference 62) and at the trailing edge of the panels. A cross-sectional drawing with all the panels assembled is given in Figure 51b. The length of each of the panels in terms of the chord and their relationship to the wing box is given in Figure 43.

The five segments of the trailing edge flap were attached by screws to the aft panels; flap angle changes had to be made with the aid of jacking screws. Segmentation of the flap allowed adjustments to the spanwise pressure distribution to make it more two-dimensional. It was known from calculations that the turbulent zones at either end of the airfoil caused a loss of lift and that flap deflections of several degrees would be needed to recover the design pressure distribution in these regions. The outboard flaps were used primarily for this purpose.

Once assembled the airfoil model at the bottom end had the appearance of the drawing in Figure 51b. The bolts shown near the quarter chord were for attachment to a fixture welded to the original test section floor. The middle bolt was fixed relative to the floor but the others, including the ones at the rear attachment points, were attached through floor pieces with elongated holes so that the angle of attack of the airfoil could be changed several degrees.

During the design of the airfoil model, a finite-element math model of the structure was formulated to predict stresses and model deformation. The airfoil contours were adjusted for these deformations and the resulting coordinates were then input to the NYU code and an "aeroelastic" pressure distribution determined. These calculations indicated that the deformations were small overall but were larger near the trailing edge than near the leading edge (see Figure 52). In any case, the effect of these predicted deformations could be easily countered by small changes in flap settings. Once the test got underway and the pressure distributions were found to be further from the design distribution than expected, a second look was taken at the aeroelastic behavior of the model.

The model contour was first measured in its unloaded condition and it was found that a small permanent deformation of the model had already occurred (see Figure 53) due to repeated loadings. Hydraulic jacks, two in front and two in back, were then used to apply simulated aerodynamic loads to cushioned bars in the forward and aft cusp regions. The deformations obtained due to this load are compared to those of the unloaded model in Figure 53. Subsequently, shims were installed between the panel's substructure and the wing box to correct for these deformations. Figure 54 shows model centerline data for the shimmed model both unloaded and loaded. Comparison of the shimmed with the unshimmed model shows a substantial improvement over the first 20% of the chord but only a slight improvement over the rear 20%. Measurements made on the model above and below the centerline of the tunnel showed similar results.

Perforated Model

The perforated model was designed and built by the McDonnell Douglas Corporation in Long Beach, California (Reference 63). It utilized an integral combination of composite materials and metals to provide the perforated surface with plenum ducts and associated metering holes. Figure 55 shows the basic construction features of the surface and the large main suction ducts below. Every other "plenum" duct was inactive since the perforated surface material was blocked by the tops of the corrugated substructure. The inactive plenum ducts on the top of the sides of the large aluminum ducts were reinforced with an aluminum insert (see Figure 55). Metering holes were located along the bottom of the plenum ducts.

Materials used in the fabrication of the suction surface are listed in Figure 56. Fiberglass and carbon fiber cloth were laid as shown around mandrels and each panel, including the perforated surface, cured in an autoclave. The surface material as noted was titanium and the holes were drilled with an electron beam. They were nominally 0.0026 inches in diameter at the outer surface and slightly large on the bottom surface. This feature reduced the possibility that the holes would become clogged. The titanium material was 0.025 inches thick and the hole's spacing was also 0.025 inches. The outside surface was smooth to the touch and was expected to be aerodynamically smooth as long as the suction mass-flow rates were not excessive.

The forward perforated panel is illustrated in Figure 57 to show some of the special construction features. Note that the first two plenum ducts that would normally be active were sealed due to the lack of access to the main duct (the most forward plenum duct) or to clogging problems associated with the bonding materials (the second plenum duct). The upper and lower forward panels were "butted" to each other and the leading edge hand smoothed. The final assembled perforated model is shown schematically in Figure 58 along with the dimensions of each panel in terms of the chord of the model. Note that the panel lengths on the top side are slightly different from those of the slotted model (Figure 43).

Another change instituted on the perforated model was the replacement of the five trailing edge flaps with three. The center flap was also designed to be controlled remotely (from the control room) to enable the attainment of the "design" pressure distribution in a shorter time. In addition, the middle flap did not contain any suction ducts. Once the perforated model tests began it was found that the center flap could only be moved a few degrees due to aeroelastic effects and was ineffective in altering the pressure distribution.

It was noted in the introduction of the present chapter that a solid lower forward panel was used for the perforated model. It was also noted that the geometry was slightly different from that of the slotted lower forward panel. This difference is illustrated in Figure 59 where the ordinate has been amplified relative to the abscissa. The calculated pressure distribution that resulted from the shape change is plotted in Figure 60 along with the original slotted-surface pressure. The only noticeable difference is in the lower forward pressures from the leading edge back to about 30% of the chord. Boundary layer calculations indicated that the flow would remain attached in the forward, solid concave region but might separate near the trailing edge at the $20 \times 10^6$ design Reynolds number. As we will see in the Results and Discussion Chapter, this was indeed the case.

Once the tests on the perforated panel were underway and the best approximation of the design pressure distribution was obtained, it became clear that the extent of laminar flow was not as large as that for the slotted model. All of the available controls were exercised but no improvement was obtained. The most probable cause was that the suction-mass-flow levels were not as high as those available on the slotted model (see Figure 61). A detailed examination of the perforated suction surface, plenum ducts, and metering holes revealed that the metering holes could be enlarged to permit higher mass flows without taxing the mass flow capabilities of the perforated surface. It was decided to remove and disassemble the perforated panels to enlarge the metering holes and to make a number of other improvements at the same time. These improvements were:

- All metering holes enlarged in first two panels.
- Second suction nozzle added to each laminar main duct in first panel.
- Laminar nozzle extensions enlarged in first two panels.
- Metering holes installed in first two plenum ducts.
- Two new vacuum pumps installed.
- Lower forward solid panel replaced with original slotted panel.

All of the changes except the last were in the direction to increase the mass flow through the two top-surface forward panels. The last was included to recover the attached flow in the aft concave region and to obtain a slightly better pressure recovery. This improvement was expected to provide some benefit to the top side as well. The suction pumps added were capable of higher incremental pressures than those in use during the slotted tests and enabled higher mass flows through the two newly activated forward plenum ducts as well as the rest of the forward panel.

Aeroelastic deformations were also a problem with the perforated-surface model. Maximum deformations in the leading and trailing-edge regions exceeded 0.20 inches which was slightly higher than the levels encountered with the slotted surface. Figure 62 shows the distribution of deviations from the design shape for the upper surface for both unloaded and loaded conditions. Consideration was given to shimming the forward and aft panels but rejected. Due to the structural characteristics of the perforated panels, shims were expected to be much less effective than they were on the slotted panels. In addition, while the shims used on the slotted panels corrected some of the deformation, they did not have a significant effect on the pressures.

## 11. Liner Fabrication

The liner's contoured surface was provided by a hard closed cell foam material which was bonded on one side to marine plywood. It was broken down into a number of blocks which were machined to the desired shape by numerically controlled milling machines (see Figure 63). These blocks in turn were bolted to a substructure that was welded or bolted to the original wind tunnel contraction and test section walls. Photographs of the liner substructure in the test section and contraction sections are given in Figures 64 and 65 respectively. The metal substructure in the contraction region was covered with plywood plates (only one is in place in Figure 65) on top of which was mounted the liner blocks to obtain the contraction shape. Joints between the blocks were filled and smoothed. Figure 66 shows a photograph of the completed contraction section looking downstream. Note that one of the foam blocks is missing at the entrance to the test section floor. A downstream view of the test section looking beyond the model upper surface is shown in Figure 67. One of the chokes can be seen on the left side of the picture beyond the trailing edge of the model. See Reference 64 for further details of the liner's construction.

Around the model/liner junctures were a number of slotted panels flush with the surface of the liner comprising what was termed the "suction collar." These panels were serviced by one or more suction hoses depending on the number of compartments in the panel. Figure 68 shows the location of the seven suction panels along with a photograph of two of the panels. The slot locations on each panel are also drawn on the sketch of Figure 68 and they can be seen in the photograph as well.

Construction features of the chokes are illustrated in Figure 69. The surface of the choke was moved in and out of the flow by a piston driven by an electrically driven bell-crank as depicted in the figure. It had a flexible surface made of fiberglass and a perforated insert just downstream of the minimum to ventilate the test-section plenum. A photograph of one of the choke plates looking upstream toward the model is given in Figure 70. Mach numbers ranging from 0.80 to 0.84 could be achieved by retracting or extending the choke. It did provide the flow quality improvements sought as discussed in the Flow Quality Improvements Chapter.

At approximately the trailing edge of the choke, the streamline liner ended and a "faired" liner began which brought the liner depth back to zero at the 78 foot station. The effective diffuser angle of the "faired" liner was quite a bit larger than that of the original diffuser. This fact coupled with the large momentum defect downstream of the model trailing edge caused considerable unsteady separated flow in the tunnel diffuser and degraded flow quality. Consequently, two rows of vortex generators were installed with eight generators in each row (two per wall) at the 60 and 71 foot stations to energize the boundary layer. They were found to be effective in reducing the amount of separation in the diffuser and were used throughout the test. Figure 71 is a photograph looking upstream past the vortex generators at the 71 foot station toward the airfoil.

## 12. Suction System

Perhaps the most complicated component of the LFC test equipment was the suction system. It provided most of the difficulties in model fabrication and required a substantial array of supporting equipment. Various elements were:

- Model and liner suction surfaces
- Subsurface plenum duct
- Metering holes
- Main suction ducts
- Calibrated suction nozzles
- Nozzle extension
- Hoses to airflow control boxes (0.25 - 1.00 inch diameter)
- Airflow control boxes with needle valves
- Sonic nozzles
- 6-inch hoses to large suction manifold
- Suction manifold
- 20-inch pipe to 10,000-cfm compressor

- 10,000-cfm compressor (rated 4 to 1 compression ratio)
- Return pipe to hollow turning vanes where flow is reinjected to the tunnel
- Vacuum pumps.

In the following discussion, a few of the more important suction elements will be discussed.

With the aid of the airfoil-pressure distributions, boundary-layer, and stability calculations, suction mass-flow requirements were established for both the slotted and perforated surfaces. Figure 30b from the Suction Requirements Section shows the variation of the suction mass-flow coefficient $C_Q$ with distance along the chord for the slotted model at two Reynolds numbers. On the upper surface, the $C_Q$ level is about 0.00015 back to about 70% chord where it increases up to the 0.001 and 0.002 range due to adverse pressure gradient in this region. As Reynolds number increases $C_Q$ decreases, most notably over the aft 20% of the chord. The lower surface was remarkable for the very high local $C_Q$ values just ahead of the sharp corners in the forward and aft cusp regions. In addition, much higher suction levels were required in the rear pressure recovery region. Note again that suction on the top and bottom sides in the test region extended to 96 and 85 percent of the chord, respectively.

Figure 30c shows the design upper-surface suction, mass-flow ($C_Q$) requirements for $R_c = 10$ and $20 \times 10^6$ and the design Mach number of 0.82 for the perforated model. A comparison of the design $C_Q$ distributions for the slotted and perforated upper surfaces is given in Figure 61 for $R_c = 10 \times 10^6$. Also shown is the maximum level of suction attained with the slotted surface as well as the maximum level of suction achieved with the perforated surface prior to its overhaul to increase its mass flow capability (see perforated panel discussion in previous section).

As noted earlier, the criteria used to determine slot width and plenum-duct height were based on the research of Dr. W. Pfenninger and his associates (References 2 and 61). Figure 72 shows a plot of the variation of the ratio of plenum-duct height to slot width with critical slot Reynolds number stemming from this work. It provides the basis for choosing the plenum-duct height once the slot width is known. If plenum slot height is too small, then slot flow oscillations and viscous slot wakes can result. The desire to keep slot width and $R_s$ small and $h/s$ high led to the establishment of the Langley LFC criteria of $h/s$ values between 10 and 20 and $R_s < 150$. Two rows of metering holes were used and they were alternated on each side of the plenum to provide a more uniform spanwise suction distribution.

Metering hole diameters were less than 0.02 inches and they were spaced 0.5 inches or less apart. The slots, metering holes, and ducts along with the rest of the suction system were designed to have a 50% oversuction capability to treat off-design conditions and design uncertainties.

Slot width is usually maintained at values on the order of the boundary layer sucked height which in turn is kept to about 20% or less of the local boundary layer thickness. Slot spacing varies directly with slot width and inversely with mass flow coefficient. Ideally, one would have many small-width slots closely spaced but practical considerations limit how close slots can be placed as well as the minimum slot width obtainable. Additional details of the slot and plenum design are given in Reference 62.

The main ducts in both the mid-span laminar regions and turbulent end zones are evacuated, or pumped, by calibrated nozzles of different sizes as illustrated in Figure 73. A bulkhead separates the turbulent zone from the laminar zones so that the noise in the former does not propagate up the ducts into the laminar regions and cause premature transition. In addition, the turbulent zones require much higher suction levels as was seen in Figure 38. Due to the higher adverse pressure gradients on the lower surface, the turbulent $C_Q$'s are quite a bit higher than for the upper surface.

Figure 74 gives a schematic of the lower surface of the model showing the turbulent zones and the area they occupy relative to the laminar or "test" zone. The intended main laminar test area was in the vicinity of the tunnel centerline since only the center flap has a suction surface.

The nozzles depicted in Figure 73 are shown in detail in Figure 75. In most ducts circular nozzles were used (top part of Figure 75) but in ducts where circular nozzles could not provide the required mass flow, with duct velocities below 250 ft/sec., rectangular nozzles were used. A transition section was incorporated in the end of the latter to transform the rectangular section to a round one. Both types of nozzles were calibrated against Reynolds number based on diameter or height. This mass-flow calibration required the measurement of the static pressure in the throat and in the duct.

Hoses of the appropriate diameter were fitted to the ends of the nozzle extensions with special couplings as illustrated in the bottom of Figure 73. A photograph of the suction nozzles, hoses, and bulkheads is given in Figure 76. Once the model was installed in the tunnel, the hoses were fed through holes in the test section floor and ceiling. Figure 77 contains a picture of the underside of the test section and the large array of hoses coming from the floor end of the model. A similar picture could be provided of the top of the tunnel since nozzles were in both ends of the model and hoses were required to service them. It should be noted that some ducts had only one nozzle and it might be in either the ceiling or floor ends of the model.

Each hose attached to a suction nozzle extension was connected at its other end to a fitting at the end of an airflow control box which controlled the amount of suction applied to individual ducts in the model. A schematic of the airflow control system including an airflow control box is given in Figure 78 and a photograph of an array of five control boxes is shown in Figure 79. A close-up photograph of several of the control boxes with the hoses attached is provided in Figure 80. The flow in each hose is controlled by a motor driven needle valve. Total mass flow to each control box is controlled by a large adjustable sonic nozzle attached to its downstream end which is connected, along with the other four sonic nozzles, to a common manifold (see Figure 78). Finally, the manifold is connected to a 10,000 cfm compressor.

A cutaway drawing of a control box is given in Figure 81 showing a drive motor, needle, shaft, acoustic liner, honeycomb/screen flow treatment, and the exit pipe. Flow entered the box around the needles of needle valves such as pictured in Figures 82a and 82b.

The construction features of the variable sonic nozzle are given in the sketch in Figure 83. These nozzles were connected to the downstream end of the control boxes and were used to prevent compressor noise from propagating upstream to the model. A photograph of one of the sonic nozzles before installation is shown in Figure 84 and after installation in Figure 85. Large six inch flexible plastic hoses connected the downstream end of the sonic nozzles to the common manifold (see Figure 86).

The original suction system installation had all five suction boxes connected directly to the sonic nozzles which were connected to the common manifold. After the experiment had been underway for some time and all the various controls exercised, it became apparent that the mass flows specified were not achievable. Subsequently, the whole suction system external to the model was disassembled and modified. The most significant change was the relocation of the control boxes above and below the test section to substantially shorten the length of the hoses connecting the model to the control boxes and reduce the line losses. In addition, a large number of hoses were increased in diameter. The resulting mass flow levels were significantly increased.

### 13. Measurements/Instrumentation

Figure 87 summarizes the large number and types of instrumentation used to determine the performance of the model during testing (see Reference 62). From these measurements, surface pressure distributions, transition location, suction mass-flow distribution, suction drag, wake drag, and airfoil lift were determined. As can be seen, both static and dynamic pressure measurements as well as surface heat transfer (thin films) measurements were obtained. Wind tunnel stagnation pressure and temperature and dewpoint temperature were determined in the "big-end" or stagnation chamber while freestream static pressure was measured on the surface of the liner at the 36 ft, 8 inch location. "Free-stream" Mach number was calculated from these measurements.

Wind tunnel reference pressures were measured using sonar-sensed mercury manometers with an accuracy of $\pm 0.2$ psf. Airfoil surface, wake-rake, suction-duct, nozzle, sonic choke and liner choke-plate pressures were measured with ESP (electroscanning pressure) modules with pressure limits of 2.5, 5, and 10 psi depending on the location and precision required. ESP modules are calibrated against integral standard gages; accuracy is generally better than 0.01 psi for the 10 psi modules and better still for the lower range units. Suction nozzle throat pressures were measured by special 2.5 psi modules with each throat pressure referenced to its own duct static pressure. Tunnel, wall-liner and diffuser pressures were measured using mechanically stepped valves with 47 ports each. Four groups of six were utilized for a total of 1128 ports. The pressure range of these scanning (or stepping) valves varied from 5 to 15 psi with the accuracy quoted as 0.01 of gage limit.

Many of the quantities measured were displayed in real time on CRT's or oscilloscopes to enable rapid adjustment of the tunnel parameters and suction system. Figure 88 shows a schematic of the real time data system and the static and dynamic quantities available for display. In addition, the wind tunnel control console displayed wind tunnel static and stagnation pressures, stagnation temperature, Mach number, and Reynolds number. Airfoil pressure and $C_Q$ distributions were displayed and superimposed on the "design" pressure distributions. The output of any thin film gage could be viewed in real time to make a visual determination of whether the flow was laminar, turbulent, or transitional.

Free stream Mach number in the 8-FT TPT is normally determined using a static pressure measurement in the chamber or plenum surrounding the slotted test section. In the LFC test setup, there was really no point downstream of the contraction where conventional "free stream" flow existed. To provide a "free stream" static measurement, liner pressure distributions were examined to find the location on the liner where the local static pressure most closely matched that corresponding to the design Mach number of 0.82 and had a near-zero gradient in the stream direction. Figure 89 shows the theoretical pressure distribution on the liner starting in the contraction and extending to the model trailing edge. The point chosen (36 ft, 8 in.) to obtain free stream static pressure is in the middle of a short flat segment approximately four feet ahead of the model leading edge at the top of the tunnel. An adjacent tap was used to provide a reference pressure for pressure instrumentation referenced to free stream pressure. The difference in pressure between the tap at 36 ft 8 in. and the adjacent reference tap was on the order of $\pm$ 0.5 psf.

Instrumentation installed on the upper and lower model surfaces to measure pressure and transition are detailed in the sketches of Figure 90. As noted on Figure 87, there were 300 static pressure orifices, 50 thin film gages, and 26 acoustic gages. The pressure ports were 0.01 inches I.D. and staggered longitudinally to prevent the wake of one orifice from adding to that of a second and third downstream, thus causing transition. It was stipulated that at least two suction slots must be between two orifices in line with one another. The highest concentration of pressure orifices along a streamline were just above and below the airfoil centerline on the top and along the centerline on the bottom. Pressure distributions presented in the present paper will come from the row just below the centerline on the top surface and along the centerline on the bottom surface. Thin-film transition gages were located primarily in the mid-span region of the wing between the mid-chord area and the trailing edge. An enlarged photograph of a thin film gage is given in Figure 91 along with those for a pressure orifice and an acoustic gage.

A number of example pressure plots will be given in the "Results and Discussion" section and they are easy to interpret. The thin-film data on the other hand needs some explanation. Figure 92 depicts the mid-span region of the LFC upper surface and the 26 thin-film gages plotted at their correct location. When the flow was laminar at a gage, it was labelled with an $L$ and when turbulent with a $T$; the transitional zones were shaded or darkened. On the right side of Figure 92 are plotted the real time voltage output from the gages just above the centerline and in the rear of the model. The first gage is clearly laminar while the next three (from bottom to top) evidence turbulent bursts or, in the case of the gage output labeled 80%, i.e., 80% of the time the boundary layer is turbulent, long periods of turbulent flow. The middle of the transition zone where the boundary layer has a maximum of turbulent energy (maximum RMS voltage output) is near the gage labelled 50%. Notice that the gages at the bottom of the sketch on the left are all turbulent while the gages in the next row are all laminar, indicating for this particular run a very non-uniform transition pattern.

Pressure orifices on the liner were positioned along streamlines and heavily concentrated in the region of the wing. Figure 93a and 93b show plots of the orifices installed in the floor and ceiling, respectively, of the liner. The walls opposite the top and bottom of the airfoil were similarly equipped. Pressure taps on the top and bottom of each side wall extended about a chord length ahead of the model while those on the centerline extended well into the contraction.

Of the 1010 liner orifices 240 were located on the two choke panels between the 53 ft and 59 ft stations. With the aid of these orifices, streamwise variations of pressure could be plotted on the CRT's to monitor, in real time, the location of the shock (or shocks) so that adjustments could be made to the flexible-wall chokes on the basis of this information.

Wake drag or profile drag was determined using a wake rake mounted on the side wall opposite the bottom surface of the airfoil. It had 47 total pressure orifices referenced to free stream stagnation pressure and 6 static pressure orifices that were referenced to free stream static pressure. The rake, pictured in Figure 94a, had a 6.2

inch span (Reference 62) and was aligned with the flow downstream of the model trailing edge. As seen in Figure 94b, some staggering of the tubes was required to obtain the desired resolution in the region where the airfoil boundary layer impacted the rake.

A schematic of a wake profile is given in Figure 95a where the measured stagnation pressure deficit, $\Delta p_t/q_\infty$, is plotted as a function of the normalized distance along the rake. Broadening and deepening of the profile will occur where flow separation, shocks or early transition occurs. Measured static and total pressure profiles along with the deduced velocity profiles for the slotted model for a range of Reynolds numbers are plotted in Figures 95b and 95c, respectively. A gradual movement of the transition location on the top surface was the cause of the gradual broadening of the total pressure and velocity profiles shown in the figure. There is much less change on the bottom surface until a Reynolds number of $15 \times 10^6$ is reached where separation in the aft cove region and a rapid forward movement of transition caused a rapid and large increase in the losses. The fact that the total pressure ratio at the ends of the rake goes to one (1.0) for all the Reynolds numbers shown indicates all the losses were captured by the rake for these conditions.

The primary variables which the experimenters had under their control were:

- Mach number
- Reynolds number
- Suction level
- Suction distribution
- Five trailing edge flaps on slotted model, three on perforated
- Angle of attack
- Liner shape.

The first four of these could be adjusted during tunnel operation while the others required access to the test section. Liner shape is included even though the liner had a “fixed” contour since both groves and raised wooden strips were used to modify the liner wall, opposite the model upper surface and on the tunnel floor, to change the area distribution of the test section and the pressure distribution on the model. A number of small changes in angle of attack were made at the start of both the slotted and perforated model tests to get a pressure distribution close to the design one. The flaps were used along with small Mach number changes to make the "vernier" adjustments in the pressure distribution.

## 14. Results and Discussion

It has been pointed out numerous times that both slotted and perforated suction surfaces were employed on the top side panels of the LFC airfoil. Experiments on both included variations in Mach number, Reynolds number, suction level and the extent of suction. A limited amount of experimentation was conducted to examine the effect of different suction distributions. The effect that these quantities had on the chordwise extent of laminar flow and drag was of primary interest in this investigation. In subsequent paragraphs, illustrative results from most of the above sensitivity studies will be discussed. The slotted surface data will be treated first followed by a section concerning the stability analyses conducted using the experimentally determined pressures and transition locations for the slotted model. A discussion of results from the perforated-surface model tests will conclude this chapter.

### Slotted Model

#### Pressure Distributions

Figures 96 and 97 show a comparison of the measured and theoretical pressure distributions on the slotted LFC airfoil model at $M = 0.82$ for chord Reynolds numbers of 10 and $20 \times 10^6$. The measured pressure distributions are essentially shock free at $R_c = 10 \times 10^6$ with full-chord laminar flow (as evidenced by surface thin-film gages used for detecting transition) on both the upper and lower surfaces. The higher, and less uniform than design, velocities on the upper surface were attributed to the nonuniform flow in the channel between the model upper surface and the liner caused by model deformation and small inaccuracies in liner design (References 65, 66). Local

Mach numbers, as noted earlier, measured on the test section wall opposite the upper surface (Figure 42) indicated that the supersonic bubble on the airfoil was larger than design and extended almost to the wall.

With increases in Reynolds number, transition moved forward gradually on the upper and rapidly on the lower surface. Figure 97 compares the measured and theoretical pressure distributions at $R_c = 20 \times 10^6$ and indicates higher experimental velocities on the upper surface and development of a weak shock at $x/c \approx 0.75$. The laminar boundary layer on the lower surface was unable to withstand the adverse pressure gradient leading into the trailing edge cusp region above $R_c \approx 14 \times 10^6$ where transition moved rapidly forward and the flow separated in the rear cusp. With decreasing extents of laminar boundary layer and the appearance of separation on the lower surface, local displacement thickness over the airfoil increased and consequently, higher free-stream Mach numbers were required to achieve the design plateau pressure distribution when Reynolds number increased. A loss in model lift and an increase in drag also occurs with decreasing extents of laminar flow.

Analysis of spanwise pressure distributions and transition patterns at the design Mach number showed that the flow was nearly two-dimensional over the model but that the leading edge, peak-pressure coefficient tended to vary along the span. In addition, transition tended to move forward with increases in Reynolds number in a non-uniform fashion. In general, transition tended to be more rearward toward the ceiling end of the model where leading edge pressure peaks were lowest, and more forward toward the floor where they were higher. At off-design subsonic Mach numbers, a peak in the nose region occured followed by a gradual decrease to a flat pressure distribution (Reference 66). An example of this type of measured off-design pressure distribution for the slotted model is shown in Figure 98 compared with that predicted for $M_\infty = 0.6$. For this case, essentially full-chord laminar flow was maintained on the upper, and most of the lower, surface by slightly increasing suction in the nose region above that required for the design pressure distribution (Figure 25).

## Suction Distribution

The measured and theoretical suction coefficient distributions required to maintain full-chord laminar flow over the slotted upper surface at $M_\infty = 0.82$ and $R_c = 10 \times 10^6$ are shown in Figure 99. The design suction requirements were based on shock-free flow, a smooth pressure distribution, and was the minimum level required to maintain laminar flow. The measured suction required for full-chord laminar flow on the slotted model was higher than that theoretically predicted; in some cases by as much as 50 percent.

There was, of course, an infinite combination of individual duct suction levels and overall suction distribution possibilities. The distributions shown (Figure 99) represent the suction coefficient level required to obtain the maximum extent of laminar flow. Small local variations were permissible within these overall distributions without an adverse effect on the extent of laminar flow. Reductions in the overall suction level of the distributions, by varying compressor controls in amounts large enough for the sum of the suction drag over the entire upper surface to be measurably different, generally resulted in either a detrimental effect on the laminar flow pattern (transition behavior) or an increase in wake drag. The higher than predicted suction requirements on the upper surface are attributed to uneven velocities and associated wavy pressure distributions on the upper surface (discussed above) and the wind tunnel environment.

## Drag Summary

The contributions to the total drag coefficient for the slotted model at $M_\infty = 0.82$ are shown in Figure 100 over a chord Reynolds number range of 10 to 20 million. The division of suction drag contributions between the upper or lower surfaces was possible (Reference 67) since the suction drag was computed duct-by-duct and integrated over each surface independently. The wake drag was separated into upper- and lower-surface components on the basis of the assumption that the wake could be divided between the upper- and lower-surface at the point on the measured wake-rake profile where the stagnation pressure loss was the greatest. The data (Figure 100) indicate that the largest contribution to the total drag was from the lower surface. With full chord laminar flow over the upper and lower surfaces for $R_c < 12 \times 10^6$, the contribution to the total drag was about 1/3 due to wake and 2/3 due to suction drag. The sharp rise in wake drag on the lower surface between 14 and 15 million Reynolds number was associated with the rapid forward movement of transition and separation of the boundary layer in the lower aft cusp. The contributions to the suction drag of the upper and lower surfaces was about 40 percent for the former and 60 percent for the latter.

Wind tunnel tests were conducted a number of years ago on several swept LFC airfoils at low speeds (References 2 and 4) with extensive laminar flow and low drag. These airfoils had suction applied only on the upper surface. Figure 101 shows the upper-surface pressure distributions and minimum total drag measured at the design lift condition ($c_l = 0.3$) on earlier low-speed ($M_\infty = 0.25$) LFC airfoils. They not only had a different sweep, compared to the present swept supercritical LFC airfoil (SCLFC(1)-0513F) but had lower design Mach numbers and lift coefficients as well. The low-speed LFC airfoil designs (based on standard NACA airfoil profiles) have favorable pressure gradients over the first 40 percent or more of the chord, and much less severe adverse pressure gradients aft, than the supercritical design. The higher design $c_l$ of the LFC airfoil (SCLFC(1)-0513F) does not permit a long run of favorable pressure gradient. Figure 101 also shows the measured minimum profile drag coefficients corresponding to the upper-surface pressure distributions at $R_c = 10 \times 10^6$. As might be expected, the present supercritical design has a larger suction drag penalty than the NORAIR and University of Michigan models, which have large extents of favorable pressure gradient (Figure 101). The wake drag contribution to the total drag is similar for the three airfoils. An increase in wake drag for $M_\infty = 0.70$ is observed on the slotted LFC airfoil (Figure 101) and attributed to the formation of a weak shock wave near the leading edge as the supersonic bubble begins to develop. As the supersonic bubble develops near the leading edge for $0.78 < M_\infty < 0.80$ full-chord laminar flow still exists, but periodic turbulent bursts occur over the upper surface causing an increase in time averaged wake drag. It may be concluded that the basic phenomenon of applying suction laminarization over an extensive supercritical zone with full-chord laminar flow at $R_c = 10 \times 10^6$ is feasible as demonstrated on the swept LFC airfoil for high lift coefficients. It is clear, however, that excessive model surface deformation at high $R_c$'s, irregular pressure distributions and a more severe environment can cause a loss in the extent of laminar flow and an increase in drag at higher Reynolds numbers.

Transition

Figure 102 shows the variation of transition with chord Reynolds number at $M_\infty = 0.82$ for the upper surface only. For constant $R_c$, both upstream and downstream $x/c$ locations are defined based on the last thin-film sensor indicating laminar flow followed by the first sensor downstream indicating fully turbulent flow, respectively. Transition occurs in between these two boundaries. In general, the results shown indicate that the extent of laminar flow was limited by the occurrence of a weak shock $(0.65 < x/c < 0.75)$ as $R_c$ is increased at $M_\infty = 0.82$. The rather large transition zone indicated in (Figure 102) is typical of airfoil sections with flat pressure distributions.

Figure 103 shows a summary of the chordwise extent of laminar flow achieved on the slotted upper surface for several Mach numbers. As Reynolds number was increased at a constant Mach number, transition moved gradually forward on the upper surface. The Reynolds number at which the forward movement began was dependent on Mach number and occurred at progressively lower Reynolds numbers as Mach number increased. For the design Mach number of 0.82, the forward movement began between $11 \times 10^6 < R_c < 12 \times 10^6$ and reached about 65 percent chord at $R_c = 20 \times 10^6$. Transition on the lower surface (not shown) moved more rapidly than on the upper surface and occurred near the leading edge for $M = 0.82$ and $R_c = 20 \times 10^6$. It may be concluded that suction laminarization over a large supersonic zone is feasible to high-chord Reynolds numbers even under non-ideal surface conditions on a swept LFC airfoil at high lift.

Figure 104 shows the maximum transition Reynolds number with Mach number for LFC experiments in wind tunnels and flight with slotted upper surfaces. Maximum values of $R_{TR}$ range between $10 \times 10^6$ to $40 \times 10^6$ for $0.1 < M_\infty < 0.82$. The present LFC airfoil results range between $10 \times 10^6$ to $22 \times 10^6$ and are in line with other wind tunnel data (References 2, 4) but lower than the flight results (References 5, 6) as might be expected. The overall data trend indicates no strong dependency on Mach number. However, transition for any given data point may be influenced by environmental disturbances and/or surface conditions.

## Effect of Suction Level on Transition

The example suction distribution and level shown in Figure 99 was established as that required for maximum transition Reynolds numbers. Since suction levels influence the amount of laminar flow obtained at a given Reynolds number, tests were conducted to assess the effect of suction level on transition location. This was accomplished by establishing the desired Mach and Reynolds number in the test section with the overall chordwise suction distribution typical of that for design but at the maximum level that the suction system was capable of, and then, recording the transition pattern as the overall suction level was incrementally decreased. A reference suction level (RSL) was defined which involved the ratio of the integrated suction coefficient at the conditions of interest to that at the design conditions, all multiplied by the square root of the Reynolds numbers at the same two conditions, i.e.,

$$\mathrm{RSL} = \frac{\hat{C}_Q \sqrt{R_c}}{\hat{C}_{Q_{DES}} \sqrt{R_{c_{DES}}}}$$

where

$$\hat{C}_Q = \int_0^1 C_Q \left(\frac{x}{c}\right) d\left(\frac{x}{c}\right)$$

$$\hat{C}_{Q_{DES}} = \int_0^1 C_{Q_{DES}} \left(\frac{x}{c}\right) d\left(\frac{x}{c}\right)$$

The reference suction level so defined allowed an examination of the dependence of boundary layer stability on suction levels and their effect on transition $N$-factors without Reynolds number being a significant factor (Reference 68).

Figure 105 shows that as the reference suction levels (RSL) are reduced for constant $M_\infty$ and $R_c$, the transition pattern moves forward and downward on the upper surface. Typically, when the suction level was reduced sufficiently, the line fairings (Figure 105) corresponding to the measured beginning of transition suddenly jumped forward. In other words, the most rearward transition location remained essentially fixed with decreasing suction until a critical level was reached. As previously discussed, suction excursions were required to estimate the correct $N$-factor corresponding to transition for each Mach and Reynolds number combination.

## Effect of Chordwise Suction Extent on Transition

Another and more practical concept for obtaining long runs of laminar flow at high Mach number is a "hybrid" configuration which combines suction over forward regions of the slotted upper surface with natural laminar flow (NLF) over rearward regions (References 66, 69, 70). An attempt was made to simulate hybrid laminar flow control (HLFC) conditions during the 8 ft TPT LFC experiments by adjusting the duct pressures to the same values as the local surface pressure to inhibit flow through the surface. This process was started at the trailing edge and extended over successively larger regions toward the leading edge; results are given in Figure 106. On the lower surface, suction was maintained for $0 < x/c < 0.25$ to minimize flow separation and, thus, control the flow between the lower surface and tunnel wall.

The results shown in Figure 106 for $M_\infty = 0.82$ and two chord Reynolds numbers indicate that laminar flow could be maintained well beyond termination of suction for suction extending up to $x/c = 0.60$ and constant $R_c$. With no suction, laminar flow was present back to about 15 percent chord for $R_c = 20 \times 10^6$ and 25 percent chord for $R_c = 10 \times 10^6$. Transition typically moves forward with increasing $R_c$ for any extent of suction. Finally, Figure 106 indicates that unless supercritical LFC airfoils are designed with adequate suction to allow laminarization through shock boundary-layer interactions, there is no reason to apply suction beyond the shock location.

The simulated HLFC conditions were, of course, carried out with the LFC airfoil flat top pressure distributions (Figure 25) which are known to be non-optimum for HLFC. Furthermore, no effort was made to smooth the unsucked downstream slotted surface.

## Boundary Layer Stability

### Stability Correlation Approach

Figure 107 illustrates the approach used in the boundary layer stability analysis for $N$-factor correlations. Measured pressure and suction distributions were used to calculate streamwise and cross-flow velocity profiles along the chord using the Kaups-Cebeci laminar boundary layer code (Reference 47). Stability analyses of these profiles gave the TS and CF disturbance amplification rates from the points of neutral stability. $N$-factors were determined based on the measured transition locations.

Solutions of the governing stability equations provide the necessary information for analyzing the disturbance growth within the boundary layer. Two of the three disturbance variables, frequency $f$, wave length $\lambda$, and wave angle $\psi$, must be specified while the third is obtained along with the disturbance amplification rate as part of the solution. The TS analysis was accomplished using a fixed frequency and fixed wave angle approach. For the incompressible stability analysis, it was assumed that the TS disturbance which amplifies most are those traveling in the local potential flow direction, $\psi = 0$. For the compressible COSAL analysis, a range of disturbance frequencies were analyzed with a wave angle of zero, and then the wave angle was varied at the maximum amplified frequency. The MARIA code was used to obtain approximate solutions to the incompressible crossflow problem for fixed wave length, stationary vortices.

## Examples of Stability Calculations

Figures 108-109 and 110-113 show examples of typical input experimental-pressure and suction distributions and subsequently computed $N$-factors using SALLY, COSAL, and MARIA codes for the transonic design and subsonic off-design test conditions on the slotted LFC airfoil. At subsonic speeds (Figures 108, 109), the pressure distributions had strong leading edge peaks. The suction distributions are typical of the design distribution for full-chord laminar flow (Figures 99, 105), however, the overall level was adjusted at a given Mach number without any attempt to optimize the distribution to achieve the maximum extent of laminar flow or absolute minimum of mass flow. For moderate to high overall suction levels, essentially full-chord laminar flow was achieved for the subsonic speed range ($0.4 < M_\infty < 0.7$) and Reynolds numbers up to about 20 million (Figure 103). Incompressible stability analyses showed that TS instability dominated the process leading to transition while the CF instability growth was so small that interaction of the two disturbances seem unlikely. A summary plot of the $N$-factors at transition due to TS instabilities is given in Figure 110 for a range of $R_c$ and relative suction levels (RSL) and $M_\infty = 0.6$. The hatched band represents the region of transition between 100 percent laminar and fully turbulent flow. The dashed lines represent trends of constant RSL with $R_c$. In general, the calculated TS $N$-factors at transition are between 9 to 11 for $10 \times 10^6 < R_c < 20 \times 10^6$ at $M_\infty = 0.6$. Stability of the boundary layer is obviously controlled by the relative amount of suction applied.

The measured transonic pressure distributions (Figures 111-113) closely follow that for design (Figures 96, 97) except for the slight waviness and higher velocities over the upper surface discussed earlier. A large supersonic zone existed over most of the upper surface culminating in a weak shock downstream for $R_c > 12 \times 10^6$. Full-chord laminar flow was maintained for the design shockless condition for $R_c \leq 12 \times 10^6$.

For the transonic case shown in Figures 111 and 112, suction was terminated at 8 percent chord and transition occurred between $0.20 < x/c < 0.28$ for $R_c = 20 \times 10^6$. Incompressible calculations (not shown) indicate that TS $N$-factors range between 10 and 13. The incompressible CF disturbances (Figure 112) grow to a maximum $N$-factor of 4.6 at $x/c = 0.12$. Further, a very weak crossflow region of opposite sign developed in the slight compression region around $x/c = 0.05$. Neither of the crossflow regions is believed likely to have significant influence on the transition process since the CF $N$-factors were less than 5. Corresponding compressible TS calculations (Figure 111) show that the $N$-factor at transition for the most critical disturbance is in the range of 5 to 7.5 for $f = 7$KHz and $\psi = -50°$.

Another transonic case is shown in Figure 113 with suction terminated at 40 percent chord and transition between $0.50 < x/c < 0.60$ for $R_c = 10 \times 10^6$. The TS $N$-factor lies between 3 to 5 using COSAL ($N$-factor between 6 to 8 using SALLY), while CF $N$-factors (not shown) reaches a maximum value of 2 using SALLY. Here, the $N$-factors were calculated for a range of frequencies at a wave angle $\psi = -50°$.

Figure 114 summarizes the effect of varying chordwise suction extent on the calculated incompressible TS $N$-factor for the slotted LFC airfoil upper surface at $M_\infty = 0.82$. The results at $R_c = 10 \times 10^6$ are for the shockless conditions and $R_c = 20 \times 10^6$ with a weak shock at the rear pressure rise region ($x/c = 0.75$). The hatched region on the figure represents the transition zone based on the correlated incompressible $N$-factors with the indicated extent of suction applied. Corresponding $x/c$ transition locations as a function of the chordwise extent of suction were discussed earlier in connection with Figure 106. For suction applied over only the first 0.25c on the slotted LFC airfoil, the incompressible transition $N$-factor is 10 to 11 at $M_\infty = 0.82$ and $10 \times 10^6 < R_c < 20 \times 10^6$. With suction applied beyond $x/c = 0.30$, the correlated TS $N$-factors are between 7 and 8 at $R_c = 10 \times 10^6$ and about 2 for $R_c = 20 \times 10^6$. The dramatic drop in TS $N$-factor at the higher Reynolds number is believed to be attributed to the formation of a weak and sometimes unsteady shock at $x/c = 0.75$. This shock may be attributed to the wavy pressure distribution due in part to model deformation under load.

## Correlation of Transition $N$-factors with Mach Number

Experimental transition data have been recently obtained and analyzed on NLF and LFC airfoils (References 71, 72) and bodies of revolution (Reference 73), in wind tunnels, and on wings or gloves in flight (References 71, 72, 74-80). The measured pressure profiles and suction distributions from a number of these experiments were used as inputs to the stability codes to calculate local disturbance-growth rates and integrated amplification ratios ($N$-factors). $N$-factors calculated at the experimentally measured transition locations for both TS and CF disturbances were compared to determine which instability was the most amplified.

Figures 115 and 116 summarize the calculated lowest $N$-factors at measured transition for TS and CF, respectively. It should be noted that there exist higher $N$-factor values than those shown in the figures; only the lower values were chosen to establish a conservative $N$-factor boundary and Mach number trend. Figure 115 shows calculated $N$-factors for situations where the TS instabilities dominated (CF negligible), whereas Figure 116 is for situations where CF dominated (TS negligible). The $N$-factors shown for the NASA SCLFC(1)-0513 wind tunnel tests with slotted suction surface were calculated, using the SALLY and COSAL codes, with identical input data for each point. All other results (References 52, 74-80) on Figures 115 and 116, were calculated using either the SALLY, COSAL, or MACK codes. The incompressible TS $N$-factors (open symbols) approach a constant lower limit of about $N = 9$, while those $N$-factor values obtained with the compressible codes (solid symbols) decrease over the indicated speed range and are about 50 percent lower at $M_\infty \cos\Lambda = 0.8$ than at $M_\infty \cos\Lambda = 0$. These trends are intended to help establish the allowable lower limit for disturbance amplifications. The limited number of incompressible CF $N$-factors (Figure 116) have a minimum value of 9 while those from the compressible codes decrease to 4 or 5. The results strongly indicate that a constant but conservative $N$-factor value ($N = 9$) can be used in applying the incompressible codes. However, knowledge of both the conservative trend and level of the $N$-factor (Figures 115, 116) is required when applying the compressible codes. The most surprising results of this correlation are the fact that the compressible codes yielded lower $N$ values than the incompressible ones and the decreases in $N$-factor with increases in Mach number for the lower-bound data.

Clearly, there is an effect of the wind tunnel environment on these numbers but the seeming larger effect on the compressible values is not understood. Examination of the noise levels and spectra of measured fluctuations for the 8-ft TPT shows (see Chapter 6, Flow Quality Improvements, etc.) that acoustic turbulence levels increase with Mach number until the wind tunnel is choked at $M_\infty = 0.8$. One might expect, based on this, that the transition $N$-factor would decrease with $M_\infty$. However, this reasoning might be questioned since the 8-ft TPT LFC experiments showed little sensitivity of transition location with increases in Mach number for $M_\infty < 0.7$ (Figures 103, 104). Transition location is believed to have been more sensitive to surface deformation, suction level, and suction extent (Figures 105, 106). Also, it is noted that lower-bound flight $N$-factors shown in Figures 115 and 116 tend to agree with the wind tunnel results at low and high speeds.

All of the $N$-factors calculated and presented in Figures 115 and 116 for the LFC airfoil were based on the beginning of transition (Reference 72). However, if the transition point were selected near the end of transition, the possibility exists that the incompressible and compressible boundary layer codes will have already predicted a laminar separation. The selection of a point in the middle of transition would have yielded amplification factors up to twice the values shown in Figures 115 and 116 for the LFC airfoil. The results shown in Figures 115 and 116 for the LFC airfoil represent calculated $N$-factors for both full chord and partial suction using the incompressible or compressible codes.

## Correlation of $N$-Factor with Turbulence Level

Figure 117 is an attempt to evaluate the influence of transonic wind tunnel turbulence level on the calculated $N$-factors at transition. Shown for comparison are calculated values of TS $N$-factors on the slotted LFC airfoil in the Langley 8-ft TPT (References 68, 70), two bodies of revolution in the NASA Ames 12-ft PWT (Reference 73), and flight results on an NLF wing glove (References 79, 80).

The N-factor data are plotted against measured turbulence levels in the facilities based on conventional single-element hot-wire measurements. The flight results are plotted against very low but unmeasured disturbance levels. Also shown for comparison is the empirical incompressible relation for the variation of TS $N$-factor with turbulence level generated by MACK (Reference 35) for $\tilde{u}/U_\infty > 0.1$ percent. All of the data tend to fall in a band indicating little influence of $\tilde{u}/U_\infty$ on $N$-factor at transonic Mach numbers (see data fairing). In fact, the LFC results shown for $0.17 < \tilde{u}/U_\infty < 0.2$ percent ($M_\infty = 0.7$) and $\tilde{u}/U_\infty = 0.07$ percent ($M_\infty = 0.6$) are for essentially full-chord laminar flow on both surfaces up to $R_c = 22 \times 10^6$. For $\tilde{u}/U_\infty = 0.05$ percent ($M_\infty = 0.82$), full-chord laminar flow existed for TS $N$-factor equal to 5 and 8 with transition between $0.2 < x/c < 0.3$. Therefore, the present limited results (Figure 117) suggest that the sensitivity of TS $N$-factor at transition to turbulence level may not be as large at transonic speeds as at low speed. Note that the $\tilde{u}/U_\infty$ data in Figure 117 were obtained from single wire hot-wire anemometers.

## Correlation of TS Maximum Amplified Frequency with $M$

As mentioned earlier, the calculated TS $N$-factors for the present results were for a range of frequencies with the wave angle fixed at zero. $N$-factors obtained using this approach have been shown to be constant for incompressible stability analysis and varying (with Mach number) for compressible analysis (Figure 115). Consequently, it is of interest to determine the variation of the maximum amplified frequencies at transition with Mach number. Figure 118 shows the variation of measured and calculated maximum frequencies at transition with Mach number. The calculated results were obtained with experimental input data to the incompressible and compressible codes and correspond to the $N$-factor results shown in Figure 115. In addition to the calculated results shown in Figure 118, measured wave forms and energy spectra were obtained from surface mounted hot-film gages on the LFC airfoil at $M_\infty = 0.82$, NLF(1)-0414 airfoil in the LaRC LTPT wind tunnel at $M_\infty = 0.1$ (Reference 21) and in flight on the Learjet 28-29 wing at $M_\infty = 0.8$ (Reference 74) that agreed with the most amplified frequencies predicted by theory. Also shown are frequencies obtained from energy spectra measured with microphones on the F-15 cone flight tests (Reference 81).

The hatched bands in Figure 118 illustrate the data trends due to the calculated incompressible and compressible data points. In either case, the maximum amplified TS frequency increases with increasing Mach number from subsonic to transonic speeds with a tendency to become essentially constant at low supersonic speeds. A somewhat lower level trend persists for the compressible results. This difference is attributed to a shift in maximum amplitude frequency in the stability calculations for both incompressible and compressible analyses. The dashed line (Figure 118) is an estimated trend based on the most amplified disturbance frequencies in wind tunnels using the approach of Mack (Reference 82) for the stability of laminar compressible boundary layers and linearized parallel flow. This estimated trend for TS frequencies in wind tunnels is higher than the calculated compressible data or measured spectra.

For $M_\infty > 0.4$, the results in Figure 118 indicate that the maximum amplitude TS frequency is between 10KHz to 20KHz. Spectra obtained from hot-wire probe measurements (Reference 23) in the free stream of the LaRC 8-ft TPT, over a range of Mach numbers, indicate that no significant energy spikes occur (Figure 119) for $f >$ 10KHz and $M_\infty < 0.82$. It should be noted that similar results were found from spectral measurements using fluctuating pressure probes in the same facility.

## Perforated Model

### Original Perforated-Surface Model

The model with the perforated upper surface was originally installed in the wind tunnel with a solid lower forward panel (see Model Construction Chapter). This panel had a different shape and pressure distribution from the slotted panel that it replaced as shown in Figures 59 and 60. Since it had no suction capability, the boundary layer on the bottom surface of the perforated model had an earlier transition than on the bottom surface of the slotted model and separation in the aft "cove" region was present at all Reynolds numbers. This in turn caused an effective decambering near the trailing edge and some loss of lift for a given angle of attack. Increasing the angle of attack to recover the design lift caused a slight increase in local velocities (more negative $C_p$) on the upper surface and an increase in waviness of the pressure distribution as well. Figure 120 gives a plot of the "best" chordwise pressure distributions obtained for the perforated surface model for $R_c = 10$ and $20 \times 10^6$ at Mach numbers of 0.816 and 0.820, respectively. The wavy pressure distribution on the top and separation in the aft cove region are readily apparent. Also a weak shock was always present at the downstream end of the upper surface supersonic bubble between 70% and 80% of the chord.

The extent of laminar flow on the perforated-surface model (as originally installed) was disappointingly low. Even at low Reynolds numbers, full chord laminar flow could not be achieved. Transition boundaries are plotted in Figure 121 for both the top and bottom surfaces and at 10 and 20 million Reynolds number to illustrate this point. Laminar flow at $R_c = 10 \times 10^6$ on the top surface of the model (Figure 121a) near the floor extended only to 60% chord and at the top to 80% chord. On the bottom surface of the model, Figure 121b, transition occurred at about 15% chord across the whole span. At $R_c = 20 \times 10^6$ transition on the top was around the 17% chord station and on the bottom in the region of 5% to 10% chord (see Figures 121c and 121d).

## Modified Perforated-Surface Model

One reason for the poor performance of the perforated surface was a shortfall in mass-flow capability relative to that needed. Slotted-surface experiments had shown that mass flows substantially in excess of design were required to maximize the extent of laminar flow (see Figure 99). Consequently, the model was disassembled and a number of changes, listed in the Suction-System Chapter, to the model and suction system were implemented. The first results following these changes are plotted in Figure 122 where the variation of suction coefficient along the chord is given for the original and modified perforated-surface configurations. Since the major modification was to the front and middle top-surface panels, it was expected that the most improvement in $C_Q$ would be seen in the first 60% of the chord and that is what Figure 122 confirms. The $C_Q$ in this region is roughly double that obtained by the original perforated panels.

Comparisons of the final suction distribution used for the upper surface of the slotted model with that of the perforated upper surface at the design Mach number and $R_c = 10 \times 10^6$ are given in Figure 123. Clearly, over the first 60% of the chord, the suction mass-flow capability of the perforated surface was greater than that of the slotted; over the aft 40%, the reverse is true. In any case, the increased suction capability of the perforated panels did improve the extent of laminar flow on the top surface. On the bottom surface, the replacement of the solid forward panel by the slotted one (used on the slotted model) had a beneficial effect on both the extent of laminar flow and the flow separation in the rear cove region.

A comparison of the pressure distributions on the slotted and modified perforated surfaces at $R_c = 10 \times 10^6$, $M_\infty \approx 0.82$, and $\alpha = 0.89$ deg is shown in Figure 124. The angle of attack of the perforated surface model is slightly larger than that for the slotted-surface model and slightly higher than the angle of attack required to achieve the "best" pressure distribution. While the perforated model Mach number is slightly lower than the slotted, the perforated top-surface suction pressures are higher with a substantial shock at $x/c \approx 0.7$. Moreover, the aft cove region of the perforated model is separated. The top-surface transition pattern for the perforated surface pressures of Figure 124 are shown in Figure 125. The improvement in the extent of transition over the original perforated-surface model (Figure 121) is modest.

A slight reduction in angle of attack and a small increase in Mach number from that in Figure 124 yielded the best pressure distributions on both the top and bottom surfaces (see Figure 126). The shock strength on the top surface is greatly diminished and the flow in the aft cove region is attached. The Mach number on the wall opposite the upper surface has been reduced and is slightly less than one. As can be seen in Figure 126 the perforated-surface pressure distribution is very nearly the same as the best slotted-surface pressure distribution with the suction pressures on the top-side still higher than the design level.

The transition pattern for the pressure distribution of Figure 126 is sketched in Figure 127. Near the top wall the flow is laminar to the trailing edge while transition on the lower 50 percent of the model occurs between 25 and 60 percent of the chord. One possible reason for this pattern may be the difference in the top-surface pressure distributions near the ceiling and floor (see Figure 128). While the data plotted in Figure 128 near the bottom wall (floor) is sparse, it does appear that the pressure gradient in the middle of the airfoil is more adverse than near the ceiling.

At a chord Reynolds number of $20 \times 10^6$ the best chordwise pressure distribution for the modified perforated-surface model is again very similar to that of the slotted model (see Figure 129). The shocks on the top surfaces of the models near the 75 percent chord station are slightly stronger than that encountered at $R_c = 10 \times 10^6$ (Figure 126). In addition, the flow in the aft cove region of the bottom surface is separated on both models. Suction pressures (negative $C_p$'s) in the mid-chord region of the bottom surfaces are slightly higher than at $R_c = 10 \times 10^6$ and this is also reflected in the higher wall Mach number opposite the bottom surface of the perforated model.

The transition pattern for a Reynolds number of $20 \times 10^6$ and $M_\infty \approx 0.82$ is shown in Figure 130. It is nearly symmetric about the centerline with transition much forward of that obtained at $R_c = 10 \times 10^6$. Transition on the centerline is approximately at 40 percent of the chord while near the ceiling and floor on the model it occurs near 30 percent chord. Clearly, Reynolds number has a powerful effect, but one must remember that as Reynolds number is increased so is the level of the free-stream disturbances and aeroelastic deformation. The chordwise distributions of pressure across the span of the model at $R_c = 20 \times 10^6$ vary much as they did at $R_c = 10 \times 10^6$ as can be seen by comparing the pressure at $R_c = 20 \times 10^6$ in Figure 131 with those in Figure 127 for $R_c = 10 \times 10^6$. However, as transition moves forward due to Reynolds number increases and the environment changes, the differences in the chordwise pressure distributions have less effect on transition.

A summary of the transition location data is given in Figure 132 where transition location is plotted as a function of chord Reynolds number. The trend with increasing Reynolds number has already been discussed, but the abrupt forward movement of transition at $R_c = 15 \times 10^6$ from 70 percent chord to 40 percent chord is noteworthy. It is hard to believe that this rapid change is a Reynolds-number or pressure-distribution effect. The fact that the suction levels on the modified perforated-surface model are higher than those on the slotted model back to 60 percent of the chord (see Figure 123), that the pressure distributions on the perforated-surface model are nearly the same from $R_c = 10 \times 10^6$ to $R_c = 20 \times 10^6$, and the transition locations on the slotted and perforated models are similar from $R_c = 10 \times 10^6$ to $R_c = 15 \times 10^6$ (see Figure 133) indicate that some local surface phenomena, perhaps due to model distortion, was the culprit. Since transition starts at $x/c < 40$ percent at $R_c > 16 \times 10^6$, an increase in suction over the aft 40 percent of the chord to approach the levels used for the slotted model would have had very little effect.

## Drag

A summary of the drag data for the modified perforated-surface model is given in Figure 134 along with that for the slotted-surface model and the fact that the flap had no suction (see Figure 123). The equivalent suction drag coefficient, $c_{d,s}$, wake drag coefficient, $c_{d,w}$, and total drag coefficient, $c_{d,tot}$, are all plotted versus chord Reynolds number. Suction drag for the perforated-surface model is seen to be less than that for the slotted-surface model over the whole Reynolds number range. This is due to the lower suction capability available over the aft 40 percent of the chord on the perforated-surface model and the fact that the flap had no suction (see Figure 123). However, the wake-drag for the perforated surface model is so much larger over the Reynolds number range that when the suction and wake drags are added to obtain the total drag, the levels for the perforated-surface model are still higher than those for the slotted-surface model. Beyond $R_c = 12 \times 10^6$ the difference ranges from 6 to 9 drag counts.

Despite the evidence provided by the test results just described, that the slotted surface appeared to be superior to the perforated surface, there was a general feeling that had the perforated-surface contour not been more wavy than the slotted surface and had the perforated and slotted panels been equally stiff, then the drag results would have been comparable. Further evidence supporting this conclusion is given in Figure 135. Plotted here is the simulated hybrid data from the slotted-surface tests along with those from the perforated- surface. It can be seen that when no suction is applied, the perforated surface gives a much longer run of laminar flow indicating that the perforated surface is more aerodynamically smooth that the slotted one (pressure distributions were essentially the same). As the extent of suction is increased, the extent of laminar flow does not increase as fast for the perforated-surface model as it does for the slotted so that when suction is applied over more than 33 percent of the chord, the slotted surface yields a longer run of laminar flow.

## 15. Lessons Learned

By the time most projects are complete, there are many things, in retrospect, that the participants would like to have done differently. The LFC project was no exception. Indeed, one measure of the LFC projects' success might be the fact that we do know how to do a number of things better than we did initially. The purpose of this section is to detail some of the more important lessons that were learned.

The choice of the airfoil and test conditions were the subject of several years of study. One of the main items of concern was the height of the supersonic bubble at design conditions. The predicted height was thought to be far enough away from the liner wall that uncertainties in the liner methodology, aeroelastic effects and onset flow would not cause the supersonic bubble to contact the liner. As noted in the Airfoil Design Chapter, the original free-air design conditions for the LFC airfoil were reduced to specifically satisfy this constraint. Many months of adjusting the angle of attack, the flap settings, the chokes, the suction levels and distribution, and adding and subtracting area from the liner opposite the top side of the airfoil never achieved a supersonic bubble of the size desired. The researchers that spent all this time looking for the design point often felt that they were looking for a needle in a haystack. Nevertheless, a steady pressure distribution close to the design distribution was finally obtained but pressures on the top side were more negative than design and the supersonic bubble was within a inch or two of the opposite wall, i.e., larger than intended. The Mach number on the wall itself was typically 0.99. Once the design pressure distribution was obtained, it was clear that small changes in Mach number would result in either a rapid forward movement of the upper surface shock to about 15 percent of chord or a more negative pressure level terminated by a strong shock at about 78 percent of chord. The latter condition would also cause the formation of a shock on the bottom surfaces. When strong shocks were encountered, on either top or bottom surfaces, laminar flow could not be maintained beyond them.

Most participants, as well as a peer group that was asked to review the project accomplishments at the end of the program, felt in retrospect that a slightly more conservative approach would have been better. Either a further reduction in test Mach number, lift coefficient, or chord of the model, or the use of a different, lower performance airfoil are possibilities. Another option would have been the use of an airfoil with essentially a flat bottom so that the suction pressures on the bottom would have been lower, the likelihood of a supersonic bubble or shock forming would have been eliminated, and the airfoil could have been moved closer to the wall opposite the lower surface. In this case, the experiment would have concentrated entirely on the upper surface.

The pressure distribution on the top surface at the "design" test condition was wavy (as opposed to smooth) as pointed out in the Results and Discussion Chapter. There are several possibilities for these waves but one of the most likely is the fact that the flow velocity over the top surface is higher than that for which the airfoil is designed. Consequently the streamlines and supersonic bubble are slightly different. If the propagation of the waves in the supersonic bubble are different from design, then the pressure distribution will be disturbed and waviness will result. The pressure distributions at subcritical speeds were quite smooth.

Other possible contributors to the wavy pressure distribution include waviness of the model surface, aeroelastic deformation of the airfoil, errors in the liner contours due to the use of two-dimensional type boundary layer codes to design it, and unknowns in the treatment of the juncture regions, including the turbulent zones at each end of the airfoil.

Calculations were made prior to the installation of the LFC slotted airfoil in the Langley 8-ft TPT of the deformation of the LFC airfoil at the maximum expected load. Since the trailing edge was thin, there was concern that it would deflect upward and change the shape of the airfoil. However, it appeared from the calculated displacements and airfoil analysis code computations using the predicted deformed shape, that the problem could easily be countered by small deflections of the trailing edge flaps (see Model Construction Chapter). Once the tests began and the wavy pressure distributions on the top surface were encountered, concern for aeroelastic deformation surfaced again. The tunnel was shut down and portions of the liner opposite the bottom of the wing were removed to install loading jacks. These jacks were located across the span and in the leading and trailing edge regions where 90 percent of the load was concentrated. Simulated aerodynamic loads equivalent to and lower than those experienced at the "design" test Mach number were applied and the deformed airfoil shape was measured. It was determined that substantially more chordwise bending at the joints between the panels (see Figures 52 and 53) was being experienced than had been predicted. The wing was subsequently disassembled and shims inserted to provide an airfoil shape, when loaded, that was much nearer to design. However, these modifications did not provide a significant improvement in the pressure distribution. The experience, nevertheless, did impress on us that loading tests should have been carried out prior to, or immediately after, the installation of the wing to check the structural calculations and provide fixes if needed.

The kinds of problems described above were exacerbated by the fact that an analysis code that could treat a yawed wing in a wind tunnel with arbitrary wall geometry did not exist. A code of this type would have permitted some parametric studies to acquire understanding of the effects of errors in liner contours and wing geometry, and also enable the design of corrections to eliminate these errors. An effort was initiated around 1978 to develop an analysis code that utilized a linear-panel method to determine the contraction flow and a full potential code for the flow around an infinite-span yawed wing. Each worked well by themselves but the problem of getting them to work together, i.e., match everywhere at the interface, was not overcome during the time the slotted model was in the tunnel and eventually dropped. Today's CFD and computer technologies would enable a much more elegant and uniform treatment of the contraction and the wing in the test section.

Another problem concerning the flow over the wing was the inability to determine its two-dimensionality. As shown in Figure 128, there were a number of chordwise rows of pressure taps but only two in the middle of the test region had a sufficient number of taps to define the chordwise pressure distribution with the required resolution. This affected our ability to adjust the trailing edge flaps to achieve a two-dimensional flow. Spanwise pressure gradients do have an effect on the boundary layer stability and must be minimized if comparisons with two-dimensional theory (used in design) are to be meaningful. It is likely that had we anticipated the wavy pressure distributions encountered in the tests, additional taps would have been added or the same number of pressure taps distributed on fewer rows with greater resolution.

Transition was determined by a distribution of flush thin-film gages very similar to those developed by Rubsin at the NASA Ames

Research Center. The time varying outputs from these sensors are greatly different depending on whether the flow over them is laminar, transitional, or turbulent. A signal analyzer to automatically characterize the sensor output as laminar, transitional, or turbulent would have eased the data reduction problem.

In addition to knowing whether a flow is laminar or turbulent, one would also like to determine the most amplified frequency to compare to the stability theory predictions. It was not clear whether the flush transition gages could consistently provide accurate spectra at the start of the LFC program and even now little spectral data has been obtained. It does seem that gages with the required sensitivity could be developed given sufficient time and funds. Thin-film gages deposited directly on metal or on plastic foil or other thin-film concepts explored in recent years (see References 83 and 84) have already been shown capable of providing good spectral data.

The determination of the profile drag was made using a single wake rake located about 8 inches downstream of the model trailing edge (see Measurements/Instrumentation Chapter). There was some ability to move the rake up or down in order to position it better when the trailing-edge flap angles were changed but none to move it from side to side to determine the spanwise variation in drag. Test conditions were often encountered where the longest run of laminar flow was not at the rake location and some ability to determine the profile drag at other places across the span would have been very useful. An ability to move the drag rake from side to side or the installation of a second rake would have been helpful.

Another device that would have provided additional confidence in the drag measurements and was used later in the HLFC tests was a fixed boundary layer rake located near the trailing edge. Analytical extrapolations from the rake to the trailing edge can provide the information needed to make top or bottom surface drag estimates, depending on which side the rake is located. In using the single wake rake for separate top and bottom surface drag estimates, it was assumed that the point of maximum stagnation-pressure loss divided the top and bottom wakes. When there was separated flow near the trailing edge on one side or another, this may not have been a good assumption. It is important to note again that the top and bottom surfaces were considered to be separate experiments and the ability to measure the drag contributed by each side was important.

The problem that the test section's environment poses for carrying out meaningful laminar flow experiments were detailed in the Tunnel Selection Chapter. An airfoil designed to have natural laminar flow back to, say 50 percent of its chord, in free air will have less than that in the wind tunnel. How much less will depend on the test section environment as well as some other factors. In the case of LFC airfoils, the environment would have a similar effect. For a given suction distribution and suction level, there will be less laminar flow in the wind tunnel than free air. In order to achieve the same amount of laminar flow in the wind tunnel as in free air, more suction must be used. When carrying out LFC or NLF experiments in a wind tunnel, it is highly desirable to be able to estimate how conservative the wind tunnel results are. This can be accomplished using the empirical equation developed by Mack for "N-factor" (Reference 35) methods or the curves given by Goradia (Reference 85) for a momentum-thickness-Reynolds-number approach. See also the discussion of Figure 117.

The relationships developed by Mack and Goradia are based on single, hot-wire, anemometry measurements. The fact that these techniques are not accurate at high subsonic speeds is discussed in the Flow Quality Improvements Chapter and documented in a number of papers. There is a need to formulate new empirical equations based on three-wire, hot-wire anemometry but many more measurements are needed in many more tunnels.

At the start of the LFC experiment, a number of months were spent making adjustments to the suction system to get the specified mass-flow levels. After a lot of detective work it was concluded that losses in the suction hoses running from the ends of the model to the airflow control boxes (see Suction System Chapter) were the primary culprit. Subsequently, many of the hoses and fittings were removed and replaced with larger ones. In addition, the airflow control boxes which were located in a rack remote from either end of the wing were moved to above, or under the test section, to shorten the length of the lines. These modifications had a most beneficial effect on the suction mass flow available. In hindsight, these problems would have been minimized had we carried out some bench tests with hoses of the same length (also bent and twisted) as employed in the wind tunnel.

The ability to adjust the trailing edge flaps on the LFC models was necessary to obtain the desired chordwise and spanwise pressure distributions. In order to do this on the slotted model, the tunnel had to be shut down and calibration equipment employed to manually reset the flaps. This was a time consuming job which was done dozens of times throughout the life of the slotted-model tests. If one or more of the flaps could have been remotely controlled, the time required to achieve the "design" test conditions would have been greatly reduced. However, it is not clear how the middle flap, with its suction ducts and slots, could have been designed to enable remote control. It may well have been possible to drive the outer flaps remotely. As noted in the Model Construction Chapter, the middle flap of the three flaps employed on the perforated model was designed to be remotely controlled but it did not have a suction surface and was ineffective.

## 16. Recommendations for Further Research

A number of items were recognized prior to undertaking, or during the course of the LFC project that needed additional research. Some had been problems for a long time while others arose due to the difficulties encountered in designing the experiment or analyzing the data. As in most research projects, there were as many problems uncovered as were solved. The more important research opportunities identified are listed below.

- Effects of wind tunnel environment on boundary layer receptivity and transition
- Code for assessing effects of streamline tunnel walls on airfoils and wings at off design conditions
- Relative "accuracy" of various compressible boundary-layer-stability codes
- Prediction and control of Görtler instabilities using newly developed theory (see Reference 86)
- Effects of surface waviness, roughness, and deformation under aeroelastic loads on laminarization at transonic speeds for high/low $c_l$'s
- Passive/active control to avoid shock-induced boundary-layer separation and drag rise at high Reynolds numbers

- Effect of spanwise pressure gradients on spanwise flow instability (see Reference 87)
- Control of crossflow instability using tailored pressure distributions
- Interaction of leading edge contamination with spanwise flow instability near leading edge.

Most would agree that the first item on this list is, by far, the most important. Additional ideas for transition and laminar-flow-control research may be found in Reference 88.

## 17. Summary Observations

A summary of some of the more important points made in the text along with some "editorial" comments are given below.

- Defined some very ambitious objectives and achieved most of them
- Overall task was long, time consuming, complex, and difficult to achieve
- Considerable theory/code development carried out in order to provide most accurate LFC designs
- Designed, fabricated, and tested two complex LFC surface concepts — slotted and perforated
- Aeroelastic effects larger than anticipated
- Top and bottom surfaces comprised different experiments. Concave areas on bottom surface provided impetus for additional research on Taylor-Görtler instability
- Complicated suction system developed to achieve extensive laminar flow on the model and for model/liner juncture boundary-layer control
- A large and complex instrumentation system was defined and applied for "real-time" data acquisition and test operations including flow diagnostics

- Increased understanding of wind-tunnel flow quality
- Achieved improved flow quality in 8-ft TPT required for transonic LFC/HLFC testing
- 3-wire hot-wire anemometers give a different picture of flow quality at high subsonic speeds relative to that at low speeds — requires further study
- Complicated liner design procedure formulated and applied – proved to be accurate
- Tunnel test conditions closely approached those specified in design–successful in achieving close approximation of design pressure distribution
- High sensitivity of airfoil pressure distribution to angle of attack and Mach number changes near design conditions made achievement of design pressure distribution extremely difficult
- Best test conditions yielded a supersonic bubble slightly larger than design
- Obtained significantly lower drag for LFC airfoil than an equivalent turbulent airfoil
- Full chord laminar flow maintained through large supersonic zone up to 12 million Reynolds number. Goal of full chord laminar flow at 20 million Reynolds number not achieved. Slotted and perforated surfaces maintained laminar flow to approximately 70% and 60% chord, respectively at a chord Reynolds number of $20 \times 10^6$.
- No significant difference between the slotted and perforated configurations when <u>all</u> factors considered
- Calibrated linear incompressible and compressible stability theory for T-S and C-F disturbances. Minimum $N$-factors at measured transition correlated.
- Obtained a large variety of data necessary for developing, understanding and verifying codes including:

| | |
|---|---|
| - $C_p$ vs. $x/c$ | - Surface conditions |
| - $C_Q$ vs. $x/c$ | - Environmental effects |
| - $(x/c)_{TR}$ | - Total Drag |

- Considerable amount of code verification and documentation of results completed – large data base available for laminar-flow researchers
- Large number of papers published containing information about, and results from, LFC project
- Data indicates that LFC is compatible with supercritical technology
- No technical reason identified indicating that LFC/HLFC should not be applied through transonic speeds

## 18. References


[1] Bobbitt, P. J.; Waggoner, E. G.; Harvey, W. D.; and Dagenhart, J. R.: A Faster "Transition" to Laminar Flow. SAE 851855, Aerospace Technology Conference and Exposition, Anaheim, CA, October 1988.

[2] Lachmann, G. V. ed.: Boundary Layer and Flow Control, Volume 2. Pergamon Press, 1961.

[3] Wagner, R.; Maddalon, D. V.; Bartlett, D. W.; Collier, F. S., Jr.; and Braslow, A. L.: Laminar Flow Flight Experiments – A Review. Natural Laminar Flow and Laminar Flow Control, M. Y. Hussaini and R. W. Barnwell (editors), Springer-Verlag, 1991.

[4] Pfenninger, Werner: Laminar Flow Control Laminarization. Special Course on Concepts for Drag Reduction, AGARD-R-654, June 1977, pp. 3-1 to 3-75.

[5] Pfenninger, W.; and Groth, E.: Low Drag Boundary Layer Suction Experiments in Flight on a Wing Glove of an F-94A Airplane with Suction through a Large Number of Fine Slots. Boundary Layer and Flow Control, Volume 2, G. V. Lachman, ed., Pergamon Press, 1961, pp. 981-999.

[6] Fowell, L. R.; and Antonatos, P. P.: Laminar Flow Control Flight Test Results, Some Results from the X-21A Program. Recent Developments in Boundary Layer Research, Part IV, May 1965. AGARDograph 97, pp. 1-76.

[7] Lumley, J. L.: Passage of a Turbulent Stream through Honeycomb of Large Length-to-Diameter Ratio. Trans. ASME, Ser. D: J. Basic Eng., Vol. 86, No. 2, June 1964, pp. 218-220.

[8] Loehrke, R. I.; and Nagib, H. M.: Control of Free-Stream Turbulence by Means of Honeycombs: A Balance between Suppression and Generation. J. Fluids Eng., Vol. 98, Ser. I, No. 3, September 1976, pp. 342-353.

[9] Prandtl, L.: Attaining a Steady Air Stream in Wind Tunnels. NACA TM 726, 1933.

[10] Schubauer, G. B.; Spangenberg, W. G.; and Klebanoff, P. S.: Aerodynamic Characteristics of Damping Screens. NACA TN 2001, 1950.

[11] Dryden, Hugh L.; and Schubauer, G. B.: The Use of Damping Screens for the Reduction of Wind-Tunnel Turbulence. J. Aeronaut. Sci., Vol. 14, No. 4, Apr. 1947, pp. 221-228.

[12] Collar, A. R.: The Effect of a Gauze on the Velocity Distribution in a Uniform Duct. R. & M. No. 1867, British A.R.C., 1939.

[13] Taylor, G. I.; and Batchelor, G. K.: The Effect of Wire Gauze on Small Disturbances in a Uniform Stream. Q. J. Mech. & Appl. Math., Vol. 2, Pt. 1, March 1949, pp. 1-29.

[14] Harvey, William D.; Stainback, P. Calvin; and Owen, F. Kevin: Evaluation of Flow Quality in Two Large NASA Wind Tunnels at Transonic Speeds. NASA TP-1737, 1980.

[15] Owen, F. K.; Stainback, P. Calvin; and Harvey, William D.: An Evaluation of Factors Affecting the Flow Quality in Wind Tunnels. AGARD-CP-348, Wind Tunnels and Testing Techniques, pp. 12-17, 12-22.

[16] Scheiman, James; and Brooks, J. D.: A Comparison of Experimental and Theoretical Turbulence Reduction from Screens,

Honeycomb and Honeycomb-Screen Combinations. A Collection of Technical Papers – AIAA 11th Aerodynamic Testing Conference, American Inst. of Aeronautics and Astronautics, 1980, pp. 129-137. (Available as AIAA-80-0433.)

[17] McKinney, Marion O.; and Scheiman, James: Evaluation of Turbulence Reduction Devices for the Langley 8-Foot Transonic Pressure Tunnel. NASA TM-81792, 1981.

[18] Scheiman, James: Considerations for the Installation of Honeycomb and Screens to Reduce Wind-Tunnel Turbulence. NASA TM-81868, 1981.

[19] Stainback, P. C.; Johnson, C. B.; and Basnett, C. B.: Preliminary Measurements of Velocity, Density, and Total Temperature Fluctuations in Compressible Subsonic Flow. AIAA 21st Aerospace Sciences Meeting, January 1983.

[20] Stainback, P. C.: A Review of Hot-Wire Anemometry in Transonic Flows. ICIASF '85, pp. 67-78, Stanford, CA, August 1985.

[21] Bobbitt, P. J.: Instrumentation Advances for Transonic Testing. Transonic Symposium: Theory, Application, and Experiment, NASA Langley Research Center, Hampton, VA, April 19-21, 1988, Jerome T. Foughner, Jr. (Compiler), NASA CP-3020, Vol. I, Part 2.

[22] Jones, G. S.; Stainback, P. C.; Harris, C. D.; Brooks, C. W.; and Clukey, S. J.: Flow Quality Measurements for the Langley 8-Foot Transonic Pressure Tunnel LFC Experiment. 27th Aerospace Sciences Meeting, Reno, NV, January 9-12, 1989, AIAA Paper No. 89-0150.

[23] Jones, G. S.; and Stainback, P. C.: A New Look at Wind Tunnel Flow Quality for Transonic Flows. SAE 881452 Aerospace Technology Conference and Exposition, Anaheim, CA, October 1988.

[24] Bauer, F.; Garabedian, P.; and Korn, D.: A Theory of Supercritical Wing Sections, With Computer Programs and Examples. Volume 66 of Lecture Notes in Economics and Mathematical Systems, Springer-Verlag, 1972.

[25] Bauer, Frances; Garabedian, Paul; Korn, David; and Jameson, Antony: Supercritical Wing Sections II. Volume 108 of Lecture Notes in Economics and Mathematical Systems, Springer-Verlag, 1975.

[26] Harris, Charles D.: Aerodynamic Characteristic of a 14-Percent-Thick NASA Supercritical Airfoil Designed for a Normal-Force Coefficient of 0.7. NASA TM X-72712, 1975.

[27] Allison, D. O.; and Dagenhart, J. R.: Design of a Laminar-Flow-Control Supercritical Airfoil for a Swept Wing. CTOL Transport Technology–1978, NASA CP-2036, Part 1, 1978, pp. 395-408.

[28] Allison, D. O.: Inviscid Analysis of Two Supercritical Laminar-Flow-Control Airfoils at Design and Off-Design Conditions. NASA TM-84657, 1983.

[29] Allison, D. O.; and Dagenhart, J. R.: Two Experimental Supercritical Laminar-Flow-Control Swept-Wing Airfoils. NASA TM-89073, February 1987.

[30] Pfenninger, W.; Reed, Helen L.; and Dagenhart, J. R.: Design Considerations of Advanced Supercritical Low-Drag Suction Airfoils. Viscous Flow Drag Reduction, Gary R. Hough, ed., AIAA, c. 1980, pp. 249-271.

[31] Van Ingen, J. L.; Blom, J. J. H.; and Goei, J. H.: Design Studies of Thick Laminar Flow Airfoils for Low-Speed Flight Employing Turbulent Boundary-Layer Suction over the Rear Part. AGARD CP-365, May 1984.

[32] Klebanoff, P. S.; and Tidstrom, K. D.: Evolution of Amplified Waves Leading to Transition in a Boundary Layer with Zero Pressure Gradient. NASA TN D-195, 1958.

[33] Dagenhart, J. R.: Amplified Crossflow Disturbances in the Laminar Boundary Layer on Swept Wings with Suction. NASA TP-1902, 1981.

[34] Srokowski, A. J.; and Orszag, S. A.: Mass Flow Requirements for LFC Wing Design. AIAA Paper 77-1222, August 1977.

[35] Mack, L. M.: Transition Prediction and Linear Stability Theory. AGARD CP-224, January 1970.

[36] Malik, M. R.; and Orszag, S. A.: Efficient Computation of the Stability of Three-Dimensional Compressible Boundary Layers. AIAA Paper 81-1277, June 1981.

[37] El Hady, N. M.: On the Stability of Three-Dimensional Compressible Nonparallel Boundary Layers. AIAA 80-1374, July 1980.

[38] Reed, H. L.; and Nayfeh, A. H.: Stability of Compressible Three-Dimensional Boundary-Layer Flows. AIAA 82-1009, June 1982.

[39] Reed, H. L.: Wave Interactions in Swept Wing Flows. Phys. Fluids, Vol. 30, 1987.

[40] El Hady, N. M.: Evolution of Resonant Wave Triads in Three-Dimensional Boundary Layers. Phys. Fluids, March 1989.

[41] Nayfeh, A. H.: Effect of Streamwise Vortices on Tollmien-Schlichting Waves. Journal of Fluid Mechanics, Vol. 107, 1981, p. 441.

[42] Nayfeh, A. H.: Influence of Görtler Vortices on Tollmien-Schlichting Waves. AIAA PPer 87-1206, 1987.

[43] Malik, M. R.: Wave Interaction in Three-Dimensional Boundary Layers. AIAA Paper 86-1129, 1980.

[44] Herbert, T.; and Morkovin, M. V.: Dialogue on Bridging Some Gaps in Stability and Transition Research. Laminar-Turbulent Transition, R. Eppler and H. Fasel, eds., Springer-Verlag, 1980.

[45] Saric, W. S.; and Reed, H. L.: Three-Dimensional Stability of Boundary Layers. Proceeding Perspectives in Turbulence Symposium, Göttingen, West Germany, May 11-15, 1987.

[46] Morkovin, M. V.: On the Many Faces of Transition Viscous Drag Reduction. C. S. Wells, ed., Plenum Publ., 1969.

[47] Kaups, K.; and Cebeci, T.: Compressible Laminar Boundary Layers with Suction on Swept and Tapered Wings. J. Aircraft, Vol. 14, No. 7, July 1977, pp. 661-667.

[48] Mack, L. M.: On the Stability of the Boundary Layer on a Transonic Swept Wing. AIAA 79-0264, January 1979.

[49] Pfenninger, W.: Special Course on Concepts for Drag Reduction, Chapter 3 - Laminar Flow Control, Laminarization. AGARD Report 654, June 1977.

[50] Smith, A. M. O.; and Gameroni, N.: Transition, Pressure Gradient, and Stability Theory. Proc. Int. Congress Appl. Mech., 9, Brussels, Vol. 4, 1956.

[51] Van Ingen, J. L.: A Suggested Semi-Empirical Method for the Calculation of the Boundary-Layer Transition Region. Report No. VTH 71, VTH 74, Delft, Holland, 1956.

[52] Hefner, J. N.; and Bushnell, D. M.: Application of Stability Theory to Laminar Flow Control. AIAA Paper 79-1493, July 1979.

[53] Smith, A. M. O.: On Growth of Taylor-Görtler Vortices Along Highly Concave Walls. Q. Appl. Math., Vol. XIII, No. 3, Oct. 1955, pp. 233-262.

[54] El-Hady, Nabil M.; and Verma, Alok K.: Growth of Görtler Vortices in Compressible Boundary Layers Along Curved Surfaces. AIAA-81-1278, June 1981.

[55] Kobayashi, R.: Taylor-Görtler Instablity of a Boundary Layer with Suction or Blowing. Rep. No. 289, Inst. of High Speed Mechanics, Tohoku Univ., Vol. 32, 1975, pp. 129-148.

[56] Pfenninger, W.; and Syberg, J.: Reduction of Acoustic Disturbances in the Test Section of Supersonic Wind Tunnels by Laminarizing Their Nozzles and Test Section Wall Boundary Layers by Means of Suction. NASA CR-2436, 1974.

[57] Carlson, Leland A.: TRANDES: A FORTRAN Program for Transonic Airfoil Analysis or Design. NASA CR-2821, 1977.

[58] Newman, Perry A.; Anderson, E. Clay; and Peterson, John B., Jr.: Aerodynamic Design of the Contoured Wind-Tunnel Lever for the NASA Supercritical, Laminar-Flow-Control, Swept-Wing Experiment. NASA TP-2335, September 1984.

[59] Carmichael, B. H.: Surface Waviness Criteria for Swept and Unswept Laminar Suction Wings. Rep. No. NOR-59-438 (BLC-123) (Contract AF33(616)-3168), Northrop Aircraft, Inc., August 1959.

[60] Carmichael, B. H.; and Pfenninger, W.: Surface Imperfection Experiments on a Swept Laminar Suction Wing. Rep. No. NOR-59-454 (BLC-124), Northrop Corp., August 1959.

[61] Pfenninger, W.; Bacon, J.; and Goldsmith, J.: Flow Disturbances Induced by Low-Drag Boundary-Layer Suction through Slots. Phys. Fluids Suppl., Vol. 10, No. 9, Pt. II, September 1967, pp. S112-S114.

[62] Harris, Charles D.; Harvey, William D.; and Brooks, Cuyler W., Jr.: The NASA Langley Laminar-Flow-Control Experiment on a Swept, Supercritical Airfoil, *Design Overview*, NASA TP-2809, May, 1988.

[63] Maddalon, Dal V.; and Poppen, William A., Jr.: Design and Fabrication of Large Suction Panels with Perforated Surfaces for Laminar Flow Control Testing in a Transonic Wind Tunnel. NASA TM-89011, 1986.

[64] Harris, C. D.; and Brooks, C. W., Jr.: Modifications to the Langley 8-Foot Transonic Pressure Tunnel for the Laminar Flow Control Experiment. NASA TM-4032, 1988.

[65] Newman, P. A.; Kemp, W. B.; and Garriz, J. A.: Wall Interference Assessment and Correlations. Transonic Symposium: Theory, Application, and Experiment. NASA CP-3020, Vol. I and II, April 1988.

[66] Harris, C. D.; Brooks, C. W., Jr.; Stack, J. P.; and Clukey, P. G.: The NASA Langley Laminar-Flow-Control Experiment on a Swept Supercritical Airfoil - *Basic Results for Slotted Configuration.* NASA TM-4100, June 1989.

[67] Brooks, C. W., Jr.; Harris, C. D.; and Harvey, W. D.: The NASA Langley Laminar-Flow-Control Experiment on a Swept Supercritical Airfoil – *Drag Equations.* NASA TM-4096, February 1989.

[68] Berry, S. A.: Incompressible Boundary-Layer Stability Analysis of LFC Experimental Data for Sub-Critical Mach Numbers. NASA CR-3999, July 1986.

[69] Brooks, C. W., Jr.; and Harris, C. D.: Results of LFC Experiment on Slotted Swept Supercritical Airfoil in Langley's 8-Foot Transonic Pressure Tunnel. NASA CP-2487, Part 2, May 1987.

[70] Berry, S. A.; Dagenhart, J. R.; Brooks, C. W., Jr.; and Harris, C. D.: Boundary-Layer Stability Analysis of LaRC 8-Foot LFC Experimental Data. NASA CP-2487, Part 2, March 1987.

[71] Harvey, W. D.; Harris, C. D.; Sewall, W. G.; and Stack, J. P.: Laminar Flow Wind Tunnel Experiments. NASA CP-3020, Vol. I and II, April 1988.

[72] Berry, S. A.; Dagenhart, J. R.; Viken, J. K.; and Yeaton, R. B.: Boundary-Layer Stability Analysis of NLF and LFC Experimental Data at Subsonic and Transonic Speeds. SAE TP-871859, October 1987.

[73] Vijgen, P. M. H. W.; Dodbele, S. S.; Pfenninger, W.; and Holmes, B. T.: Analysis of Wind Tunnel Boundary-Layer Transition Experiments on Axisymmetric Bodies at Transonic Speeds Using Compressible Boundary-Layer Stability Theory. AIAA 88-0008, January 1988.

[74] Croom, C. C.; Manuel, G. S.; and Stack, J. P.: In-Flight Detection of Tollmien-Schlichting Instabilities in Laminar Flow. SAE Paper 871016, April 1987.

[75] Boeing Commercial Airplane Company: Flight Survey of the 757 Wing Noise Field and Its Effects on Laminar Boundary Layer Transition. Vols. I and II. NASA CR-178216.

[76] Rozendaal, R. A.: Natural Laminar Flow Flight Experiments on a Swept Wing Business Jet - Boundary Layer Stability Analysis. NASA CR-3975, May 1986.

[77] Rozendaal, R. A.: Variable-Sweep Transition Flight Experiment (VSTFE) Stability Code Development and Clean-Up Glove Data Analysis. NASA CP- 2487, March 1987.

[78] Befus, J.; Nelson, R.; Latos, J., Sr.; and Ellis, D.: Flight Test Investigations of a Wing Designed for Natural Laminar Flow. SAE TP-871044, April 1982.

[79] Runyan, L. J.; Navran, B. H.; and Rozendaal, R. A.: F-111 Natural Laminar Flow Glove Flight Test Data Analysis and Boundary-Layer Stability Analysis. NASA CR-166051, 1984.

[80] Runyan, L. J.: Boundary Layer Stability Analysis of a Natural Laminar Flow Glove on the F-111 Tact Airplane. Symposium on Viscous Drag Reduction, Dallas, Texas, November 7-8, 1979.

[81] Fisher, D. F.; and Dougherty, N. S.: In Flight Transition Measurements on a 10 Degree Cone at Mach Numbers Form 0.5 to 2.0. NASA TP-1971, June 1982.

[82] Mack, L. M.: Progress in Compressible Boundary-Layer Stability Computations. Proceedings of the Boundary-Layer Transition Workshop, Vol. IV, Rept. No. TOR-0172(52816-16)-5, December 1971.

[83] Johnson, Charles, B.; Carraway, Debra L.; Hopson, Purnell, Jr.; and Tran, Sang Q.: Status of a Specialized Boundary Layer Transition Detection System for Use in the U.S. National Transonic Facility. Presented at the 12th International Congress on Instrumentation in Aerospace Simulation Facilities, Williamsburg, Virginia, June 22-25, 1987.

[84] Stack, J. P.; Mangalam, S. M.; and Berry, S. A.; A Unique Measurement Technique to Study Laminar-Separation Bubble Characteristics on an Airfoil. AIAA Paper 87-1271, June 1987.

[85] Goradia, S. H.; Bobbitt, P. J.; and Harvey, W. D.: Computational Results for the Effects of External Disturbances on Transition Location on Bodies of Revolution from Subsonic to Supersonic Speeds and Comparisons with Experimental Data. Aerospace Technology Conference and Exposition, Anaheim, CA, September 25-28, 1989, SAE Technical Paper Series 892381.

[86] Kalburgi, Vijay; Mangalam, S. M.; Dagenhart, J. R.; and Tiwari, S. N.: Görtler Instability on an Airfoil. Proceedings of Conference on "Research in Natural Laminar-Flow Control," NASA CP2487, Part 1, 1987.

[87] Goradia, S. H.; Bobbitt, P. J.; Morgan, H. L.; Ferris, J. C.; and Harvey, W. D.: Results of Correlations for Transition Location on a Clean-Up Glove Installed on an F-14 Aircraft and Design Studies for a Laminar Glove for the X-29 Aircraft Accounting for Spanwise Pressure Gradient. Proceedings of a "Transonic Symposium" held at NASA Langley Research Center, NASA CP-3020, Vol. II, April 19-21, 1988.

[88] Bobbitt, Percy J.: Transition Research Opportunities at Subsonic and Transonic Speeds. Instability, and Transition Proceedings. NASA Langley Research Center, May 15 - June 19, 1989, Robert G. Voigt and M. Y. Hussaini (eds.), Springer-Verlag.

## 19. Nomenclature

Symbols

| | |
|---|---|
| $a$ | local speed of sound |
| $a_\infty$ | speed of sound in free stream |
| $A$ | wave amplitude |
| $A_0$ | wave amplitude at neutral stability point |
| $B$ | $= (\beta\theta R_\theta)_{\max}$ also model span |
| $c$ | airfoil chord |
| $c_d$ | section drag coefficient, $= \frac{\text{drag force}}{\frac{1}{2}\rho_\infty U_\infty^2 c}$ |
| $c_{d,s}$ | suction drag coefficient |
| $c_{d,w}$ | total of skin friction, shock and form drags |
| $c_{d,wave}$ | wave drag coefficient |
| $c_l$ | lift coefficient, $= \frac{\text{lift force}}{\frac{1}{2}\rho_\infty U_\infty^2 c}$ |
| $C_P$ | pressure coefficient, $= \frac{p-p_\infty}{q_\infty}$ |
| $C_Q$ | coefficient of suction, $= \frac{\rho_w w_w}{\rho_\infty U_\infty}$ |
| $C_{QM}$ | magnification of suction coefficient distribution in model turbulent region over that in model laminar test region |
| $d$ | suction orifice diameter |
| $f_{\max}$ | maximum amplified frequency |
| $G$ | Görtler parameter, $= R_\theta(\theta/r)^{\frac{1}{2}}$ |
| $h$ | amplitude of surface wave also slot plenum height and rectangular suction orifice height |
| $\tilde{m}$ | RMS value of fluctuating mass flow |
| $m_\infty$ | free stream mass flow, $= \rho_\infty U_\infty$ |
| $M$ | local Mach number $U/a$ |
| $M_\infty$ | free stream Mach number, $= \frac{U_\infty}{a_\infty}$ |
| $N$ | $ln\ A/A_0$, amplification factor |
| $p$ | pressure |
| $\Delta p_t$ | $= P_t - P_{t_\infty}$ |
| $p'$ | fluctuating pressure |
| $\tilde{p}$ | RMS value of fluctuating pressure |
| $q_\infty$ | free stream dynamic pressure, $= \rho_\infty U_\infty^2/2$ |
| $r$ | radius of curvature |
| $R_\theta$ | Reynolds number based on momentum thickness ($\theta$) |
| $R_\infty$ | unit Reynolds |

| | |
|---|---|
| $R_{TR}$ | value of Reynolds number at transition |
| $R_c$ | Reynolds number based on chord, $= \frac{\rho_\infty U_\infty c}{\mu_\infty}$ |
| $R_s$ | Reynolds number based on slot width |
| $s$ | slot width |
| $t$ | maximum thickness of airfoil |
| $T$ | temperature $°R$ or $°F$ |
| $u'$ | $x$ component of fluctuating velocity |
| $\tilde{u}$ | RMS value of $x$ component of fluctuating velocity |
| $u, v, w$ | disturbance velocity components in $x, y$ and $z$ directions |
| $U$ | mean longitudinal velocity |
| $U_\infty$ | free stream velocity |
| $x$ | distance measured in streamwise direction |
| $y, z$ | cartesian coordinates normal to $x$; $z$ also used to indicate distance along wake rake |
| $X_M, Z_M$ | model plan form coordinates |
| $\alpha$ | angle of attack |
| $\beta$ | disturbance growth rate or flap deflection angle |
| $\delta u, \delta v, \delta w$ | model perturbation velocities |
| $\theta$ | boundary layer momentum thickness |
| $\lambda$ | wave length |
| $\Lambda$ | sweep angle |
| $\rho_\infty$ | free stream density |
| $\sigma$ | spacial growth rate |
| $\psi$ | angle of wave propagation |

Abbreviations

| | |
|---|---|
| ACEE | NASA's Aircraft Energy Efficiency Program |
| ARC | NASA Ames Research Center |
| ATM | atmospheric pressure, 14.7 psi |
| CF | crossflow |
| HLFC | hybrid laminar flow control |
| Hz | Hertz |

| | |
|---|---|
| LE | leading edge |
| LFC | laminar flow control |
| LRC | NASA Langley Research Center |
| LTPT | NASA Langley Low Turbulence Pressure Tunnel |
| RSL | reference suction level ($= C_Q/C_{Q_{ref}}$) |
| TE | trailing edge |
| TS | Tollmien-Schlichting |
| 8-ft TPT | NASA Langley 8-ft Transonic Pressure Tunnel |
| 12-ft PT | NASA Ames 12-ft Pressure Tunnel |

## Subscripts

| | |
|---|---|
| $CF$ | crossflow |
| $DES$ | design |
| $l$ | local |
| max | maximum value of quantity |
| $n$ | normal to the leading edge |
| $s$ | suction component |
| $SUC$ | extent of suction |
| $sonic$ | sonic value |
| $t$ | total or stagnation pressure conditions |
| $tot$ | total value |
| $TR$ | value at transition |
| $TS$ | Tollmien-Schlichting |
| $\infty$ | free stream condition |
| $w$ | wake component of drag or value at model surface |

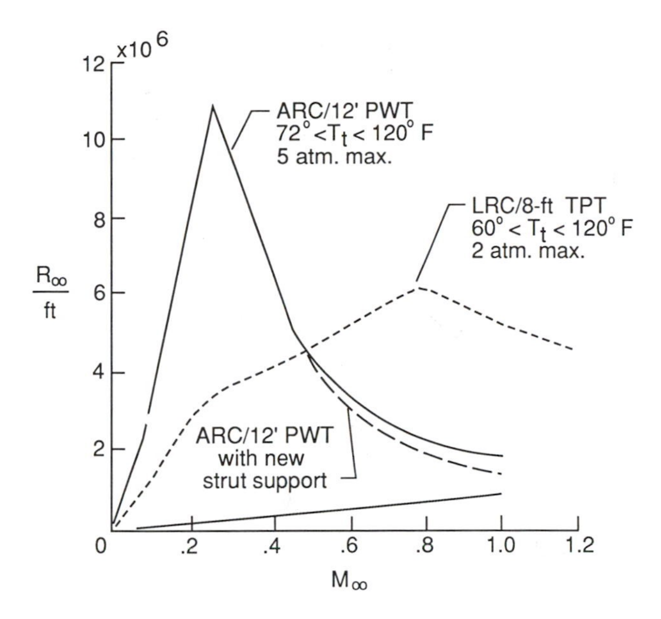


Figure 1. Reynolds number and Mach number capabilities of the LRC 8-ft TPT and the ARC 12-ft PT.

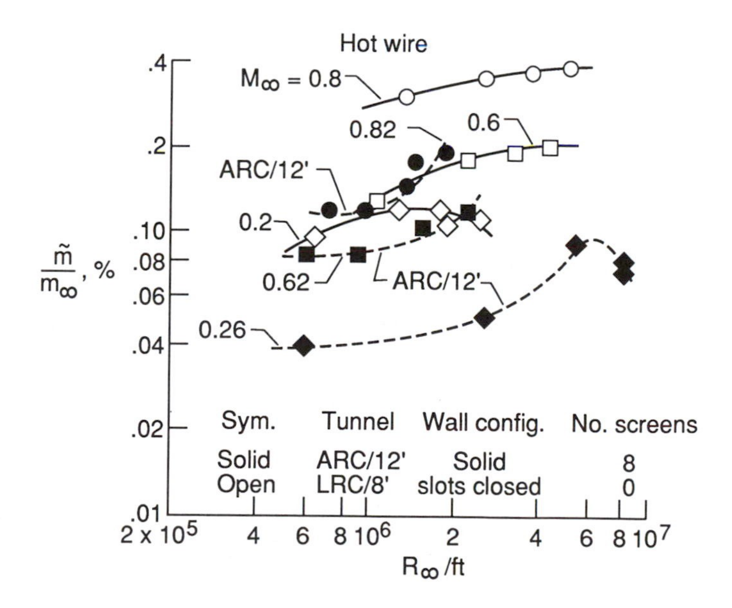


Figure 2. The variation of normalized mass-flow fluctuations with unit Reynolds number for the LRC 8-ft TPT and ARC 12-ft PT.

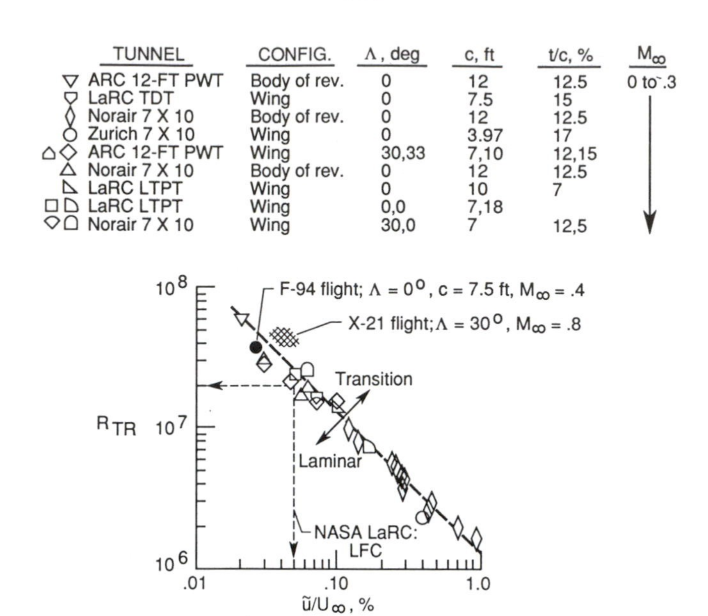


Figure 3. Influence of turbulence level on maximum transition Reynolds number for low-drag wings and bodies of revolution with laminar flow control in wind tunnels and flight.

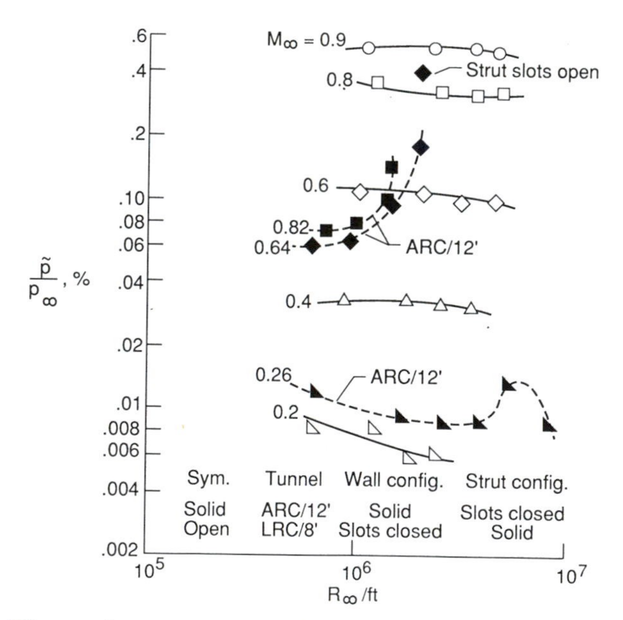


Figure 4. The variation of normalized pressure fluctuations with unit Reynolds number for the LRC 8-ft TPT and the ARC 12-ft PT.

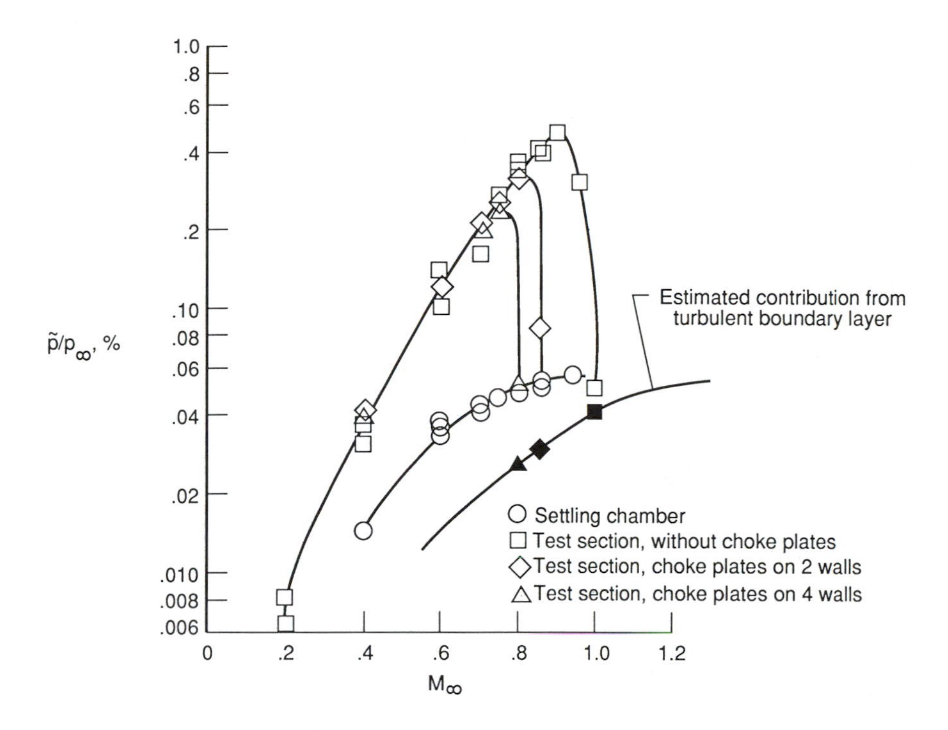


Figure 5. Normalized pressure fluctuations versus Mach number for the LRC 8-ft TPT with and without chokes.

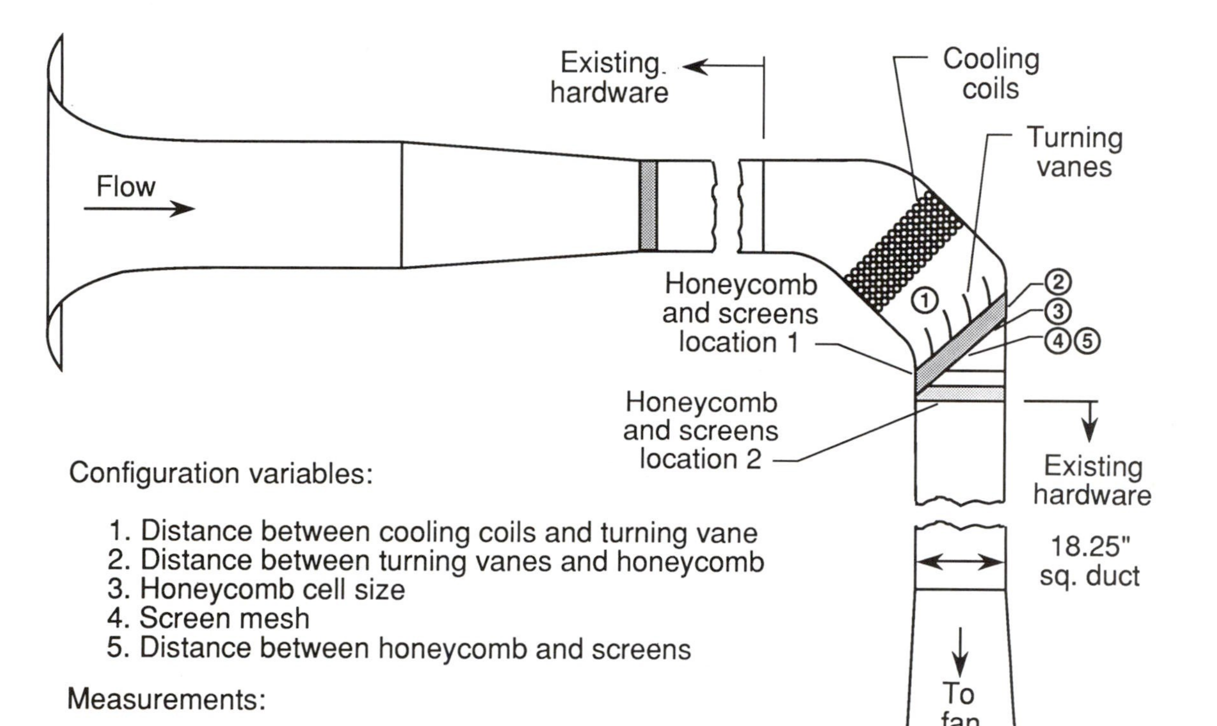


Figure 6. Schematic of the apparatus used to test various flow-quality fixes proposed for the LRC 8-ft TPT.

- Screen
  - 5 Screens @ one foot spacing
  - Solidity: 35.2% (openness: 64.8%)
    - 30 mesh - 0.0065 inch diameter stainless steel wire
- Honeycomb
  - Length/diameter: 9.3
  - Length: 3.5 inches
- Cooler
  - 8 banks of one inch diameter tubes
  - Tube spacing: 3.25 inch
  - Fin diameter: 2.25 inch
    - Spacing: 0.126 inch

Figure 7. Turbulence manipulators installed in the 8-ft TPT as well as the characteristics of the cooler situated in the corner ahead of the honeycomb and screens.

Figure 8. A photograph of the honeycomb installed in the LRC 8-ft TPT taken from just downstream of the turning vanes.

Figure 9. A photograph of the most downstream screen taken from the entrance to the test section.

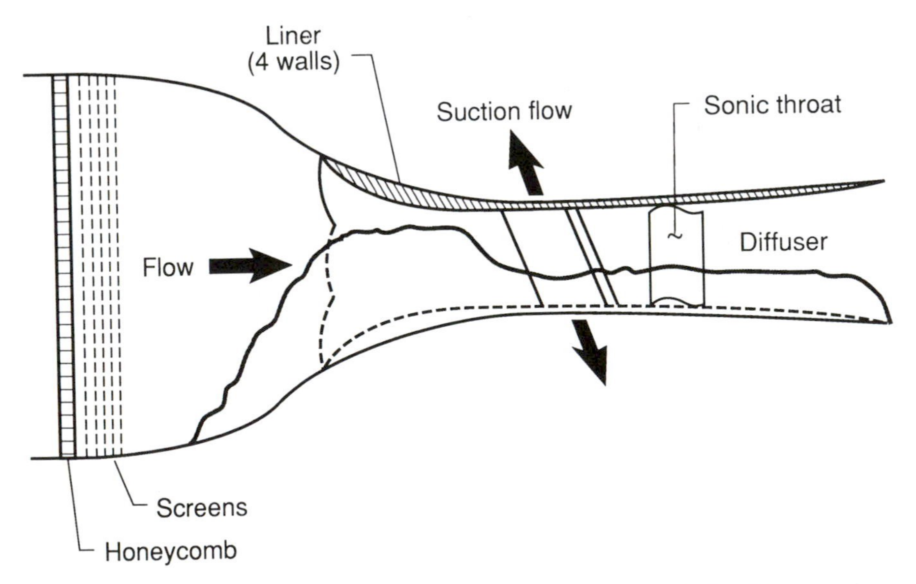


Figure 10. Sketch of the test setup in the LRC 8-ft TPT and the honeycomb and screens.

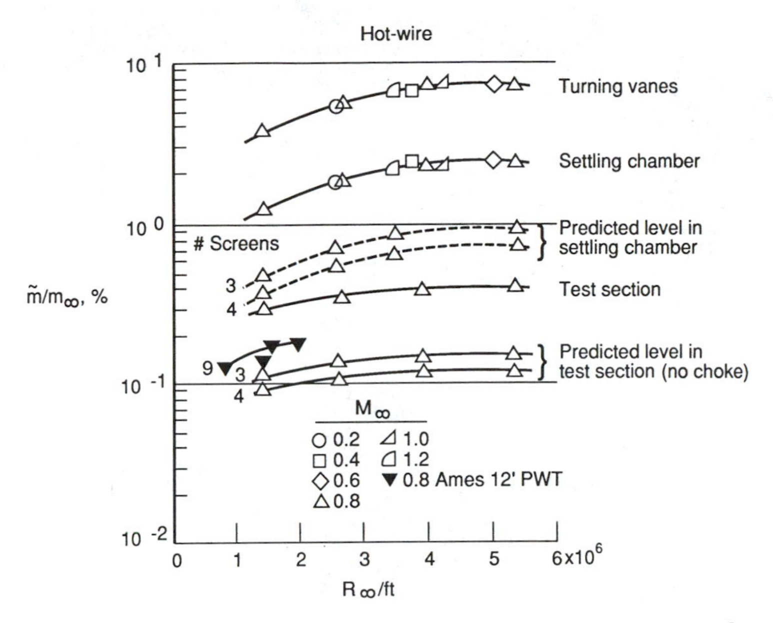


Figure 11. Measured and predicted mass fluctuations as a function of unit Reynolds number at various locations in the tunnel circuit. Both LRC 8-ft TPT and ARC 12-ft PT results are shown.

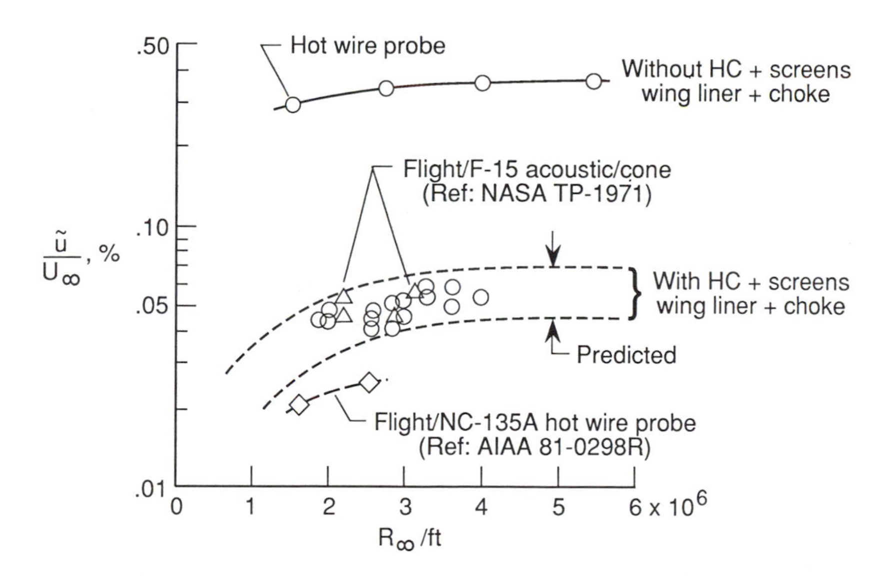


Figure 12. Single wire hot-wire-anemometer measurements of $\frac{\tilde{u}}{U_\infty}$ with and without honeycomb, screens, LFC wing, liner, and choke in the 8-ft TPT compared to flight data. $M_\infty = 0.82$ .

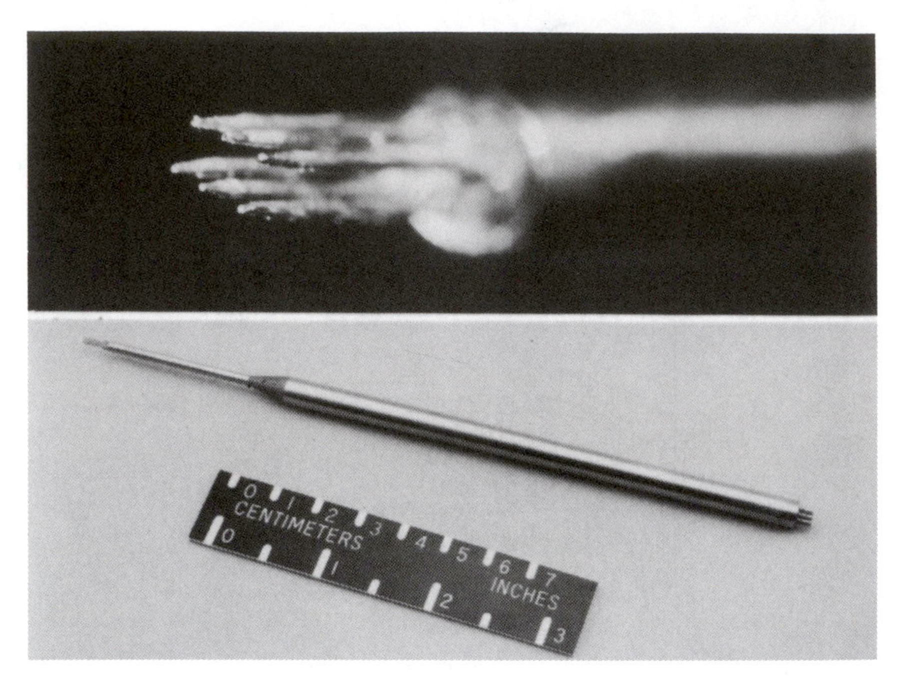


Figure 13. Enlarged photographs of a three wire hot-wire probe and probe head.

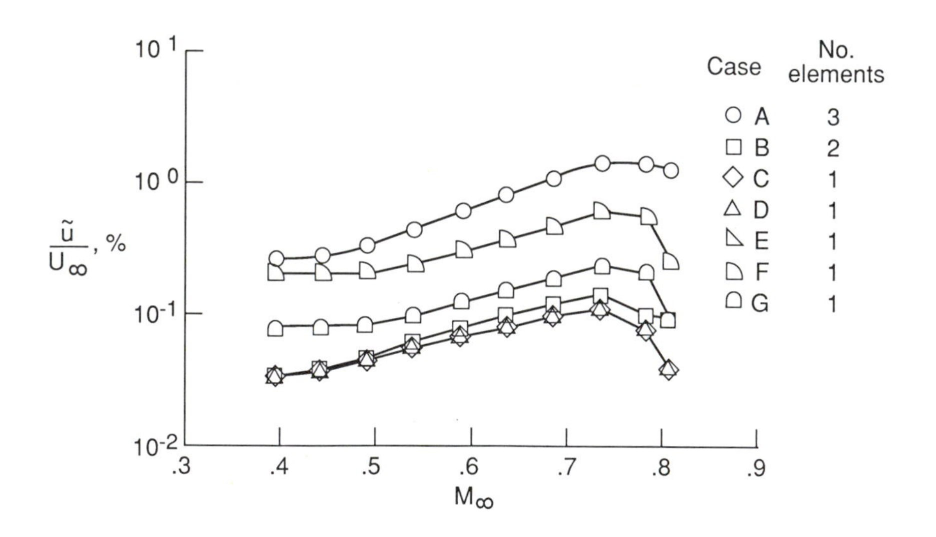


Figure 14. Comparison of single and multi-wire hot-wire-anemometer measurements of $\frac{\tilde{u}}{U_\infty}$ in the Langley Research Center 8-ft TPT.

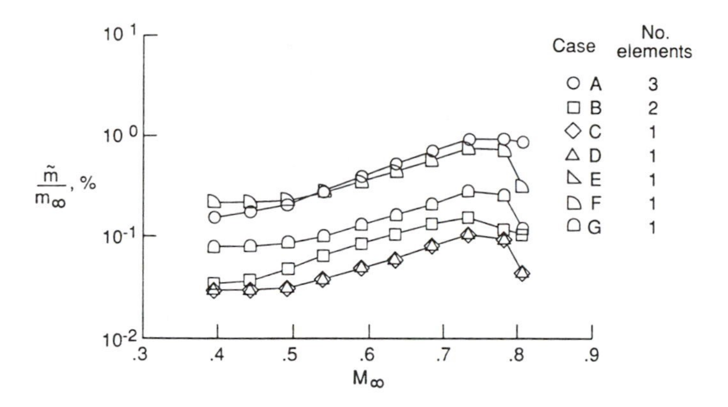


Figure 15. Comparison of single and multi-wire-anemometer measurements of mass fluctuations in the Langley Research Center 8-ft TPT.

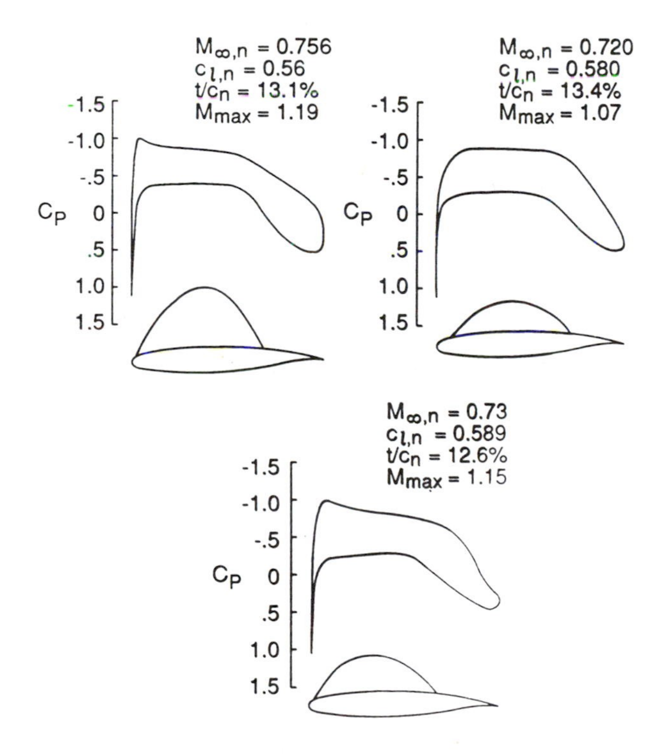


Figure 16. Evolution of the "Baseline Airfoil."

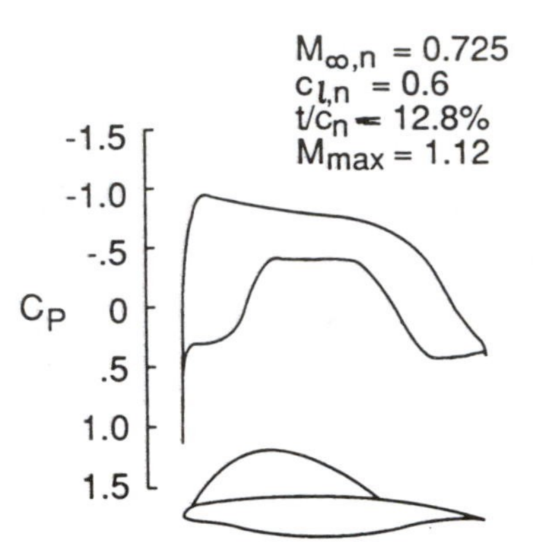


Figure 17. An early LFC airfoil with the undercut leading-edge.

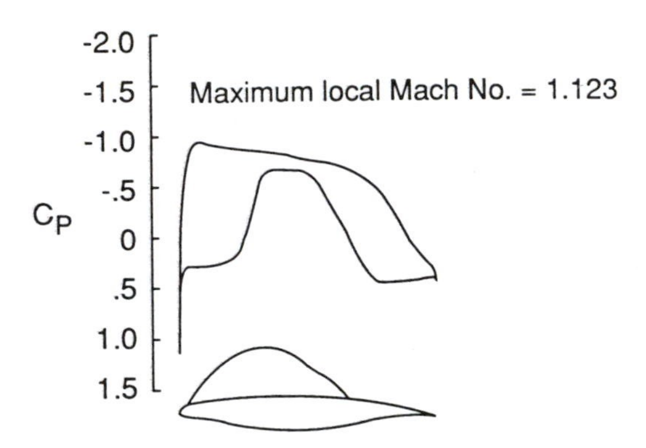


Figure 18. Geometry and pressure distributions of the "Baseline Airfoil."

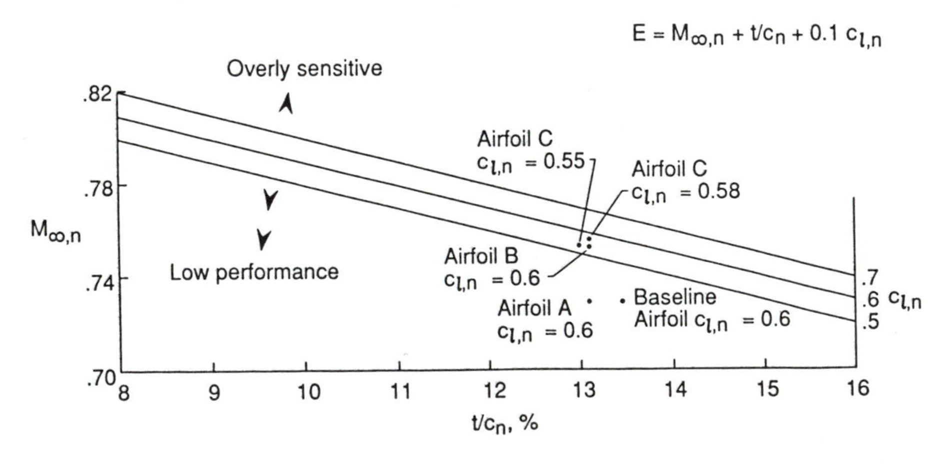


Figure 19. Plot showing the relative "efficiency" of various LFC airfoil designs.

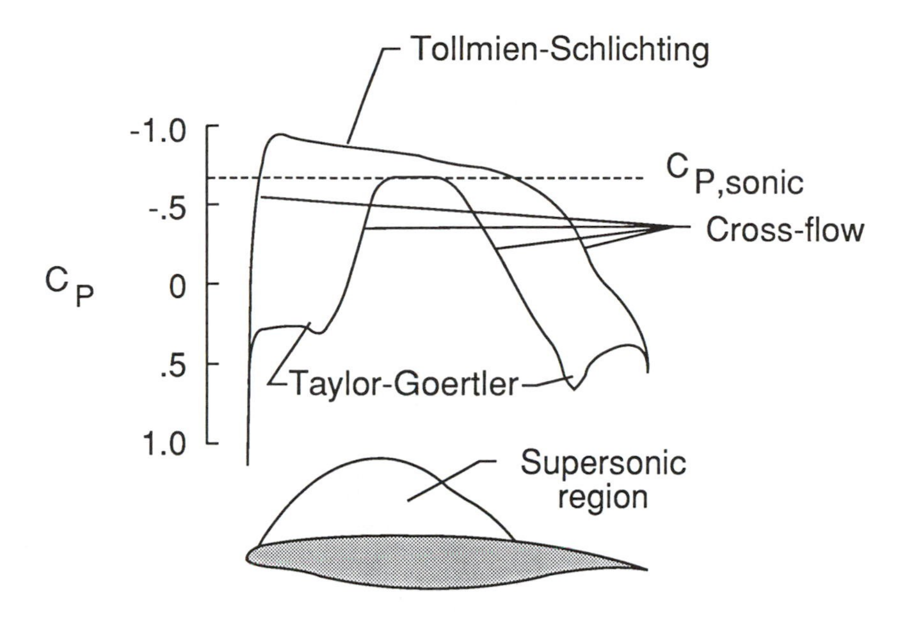


Figure 20. Pressure distribution for the "Baseline Airfoil" with modified contours in the forward and aft cusp regions on the lower surface. This airfoil is designated as Airfoil A.

| | Airfoil | $t/c_n$ |
|---|---|---|
| ——— | B | 0.131 |
| - - - - | A | .135 |

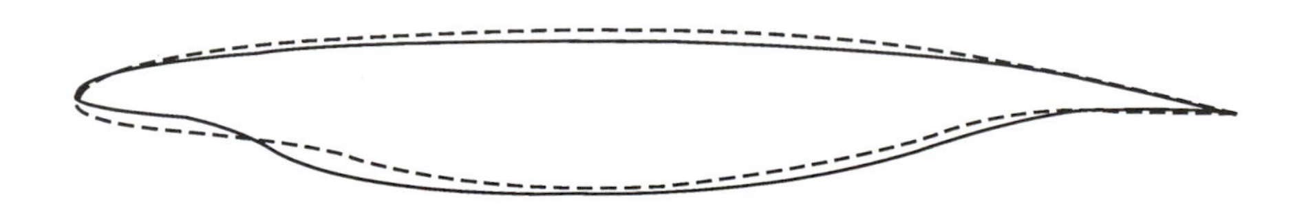

(a) Airfoil geometries.

Figure 21. Geometry, drag rise curves, and pressure distributions for Airfoils A and B.

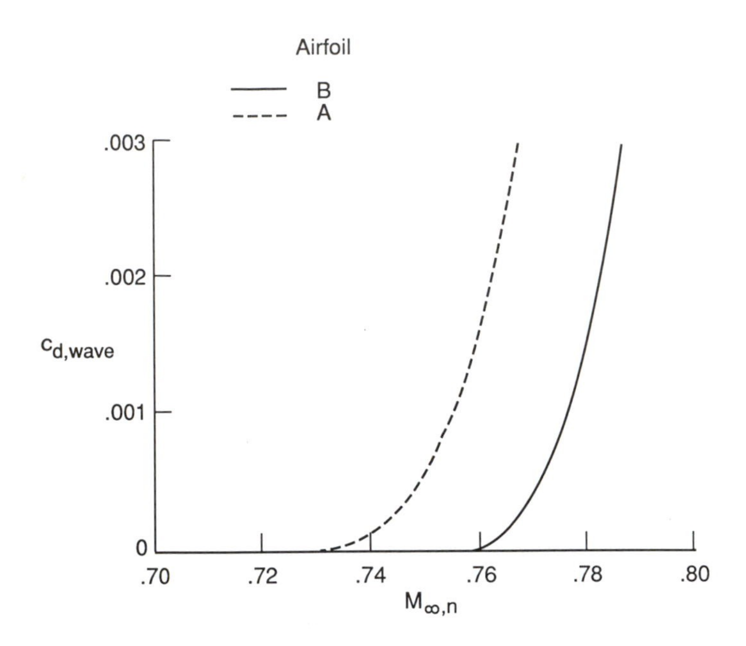


(b) Wave drag versus leading edge normal Mach number.

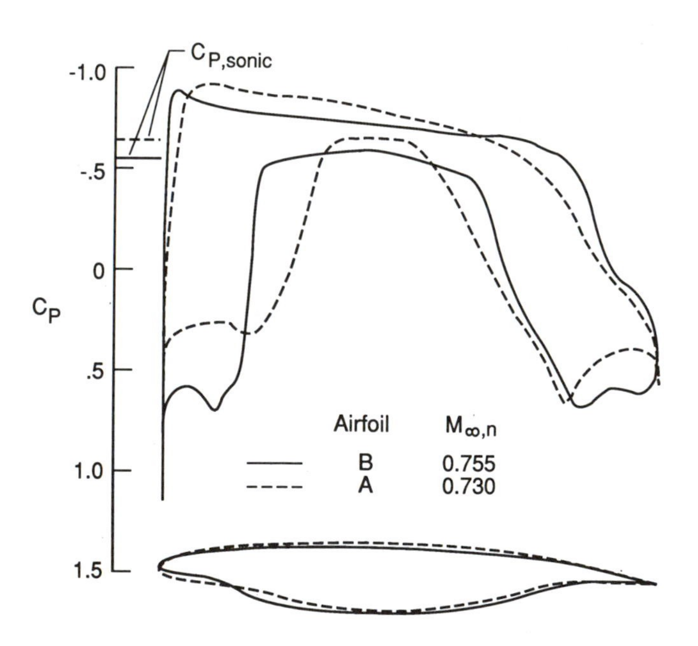


(c) Pressure distributions.

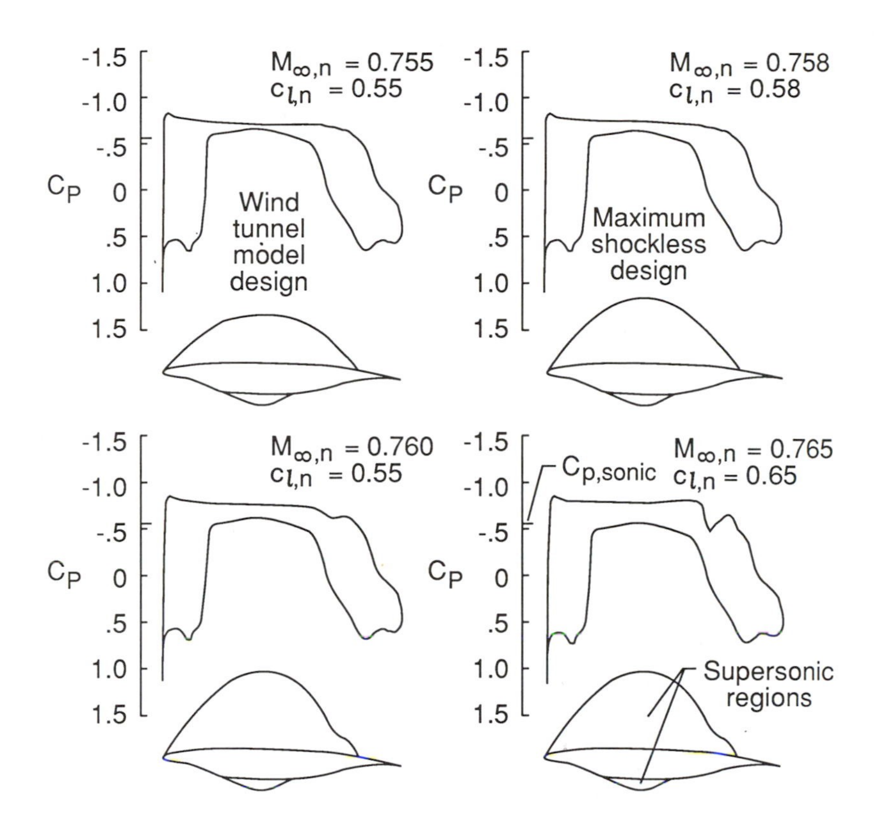


Figure 22. Pressure distributions for Airfoil B at the wind-tunnel and maximum-shockless design points as well as for two "off design" Mach numbers and lift coefficients.

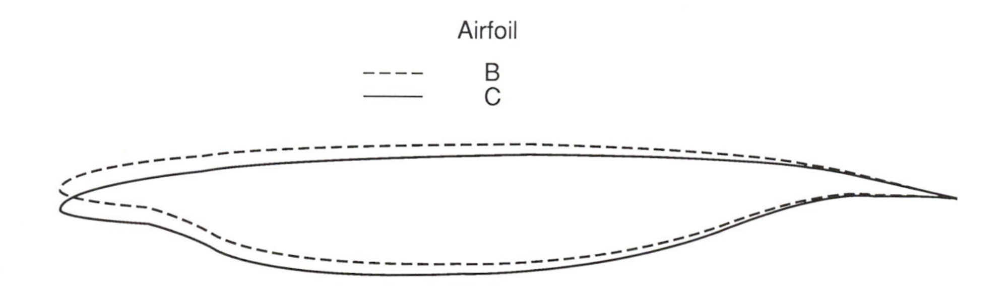


Figure 23. Comparison of the geometries of Airfoils B and C.

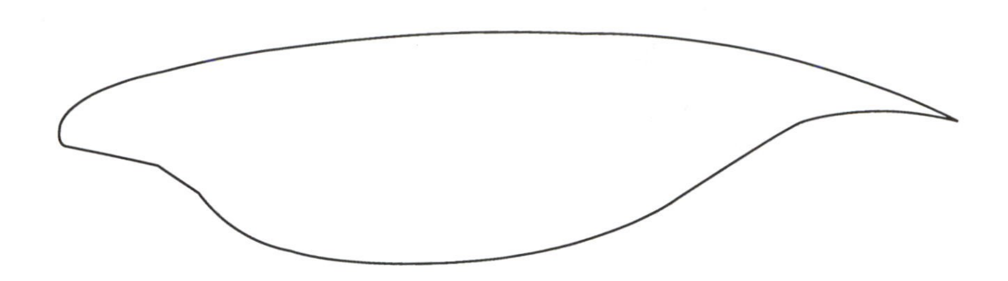

Figure 24. Geometry of Airfoil C with ordinates magnified by a factor of 2.

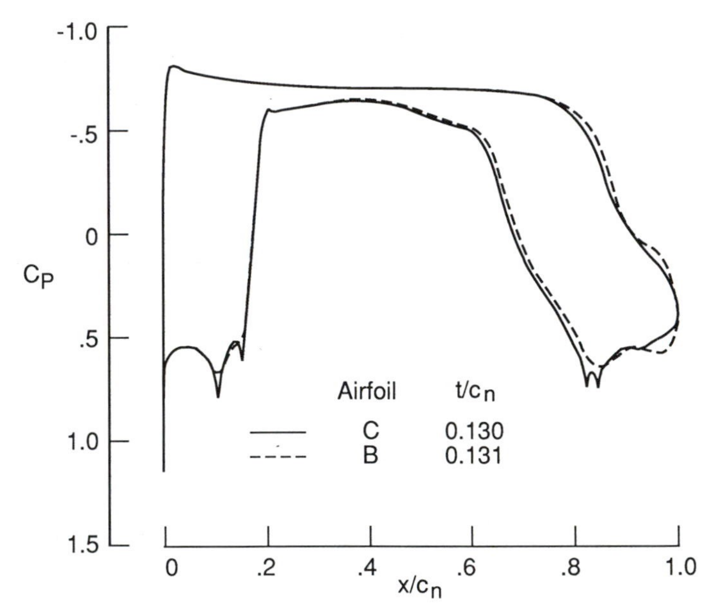


Figure 25. Pressure distribution for Airfoils B and C. $M_{\infty,n} = 0.755$ and $c_{l,n} = 0.55$.

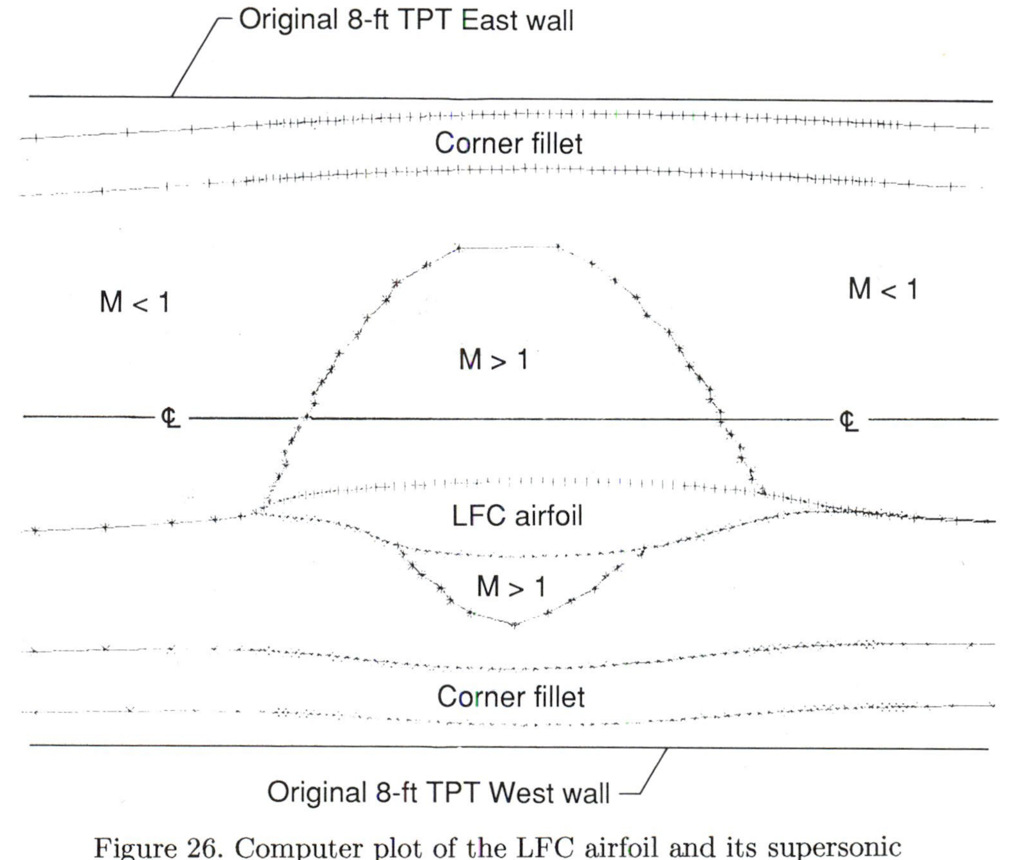


Figure 26. Computer plot of the LFC airfoil and its supersonic "bubbles" showing their relationships to the liner corner fillets. $M_\infty$=0.82, $c_2$=0.47 and $\Lambda$=23° .

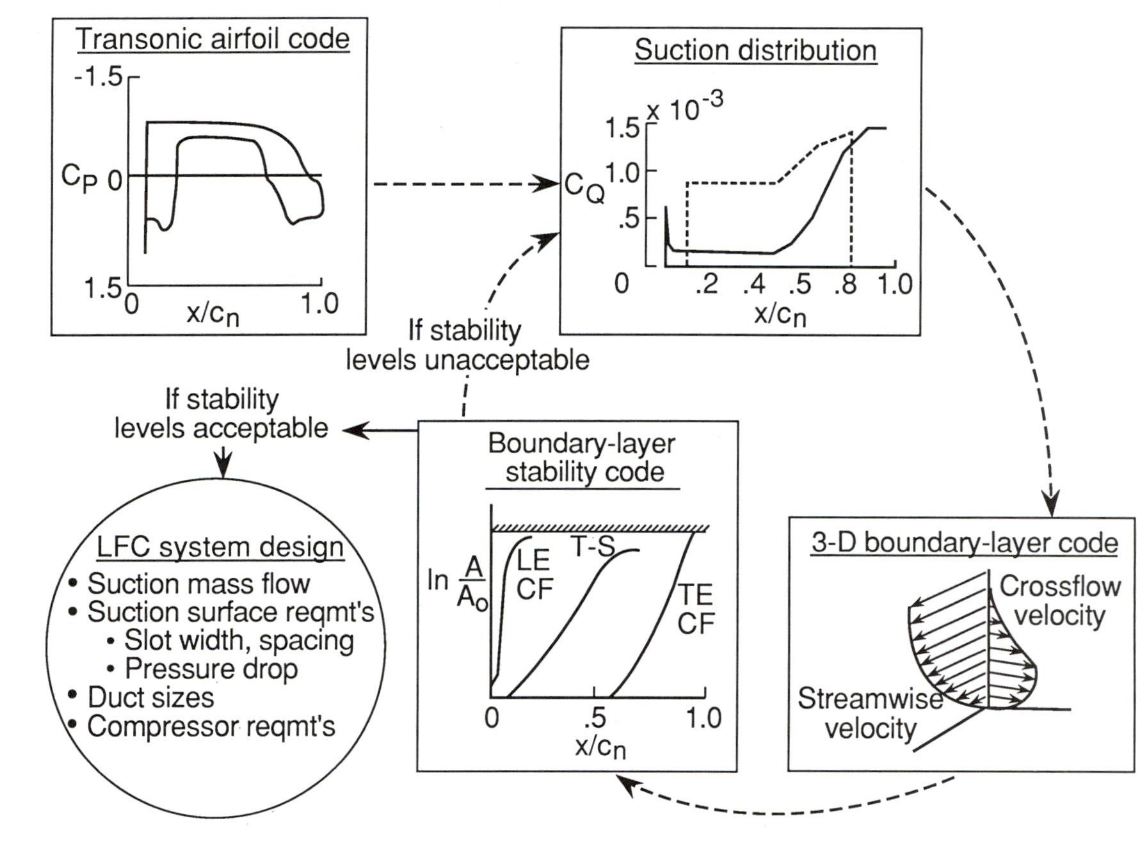


Figure 27. Block diagram of LFC airfoil design cycle.

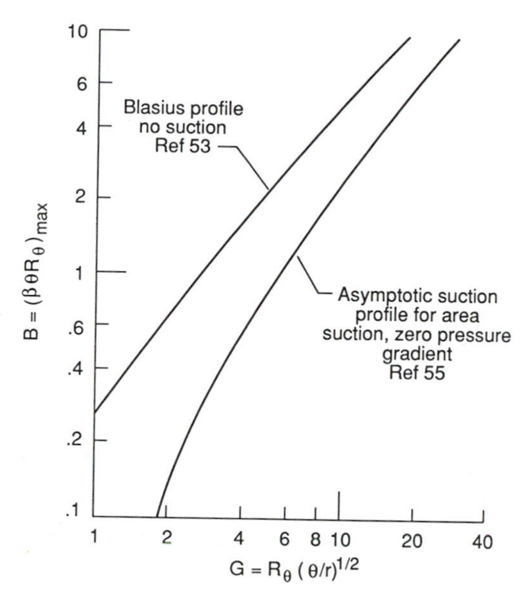


Figure 28. Linearized maximum growth factor B of Taylor-Görtler vortices versus Parameter G for incompressible flow with/without suction.

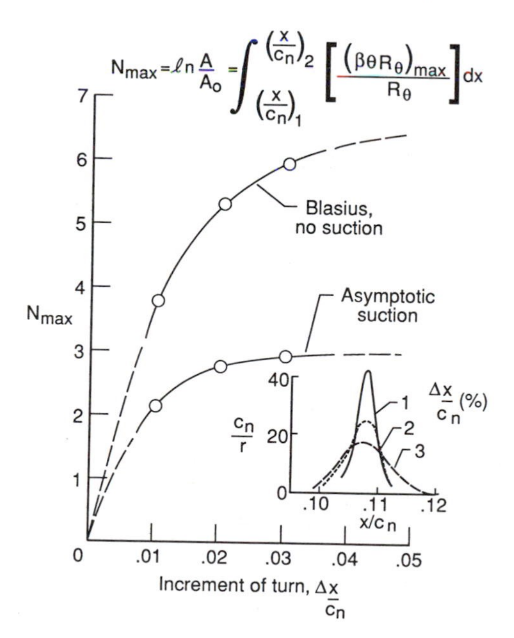


Figure 29. Maximum disturbance amplitude ratio $N_{\max}$ for Taylor-Görtler vortices versus flow turn increment for LFC airfoil lower surface design.

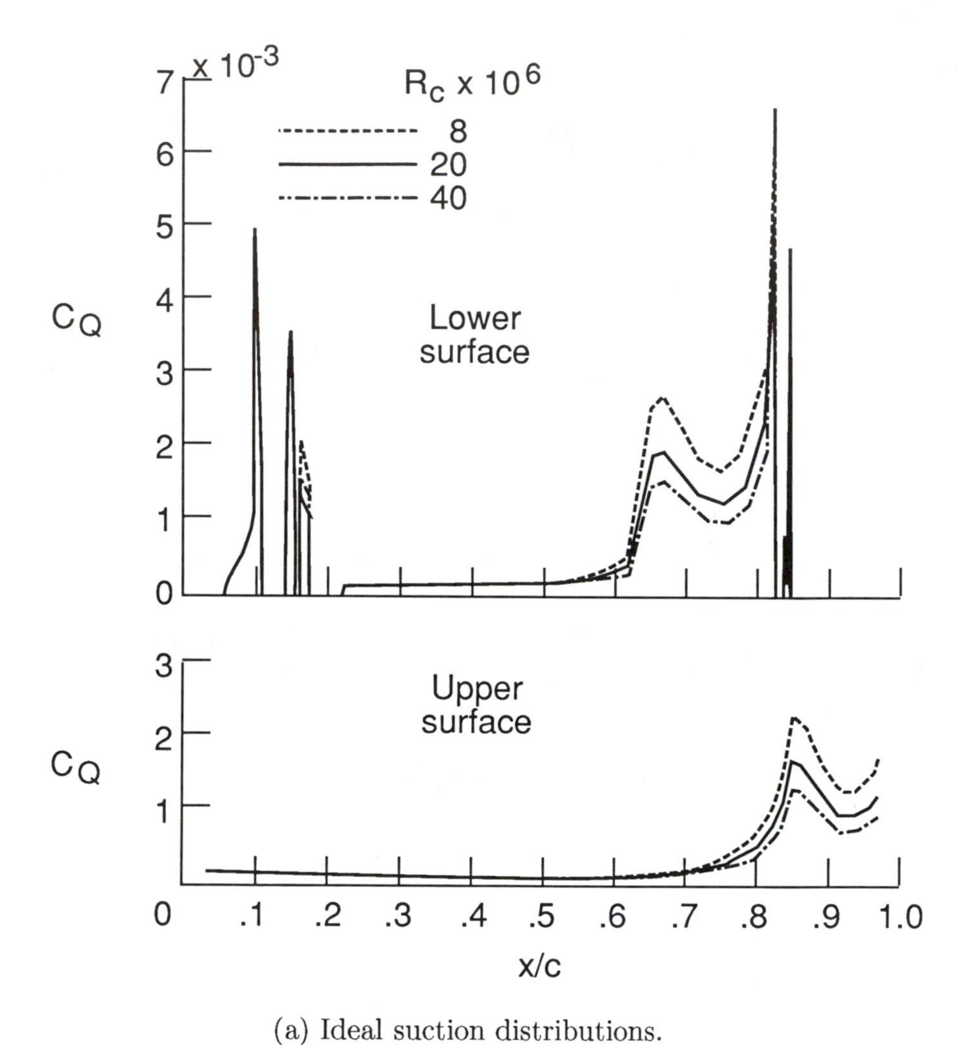


(a) Ideal suction distributions.

Figure 30. Suction mass-flow distributions for the slotted and perforated models at several chord Reynolds numbers and $M_\infty=0.82$.

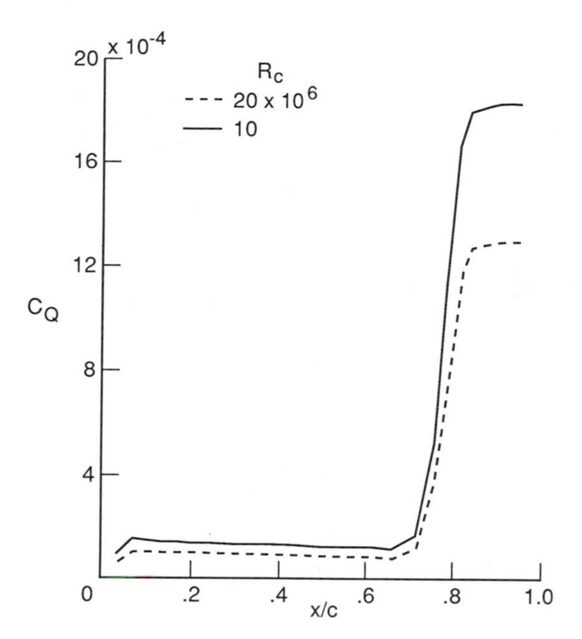


(b) Design slotted-surface suction distribution for upper surface.

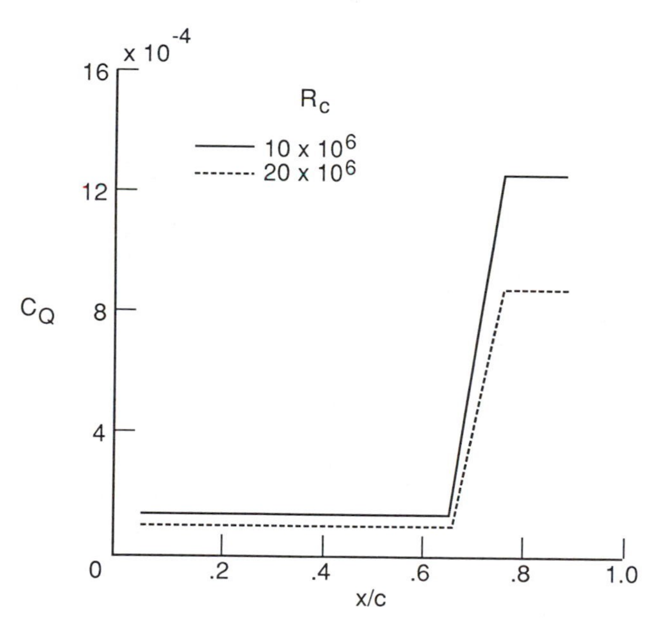


(c) Design perforated-surface suction distribution for upper surface.

Figure 30. Concluded.

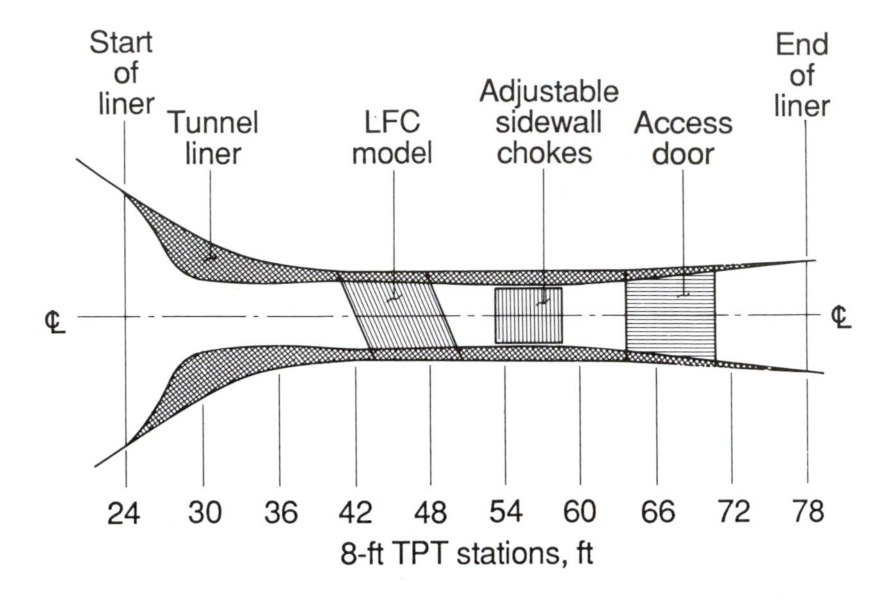


Figure 31. West-side view of LFC test setup in 8-ft TPT.

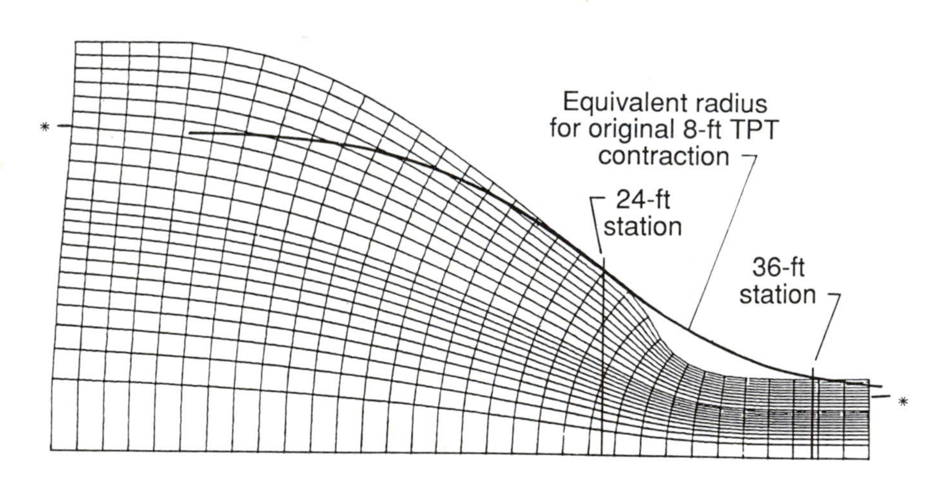


Figure 32. Stream-tube code grid for equivalent axisymmetric contraction.

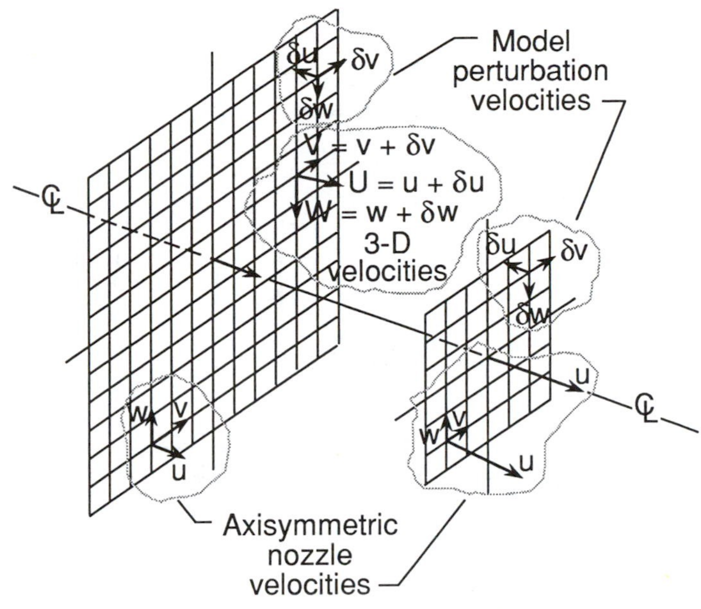


(a) Velocity field superposition.

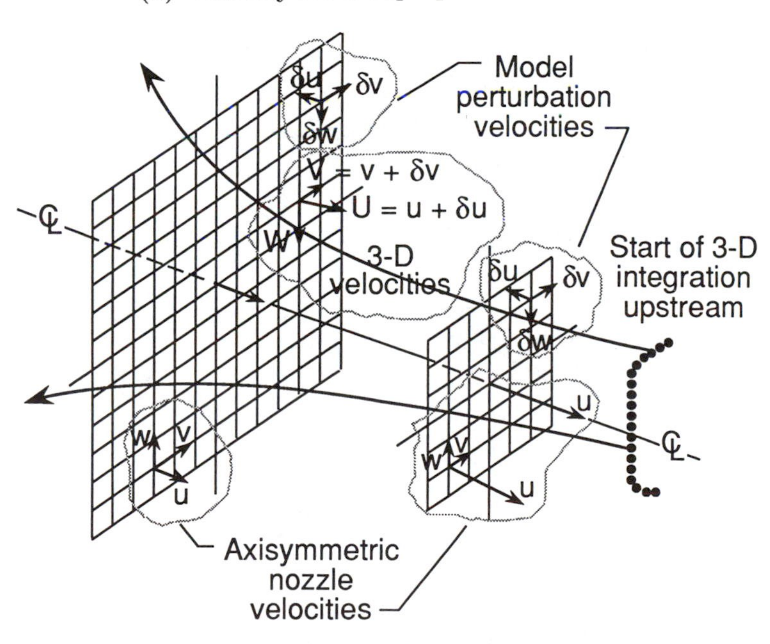


(b) Streamline filament integration.

Figure 33. Sketches depicting the superposition of wing perturbation velocities on contraction nozzle velocity field and integration of the combined velocities to obtain streamlines.

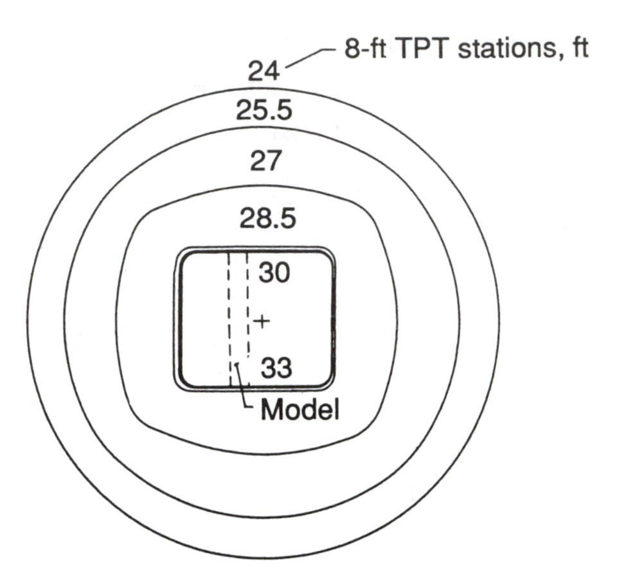


Figure 34. Cross-sectional views of contraction liner contours.

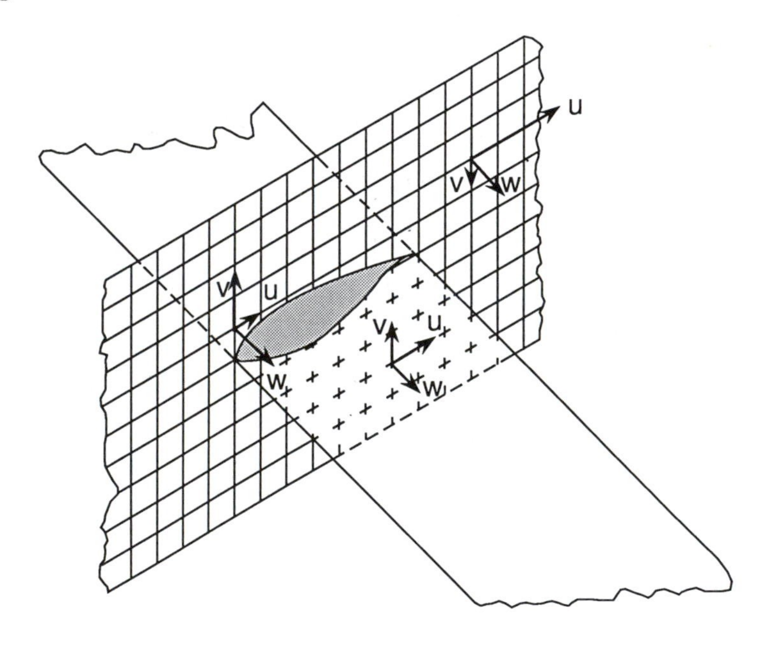


(a) Velocity field.

Figure 35. Integration of 3-D velocity field composed of 2-D airfoil plus constant sweep components.

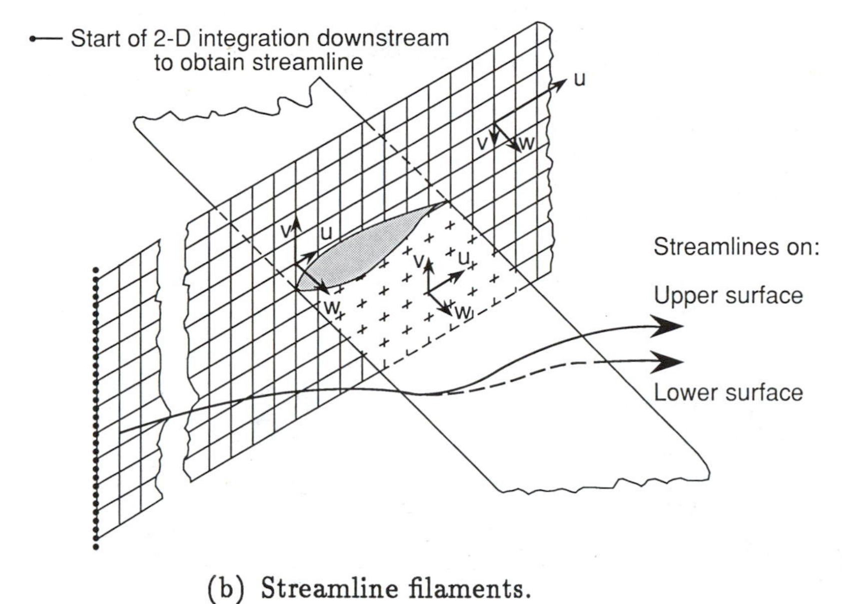


(b) Streamline filaments.

Figure 35.- Concluded

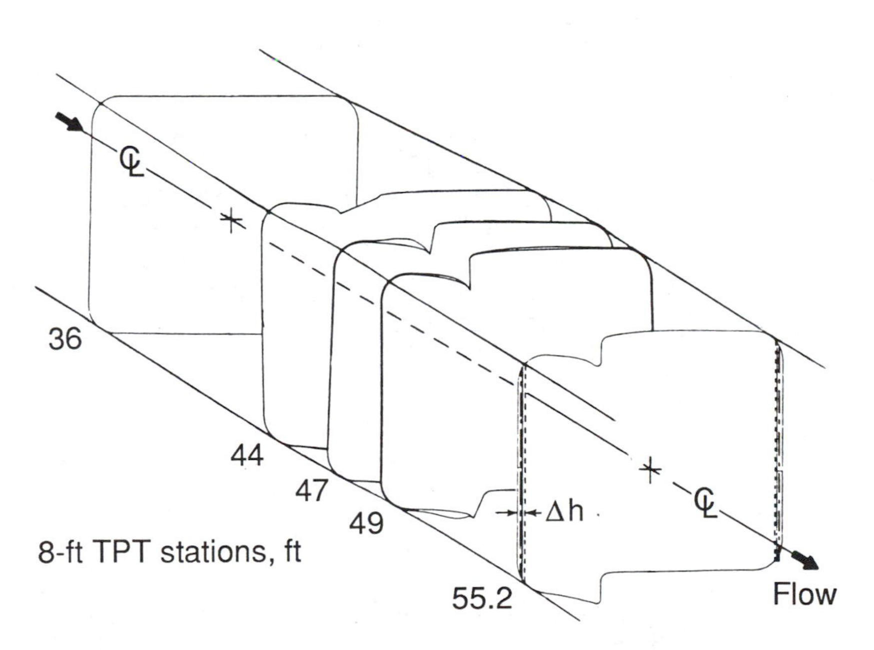


Figure 36. Streamline liner contours showing the step downstream of the trailing edge.

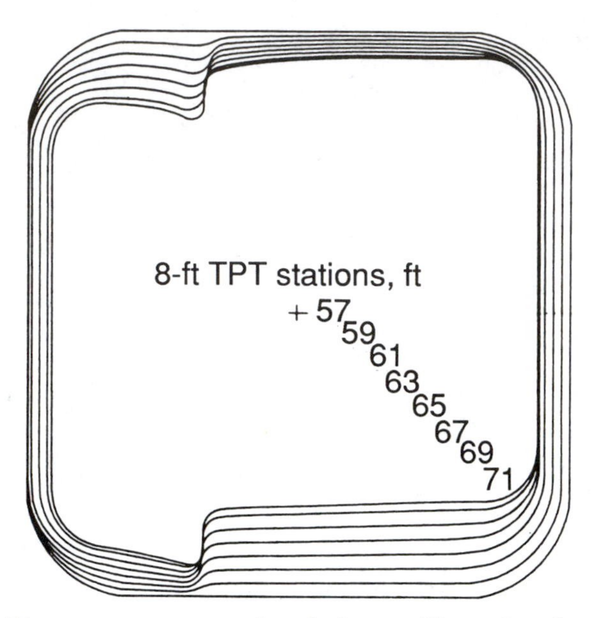


Figure 37. Upstream cross-sectional views of liner showing the "fairing-out" of the steps.

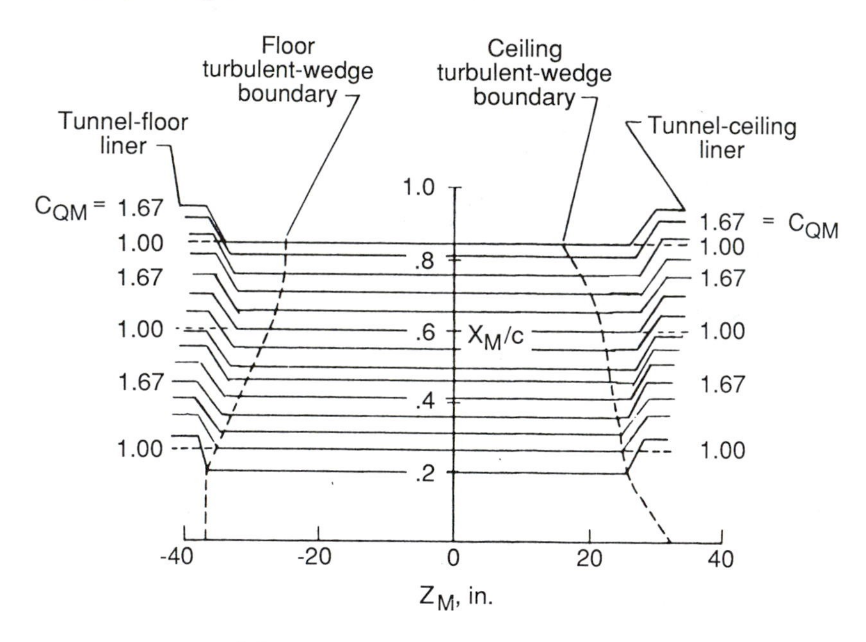


(a) Calculated at design.

Figure 38. Carpet plots of the increased suction requirements in the turbulent zones at each end of the model lower surface.

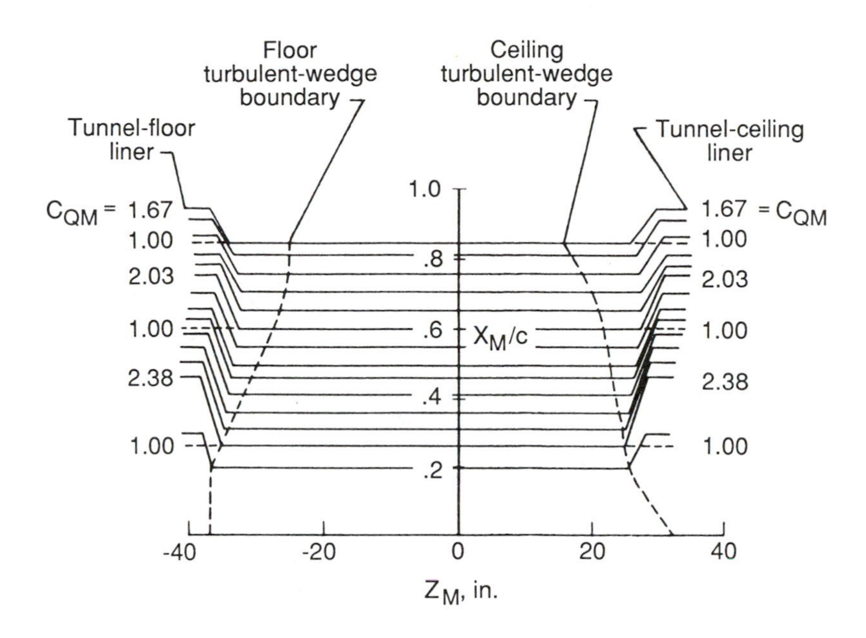


(b) Incorporated in model.

Figure 38.- Concluded

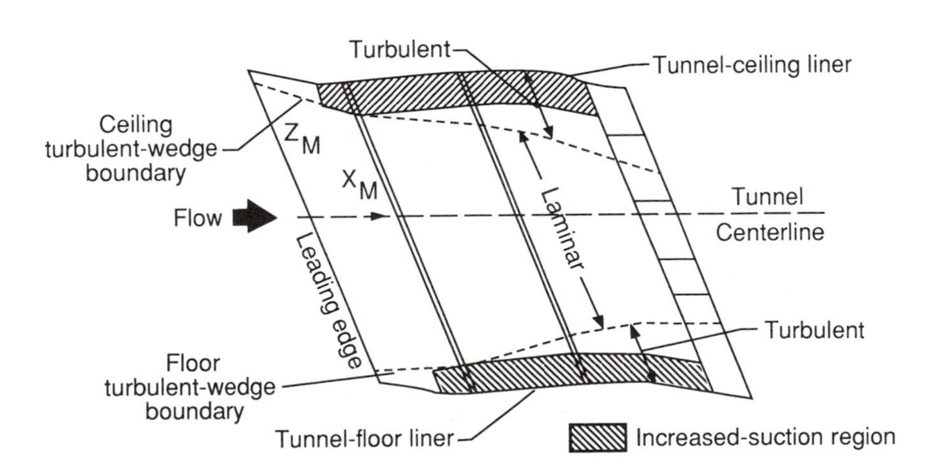


Figure 39. Planform view of model lower surface showing turbulent, laminar, and increased-suction regions.

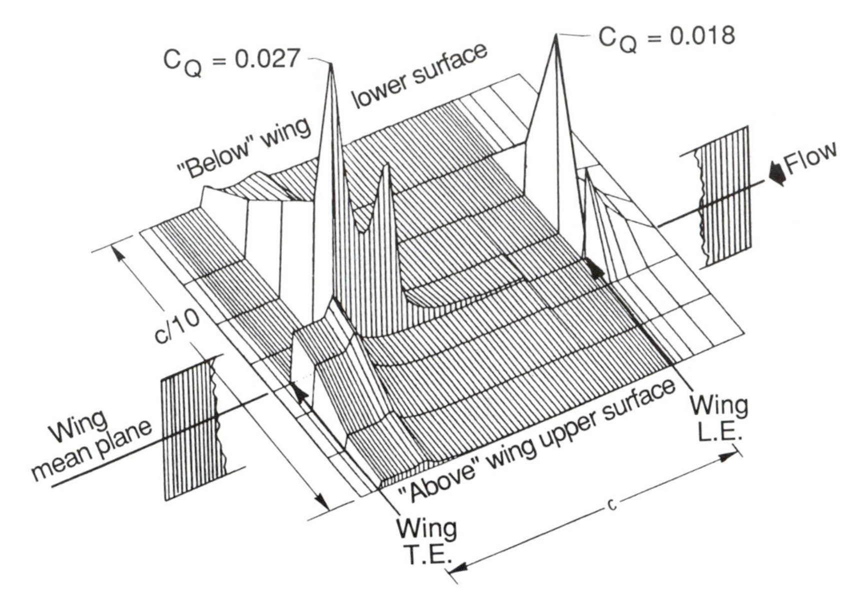


Figure 40. Calculated suction distributions provided by liner end plates at design.

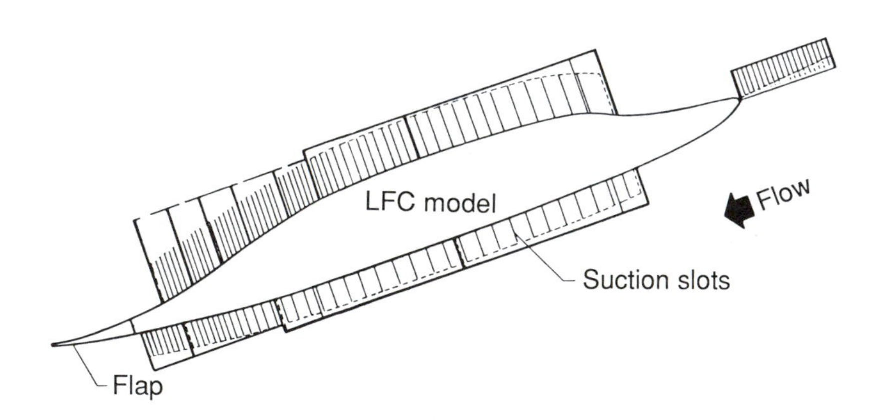


Figure 41. Suction panel-block layout on model end plate adjacent to model.

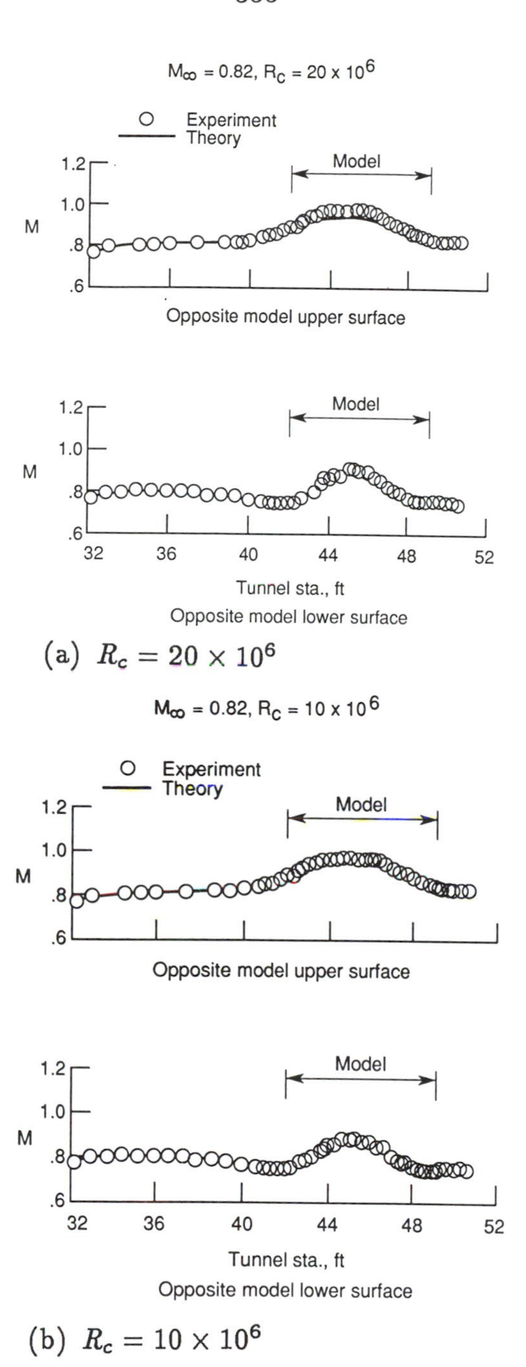


(a) $R_c = 20 \times 10^6$

(b) $R_c = 10 \times 10^6$

Figure 42. Measured and predicted Mach number distributions on 8-ft TPT liner surface centerlines.

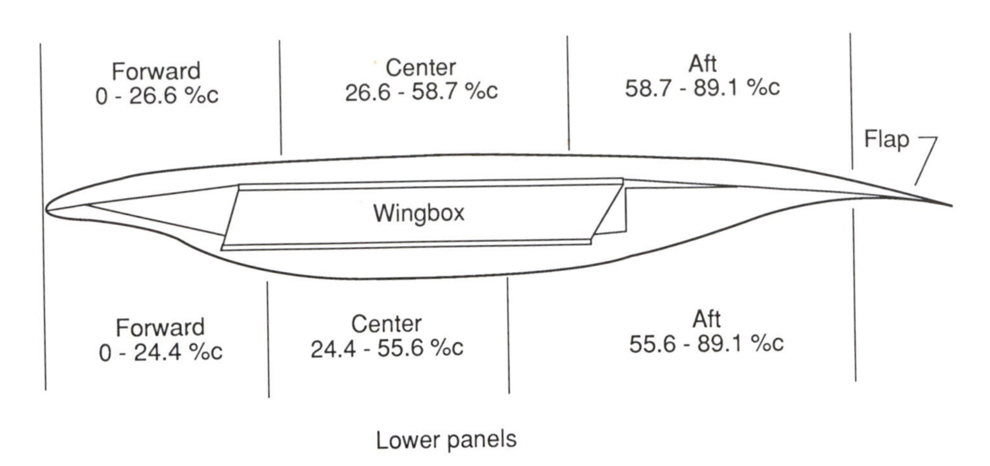


Figure 43. Cross section of slotted model showing panel arrangement.

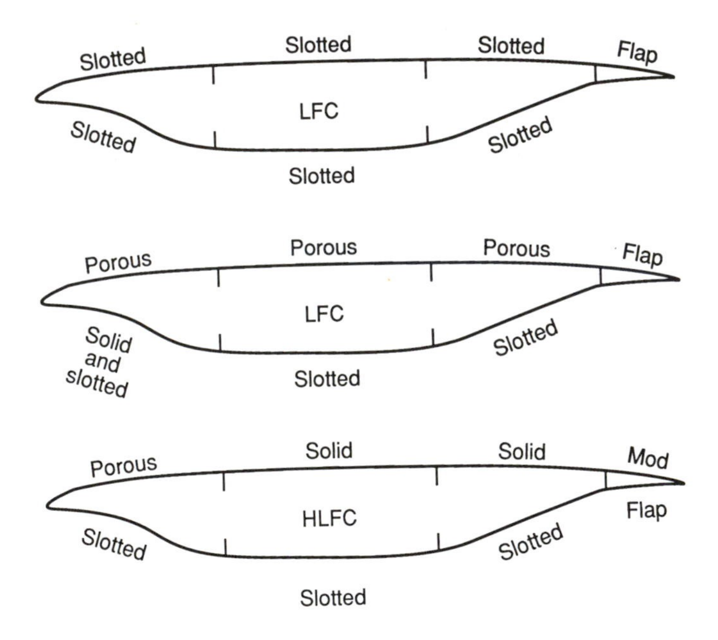


Figure 44. LFC/HLFC airfoil suction-surface configurations tested.

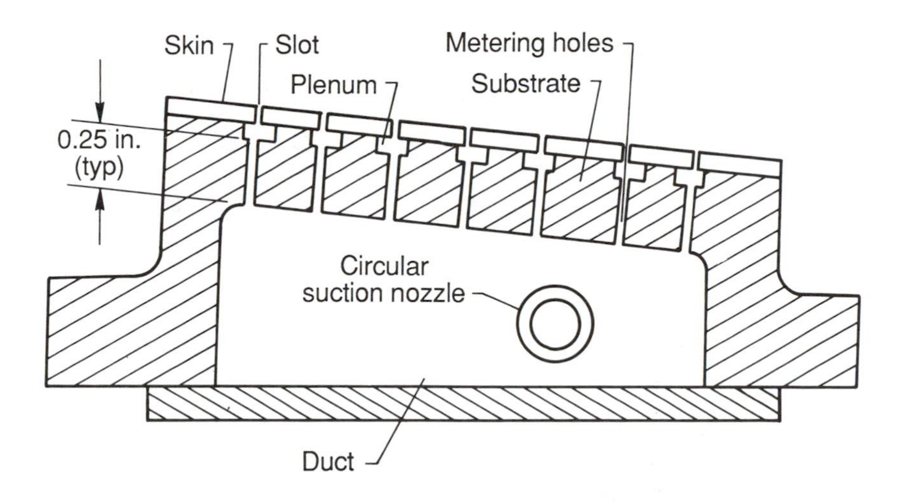


Figure 45. Cross-sectional sketch of slotted-surface suction duct.

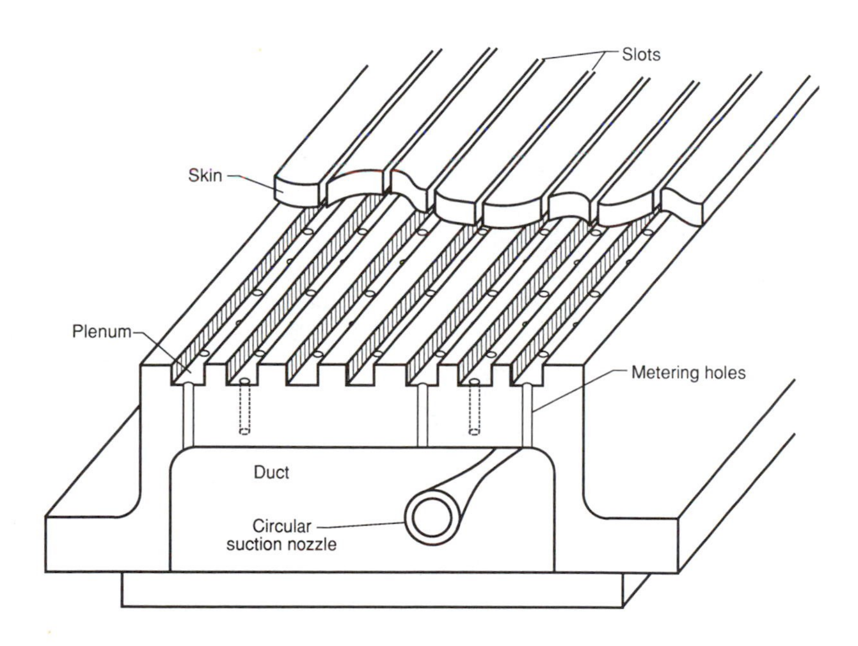


Figure 46. Isometric sketch of slotted-surface suction duct.

Figure 47. Photograph of lower aft panel mounted on numerically controlled milling machine.

Figure 48. Photograph showing plenum slots being cut on numerically controlled milling machine.

Figure 49. Photograph showing metering holes being drilled in lower forward panel.

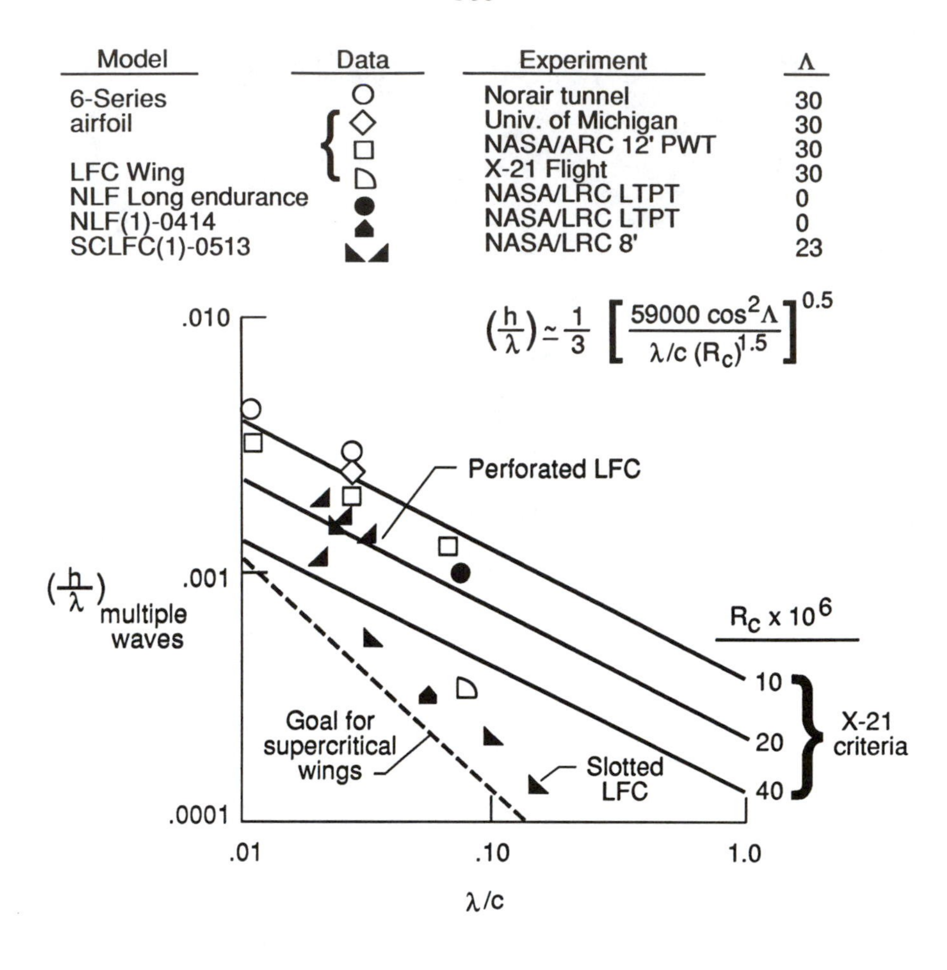


Figure 50. Max permissible amplitude ratio $(h/\lambda)$ of multiple waves for NLF/LFC wings in wind tunnels and flight.

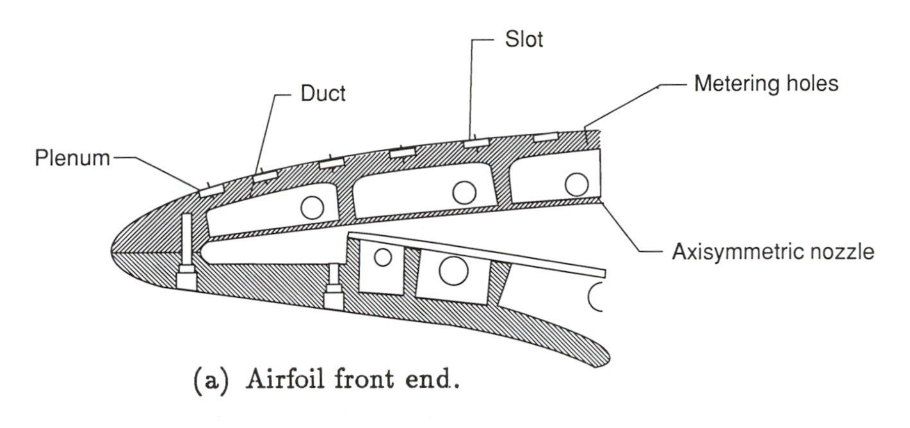


(a) Airfoil front end.

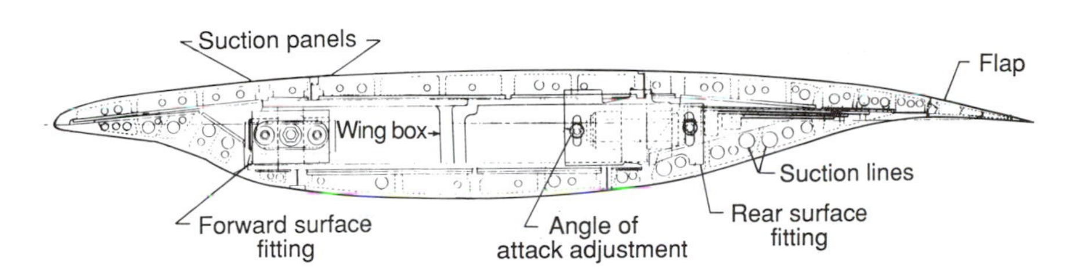


(b) Bottom end view of complete airfoil.

Figure 51. Sketches showing construction features of slotted model.

Off design
$M_{\infty,n} = 0.75$, $c_{l,n} = 0.77$
1.25 ATM/825 psf load at T.E.

T. E. Tip deflection = 0.046 in.
Spar Program Output

Loaded
No load

Figure 52. LFC airfoil stress/deflection calculations.

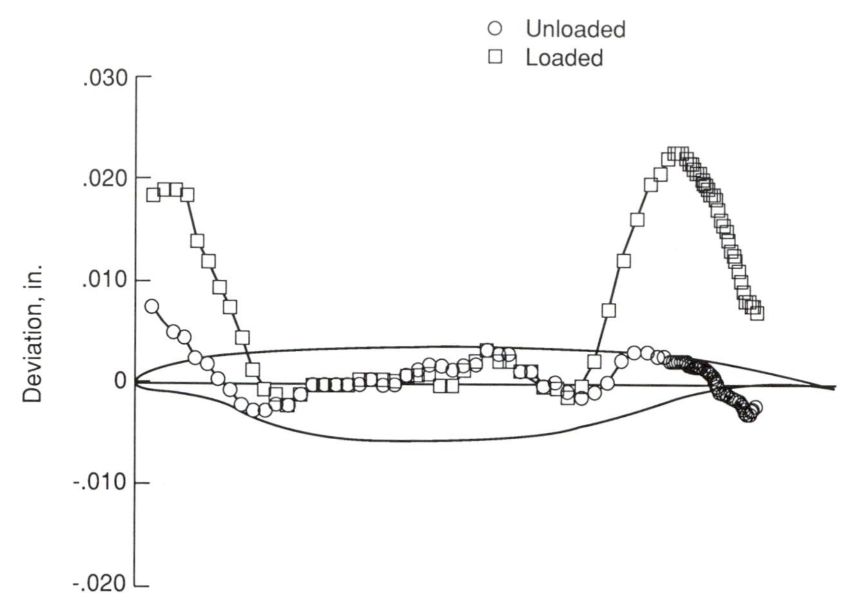


Figure 53. Effect of simulated loading on deviation of upper surface from design ordinates along model centerline without shims.

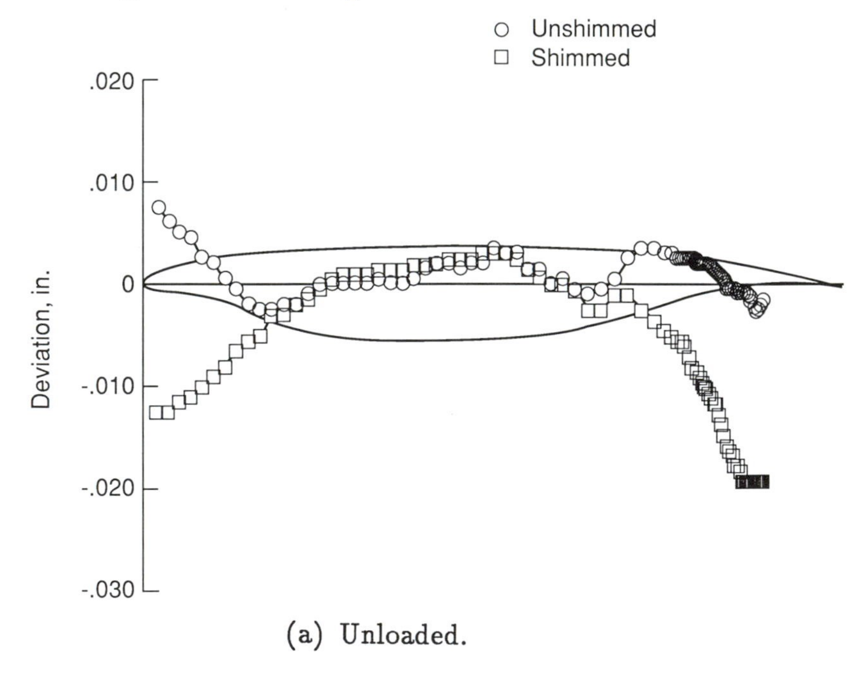


(a) Unloaded.

Figure 54. Effects of shims on deviation of upper surface from template along centerline.

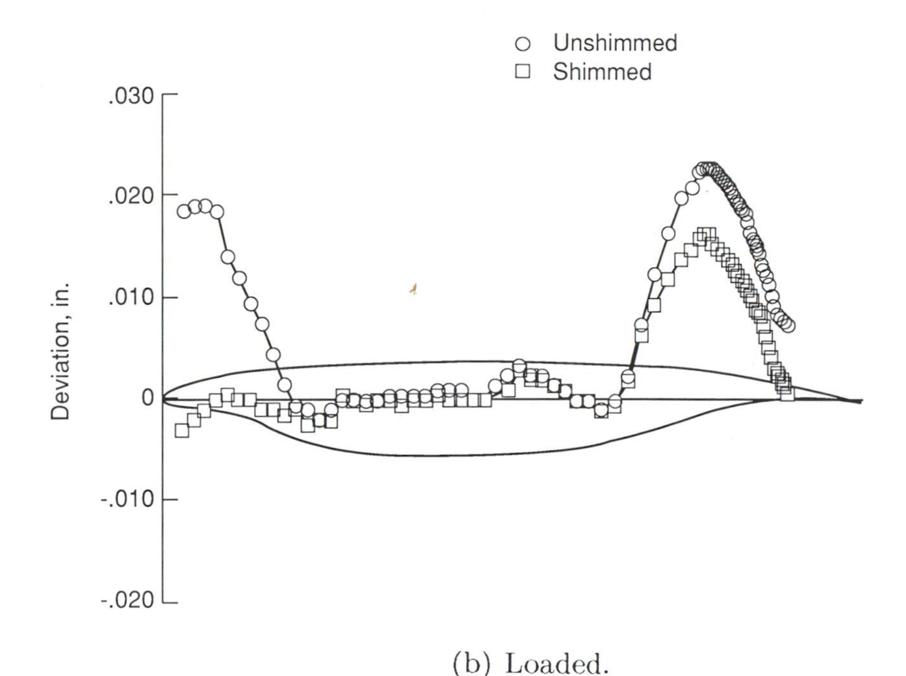


(b) Loaded.

Figure 54. - Concluded

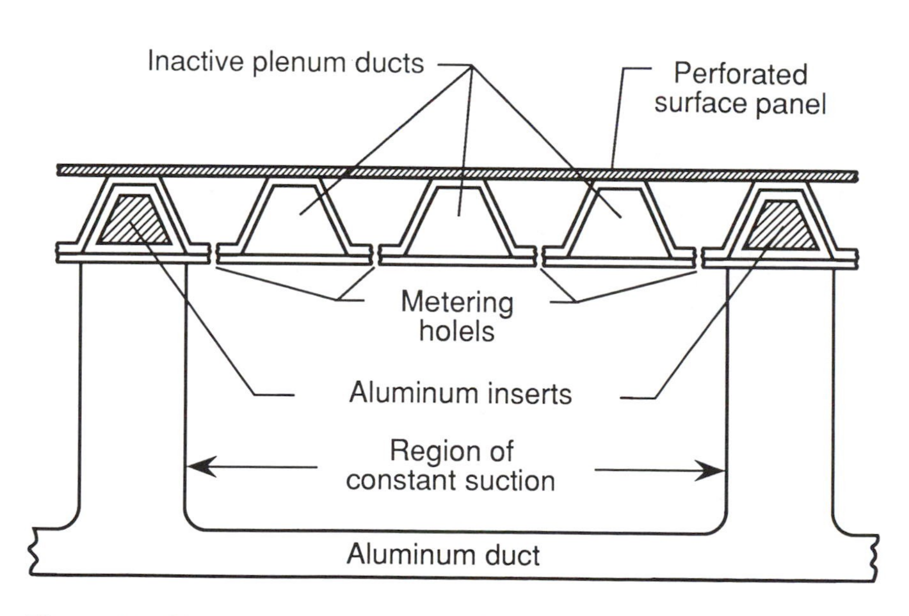


Figure 55. Sketch of cross-section of perforated surface construction features.

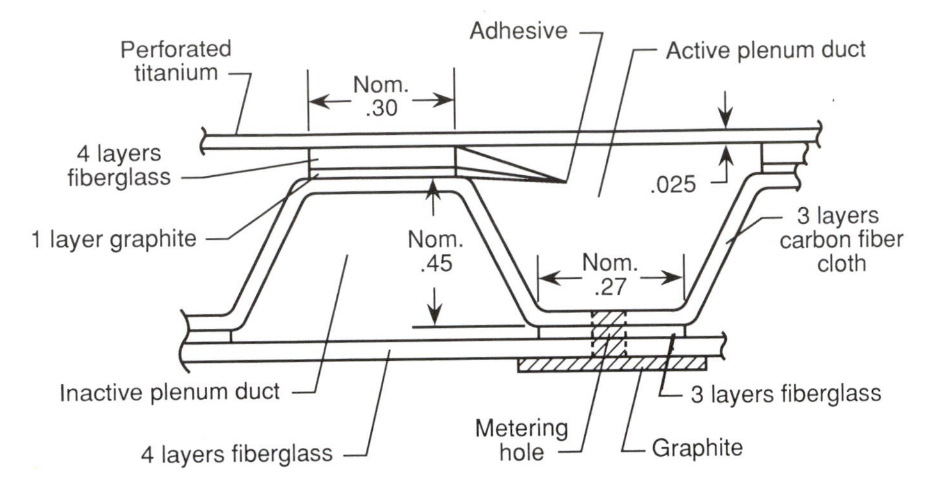


Figure 56. Detailed sketch of perforated suction surface showing dimensions and materials.

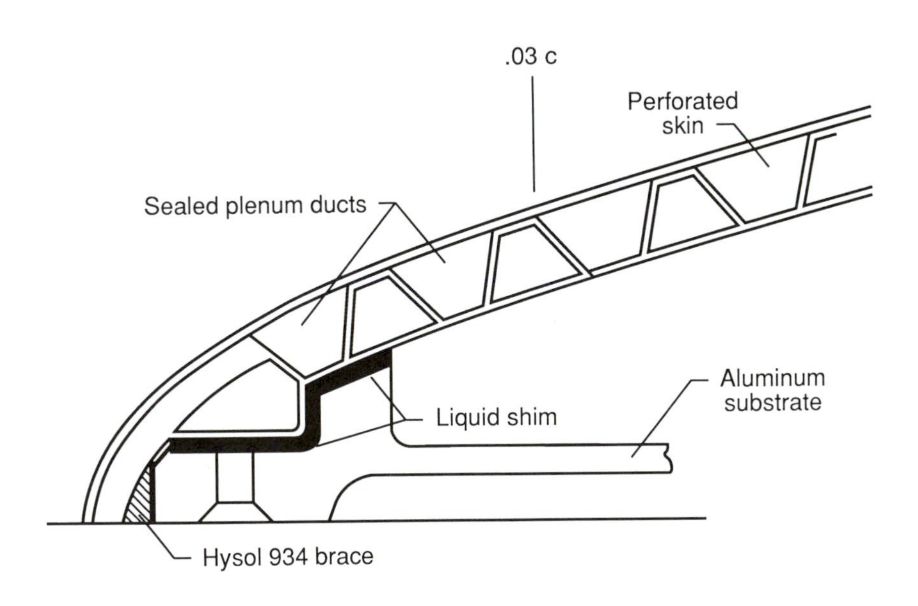


Figure 57. Details of construction features of nose region of perforated model.

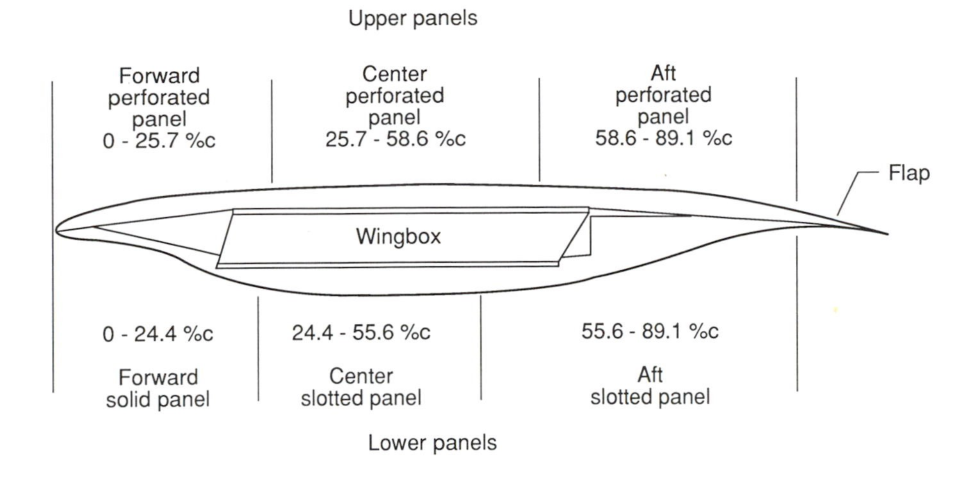


Figure 58. Cross-section of perforated model showing panel arrangements.

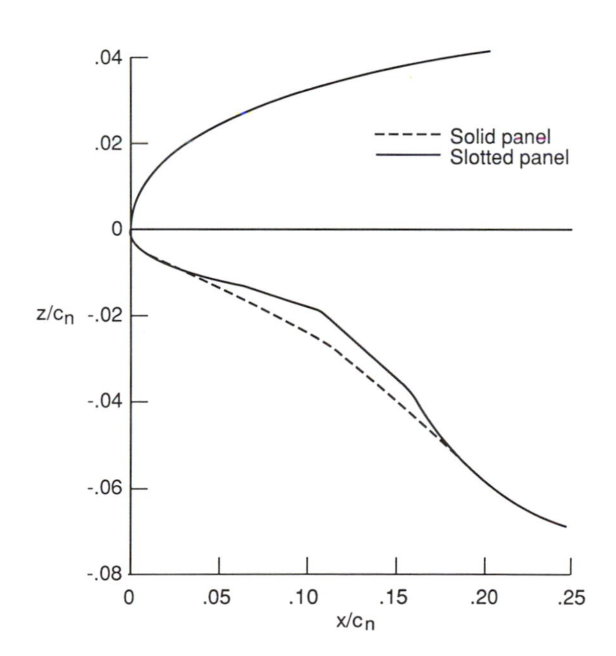


Figure 59. Comparisons of the geometries in the lower forward region of the slotted and solid panels.

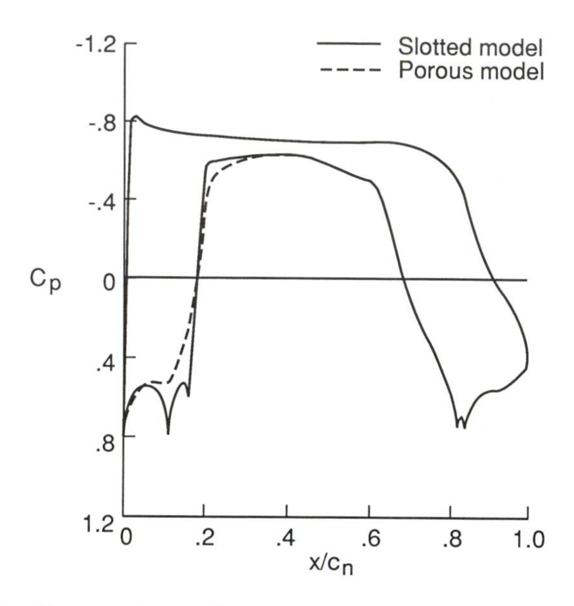


Figure 60. Comparison of design pressure distributions for slotted and porous models, $M_{\infty,n} = 0.755$; $c_{\ell,n} = 0.55$.

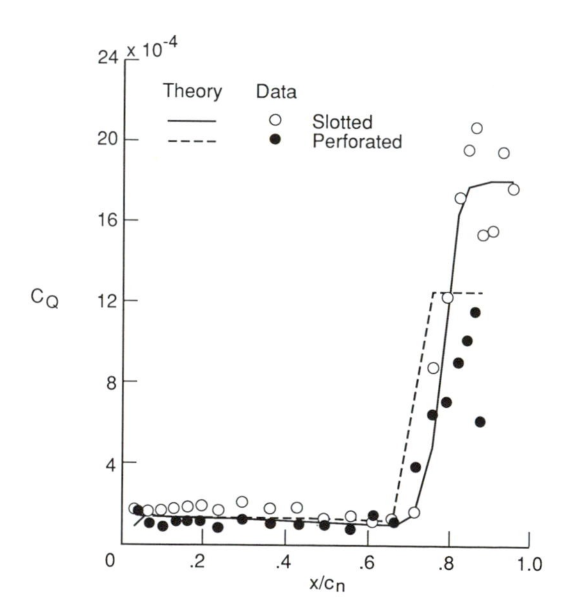


Figure 61. Comparison of maximum suction distributions on upper surface of perforated and slotted models, $R_c = 10 \times 10^6$.

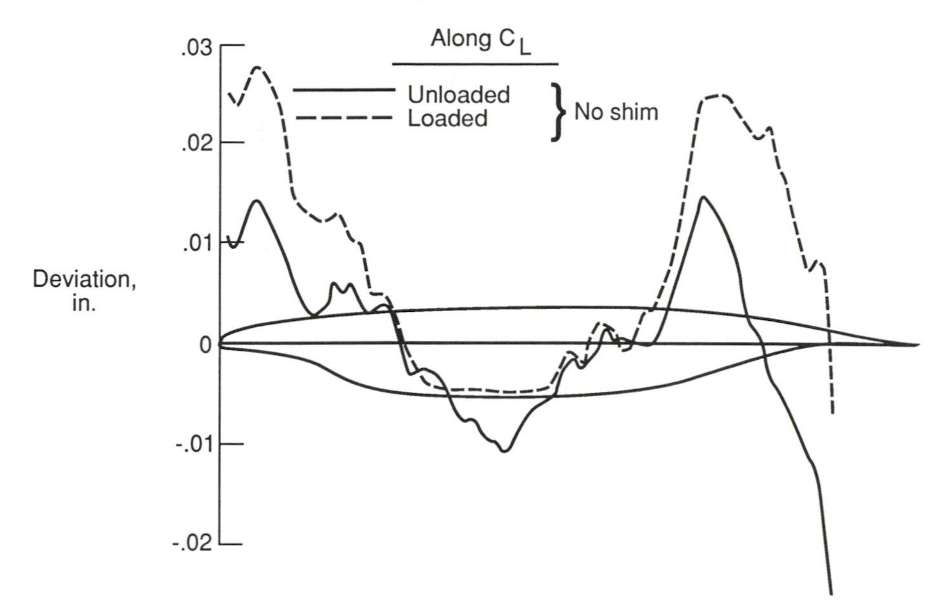


Figure 62. Deviation of upper surface of perforated model from the design ordinates in both the loaded and unloaded condition.

Figure 63. Photograph of liner-block contour being machined.

Figure 64. Photograph of liner substructure being installed in the test section of the 8-ft TPT.

Figure 65. Photograph of contraction liner substructure.

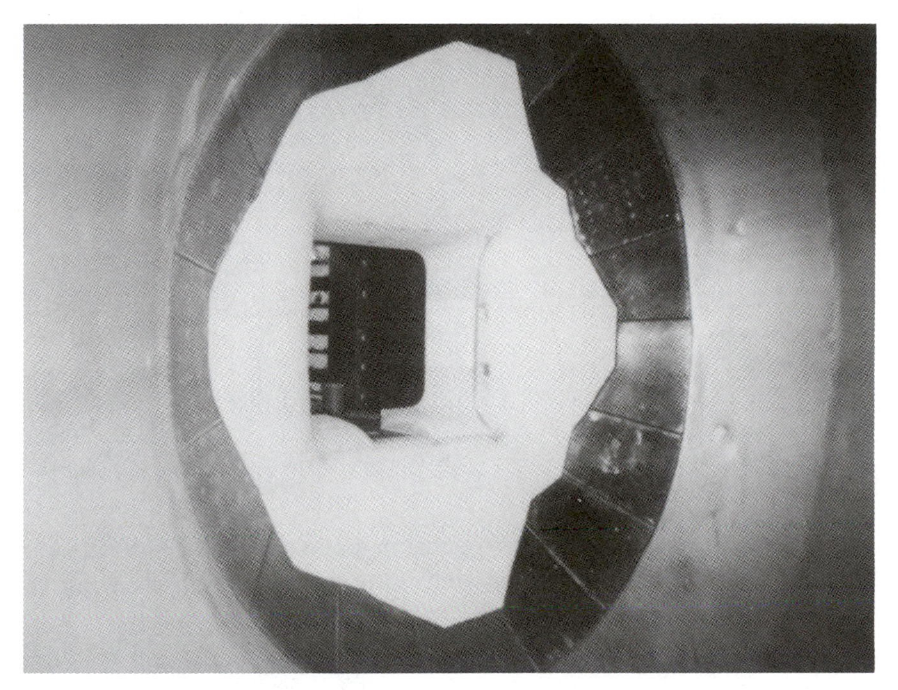

Figure 66. Photograph of contraction liner after installation. Note that one liner block is missing at the entrance to the test section.

Figure 67. Photograph looking downstream of LFC slotted model and liner after installation.

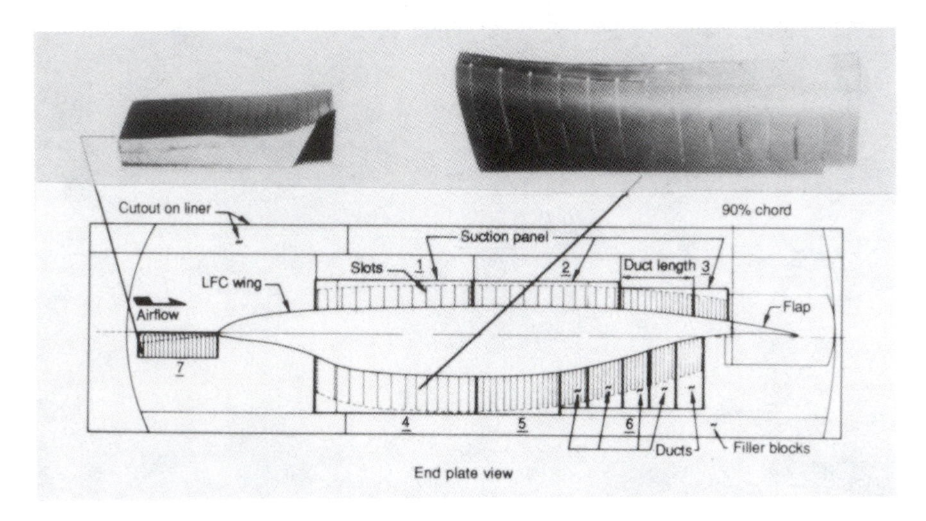


Figure 68. Photograph of two suction panels used in juncture region adjacent to ends of the model along with a sketch showing the location of seven panels.

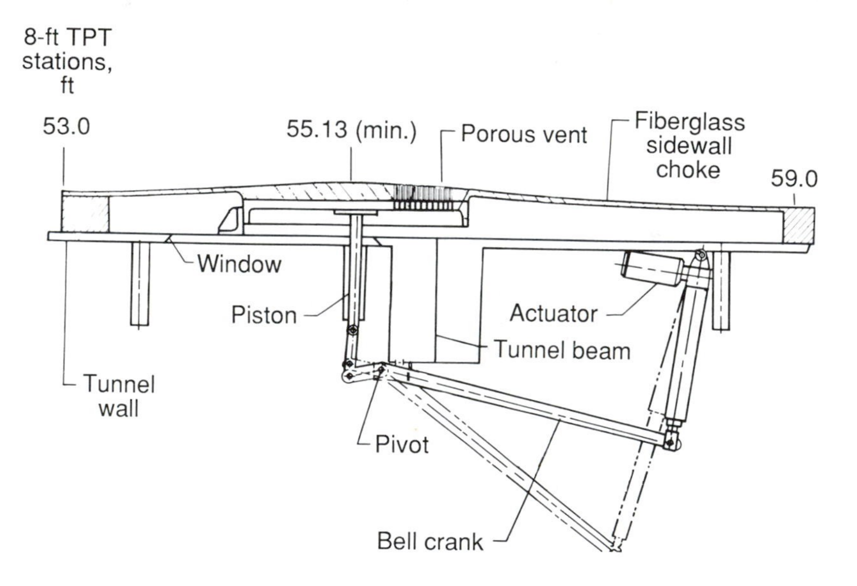


Figure 69. Sketch of adjustable choke plate.

Figure 70. Photograph of one of the choke plates looking upstream toward the model.

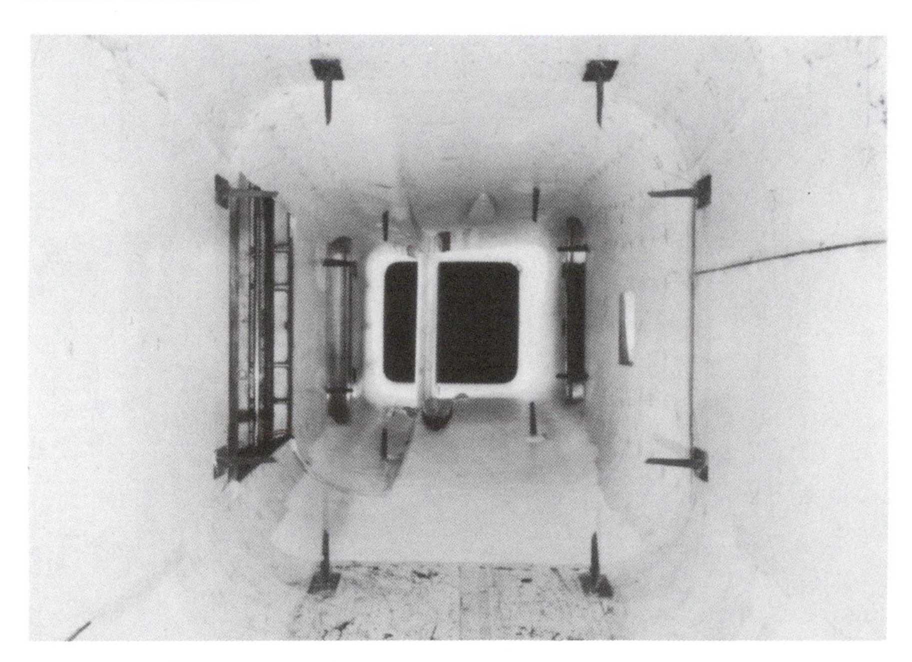

Figure 71. Photograph of vortex generators looking upstream toward the model.

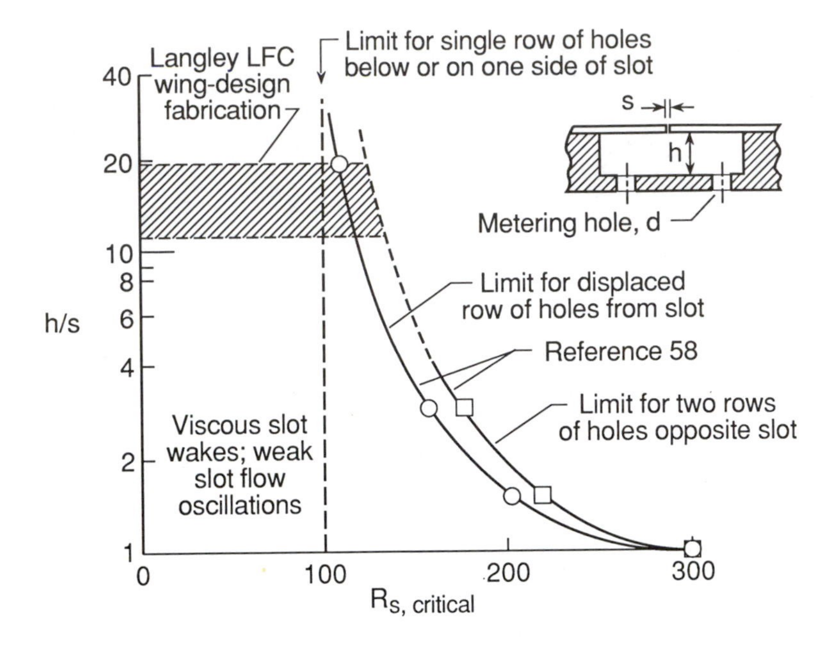


Figure 72. Slotted suction surface metering hole design criteria.

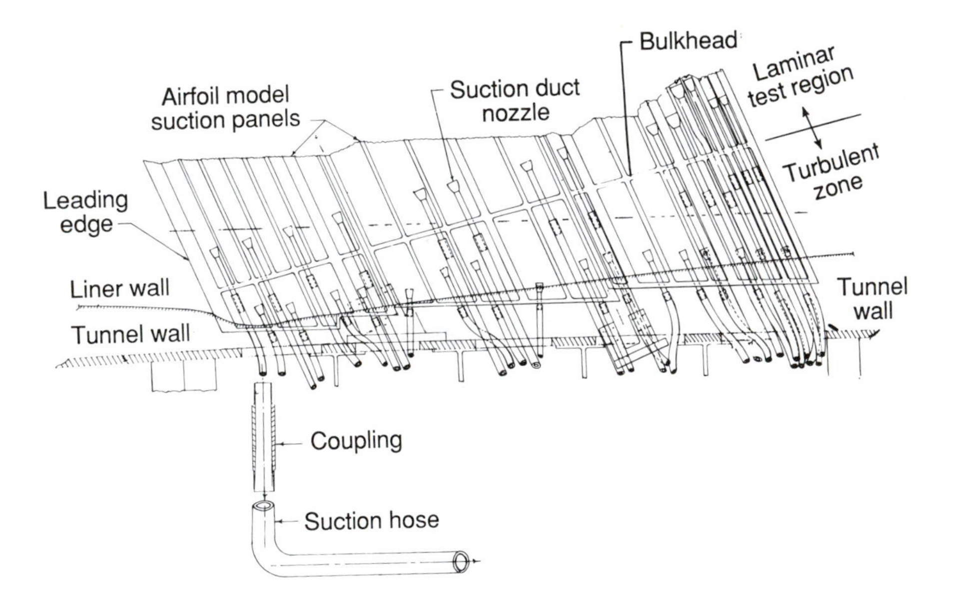


Figure 73. Sketch of end of model nearest the test suction floor showing arrangements of section nozzles.

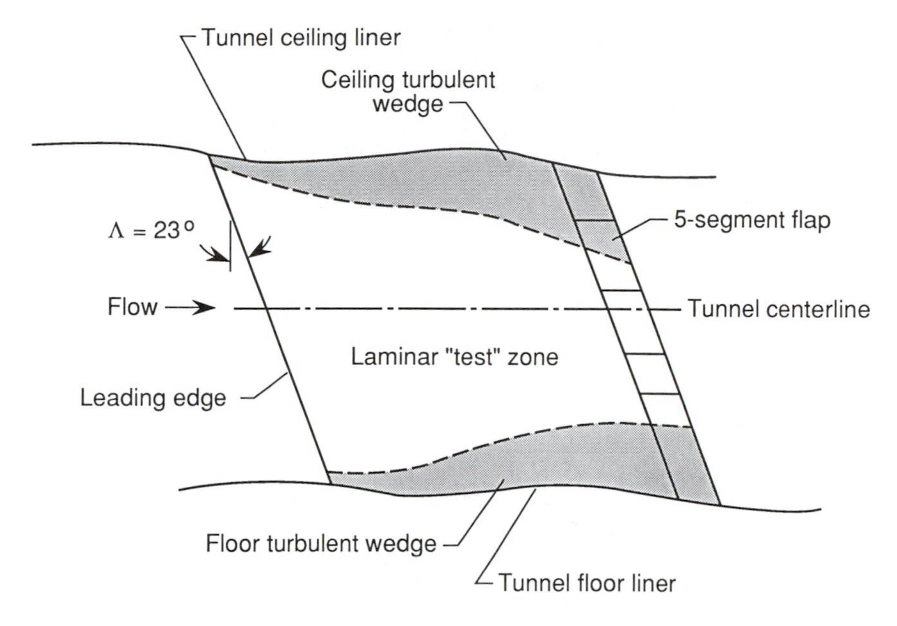


Figure 74. Planform view of model lower surface showing turbulent wedges at each end of model and laminar "test" zone in the middle.

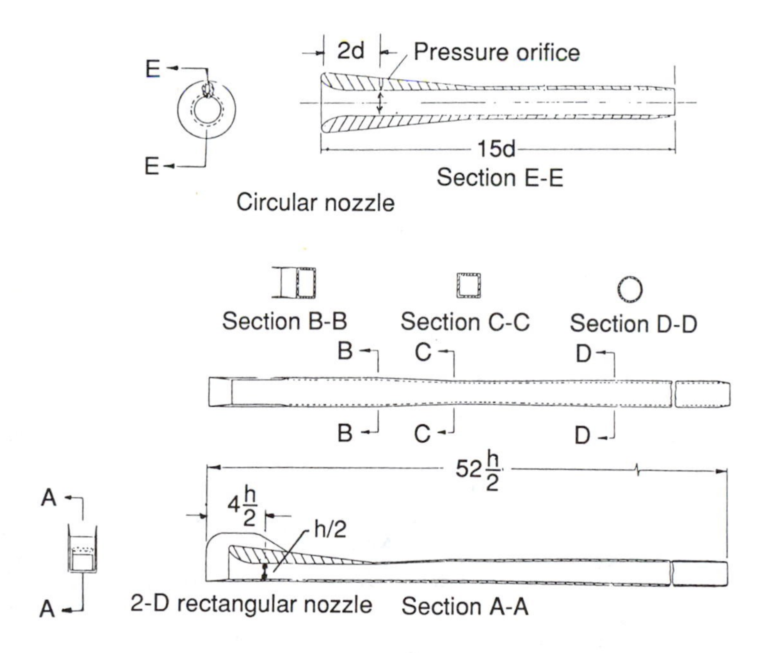


Figure 75. Details of circular and rectangular suction nozzles.

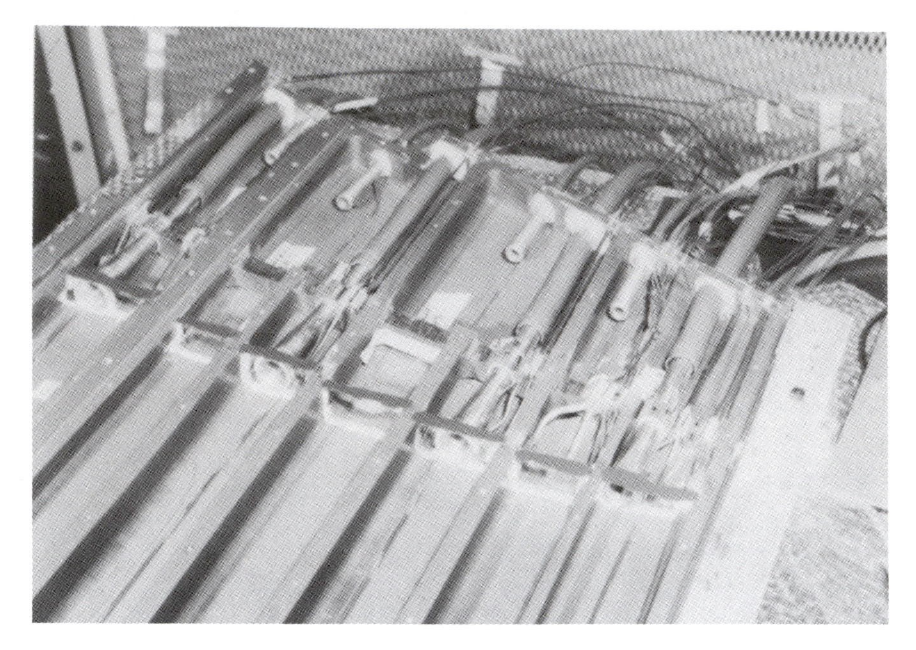

Figure 76. Photograph of underside of suction panel showing suction nozzles, hoses, and bulkheads.

Figure 77. Photograph of external connector hoses between end of model nearest test section floor and airflow control boxes.

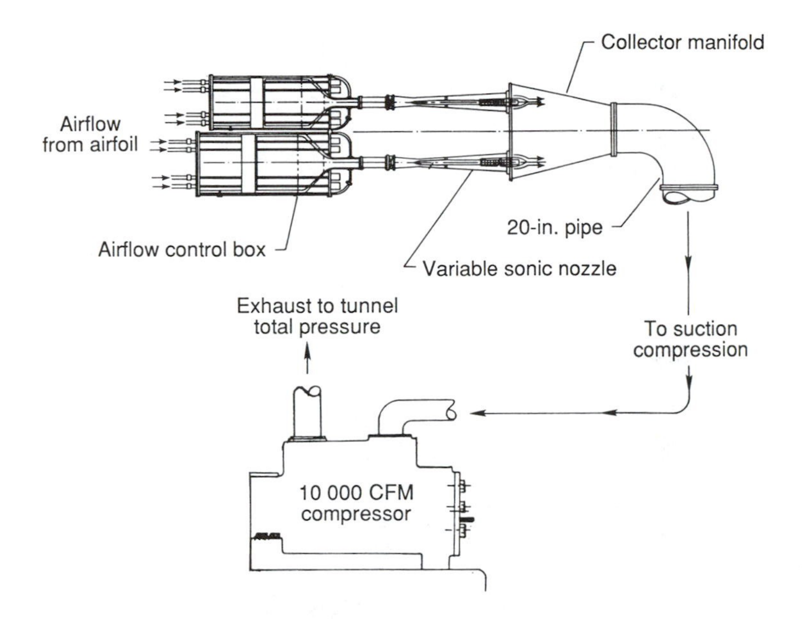


Figure 78. Schematic of suction airflow system.

Figure 79. Photograph of an array of five control boxes.

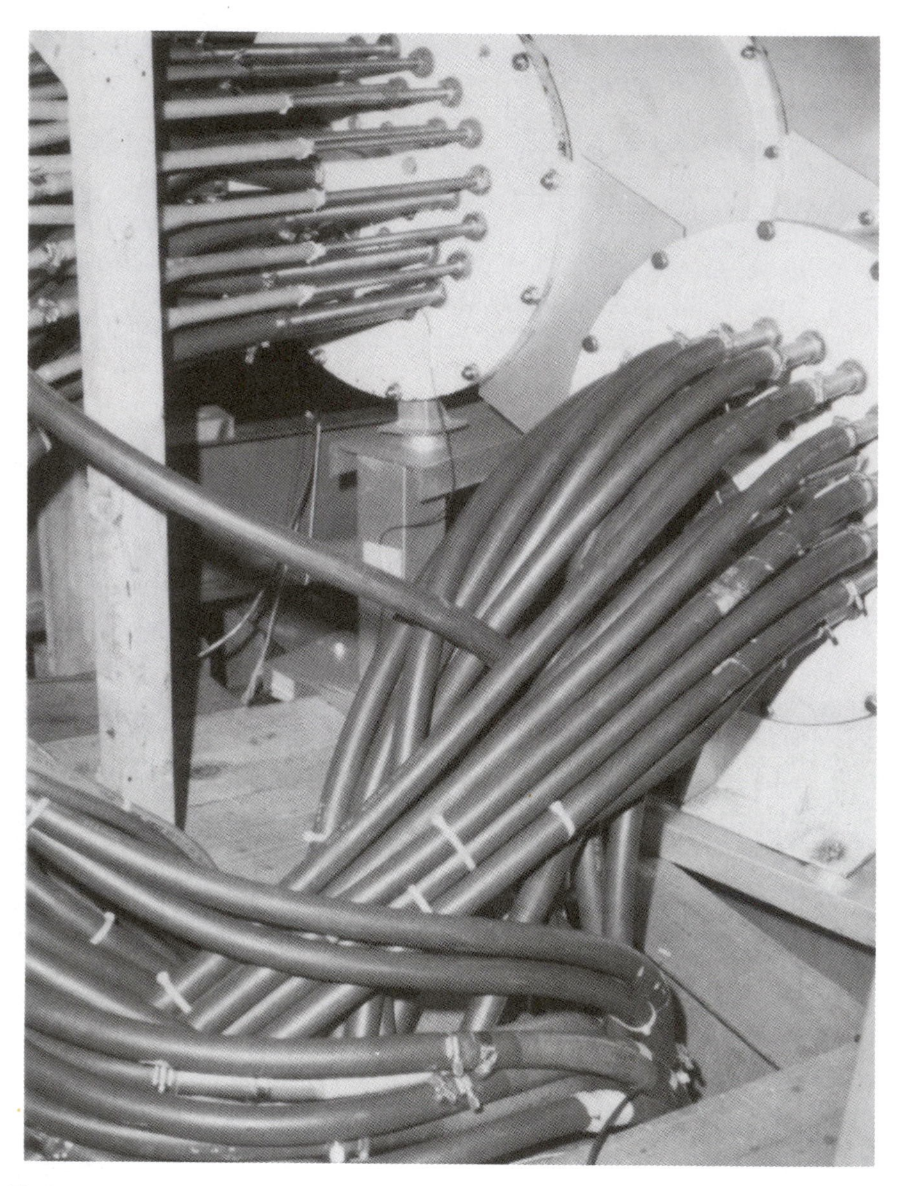

Figure 80. Close up photograph of several control boxes with hoses attached.

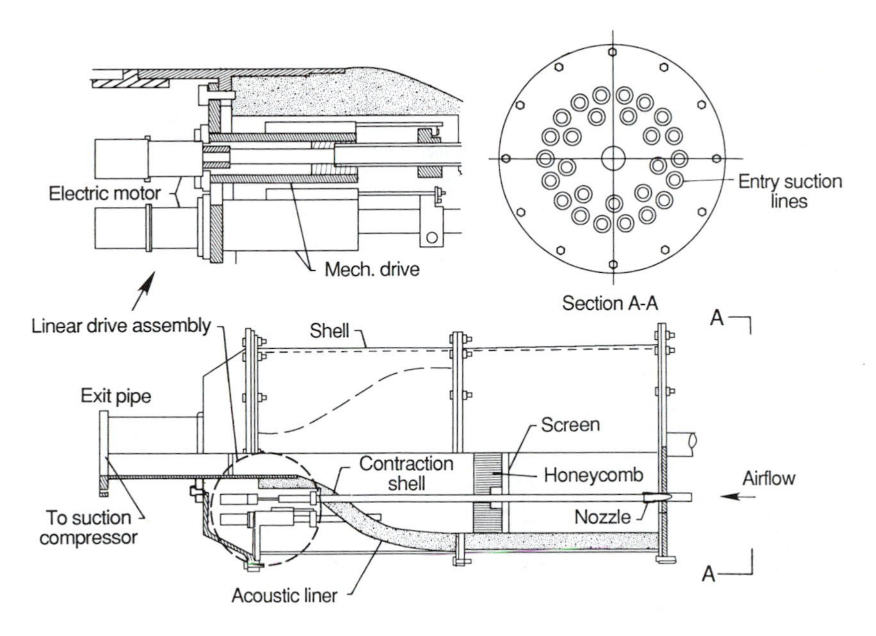


Figure 81. Details of airflow control box.

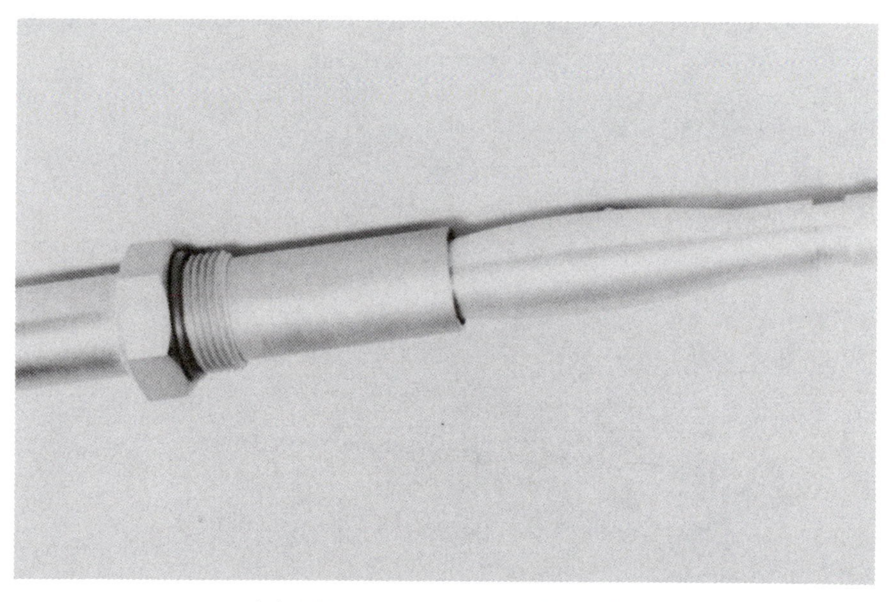

(a) Photograph of needle valve.

Figure 82. Photographs showing details of airflow control box needle valve and "needles" installed in an airflow control box.

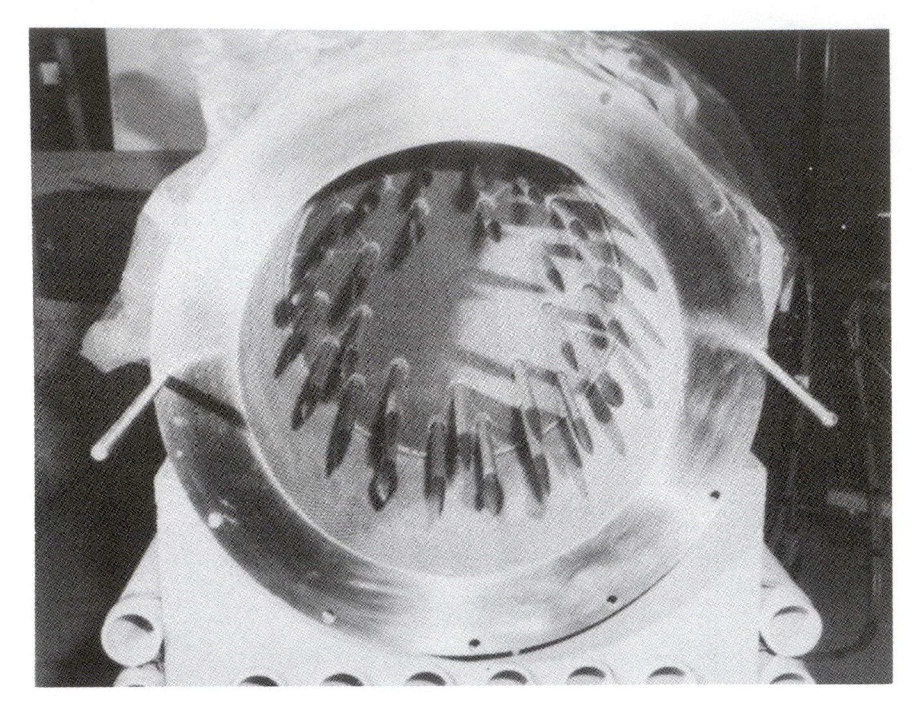

(b) Photograph of airflow control box with end removed to show "needles" of the needle valves.

Figure 82.- Concluded

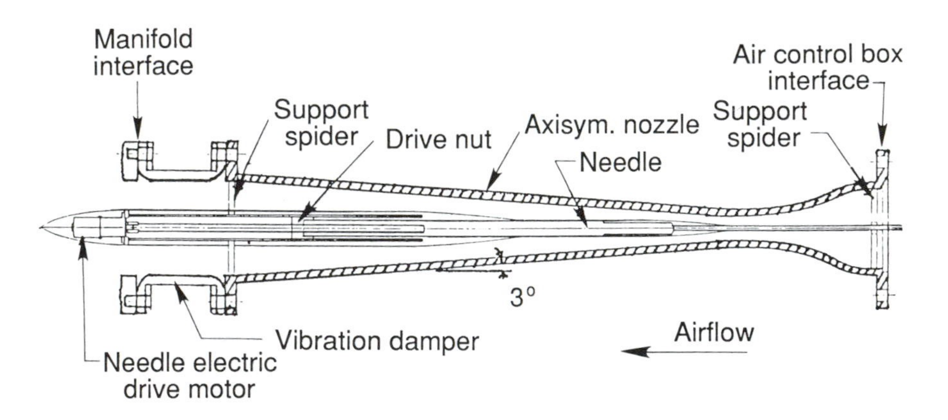


Figure 83. Sketch of variable sonic nozzle.

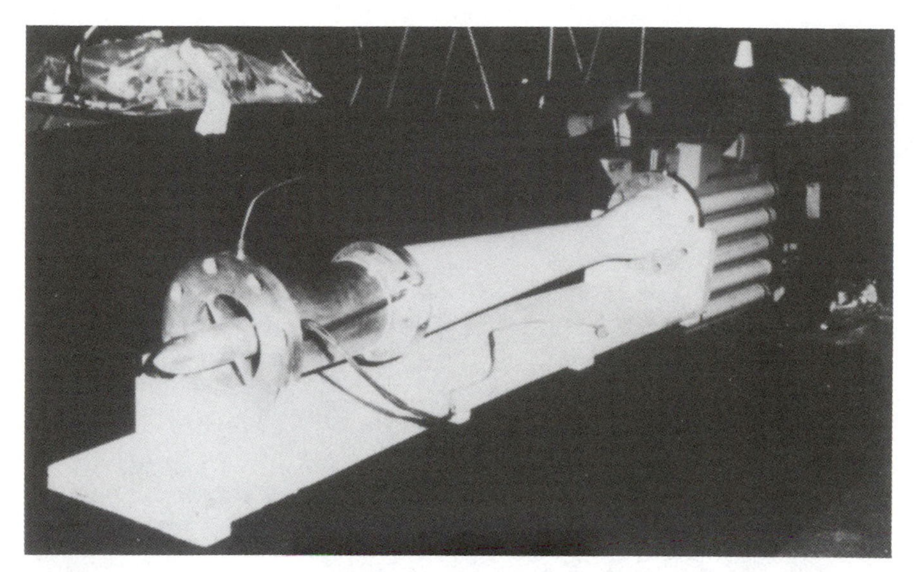

Figure 84. Photograph of single sonic nozzle before installation.

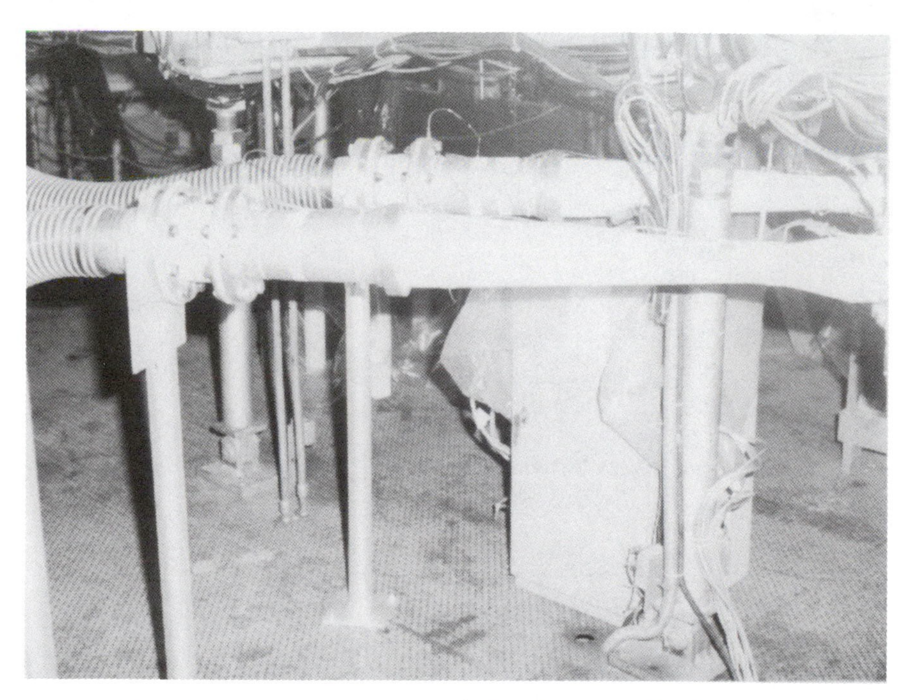

Figure 85. Photograph of sonic nozzle after installation.

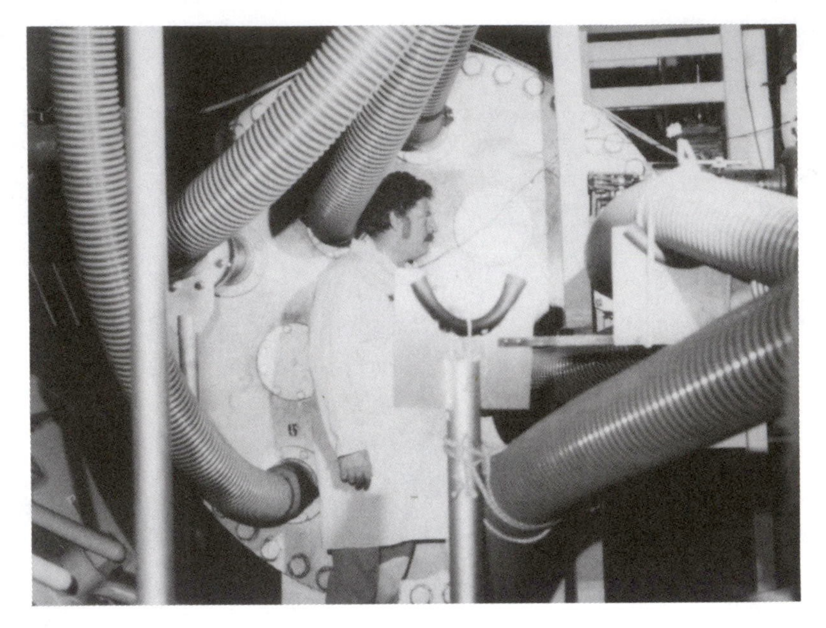

Figure 86. Photograph of large flexible hoses connecting sonic nozzles to common manifold.

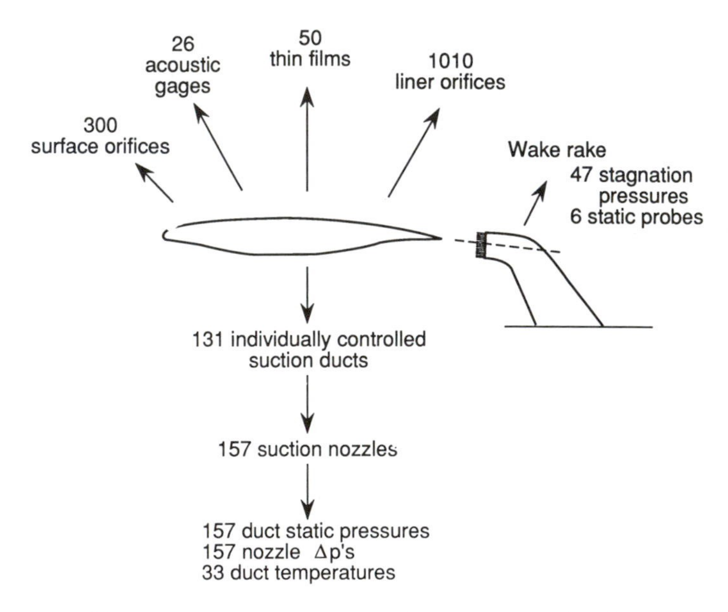


Figure 87. Sketch showing types and quantity of primary instrumentation.

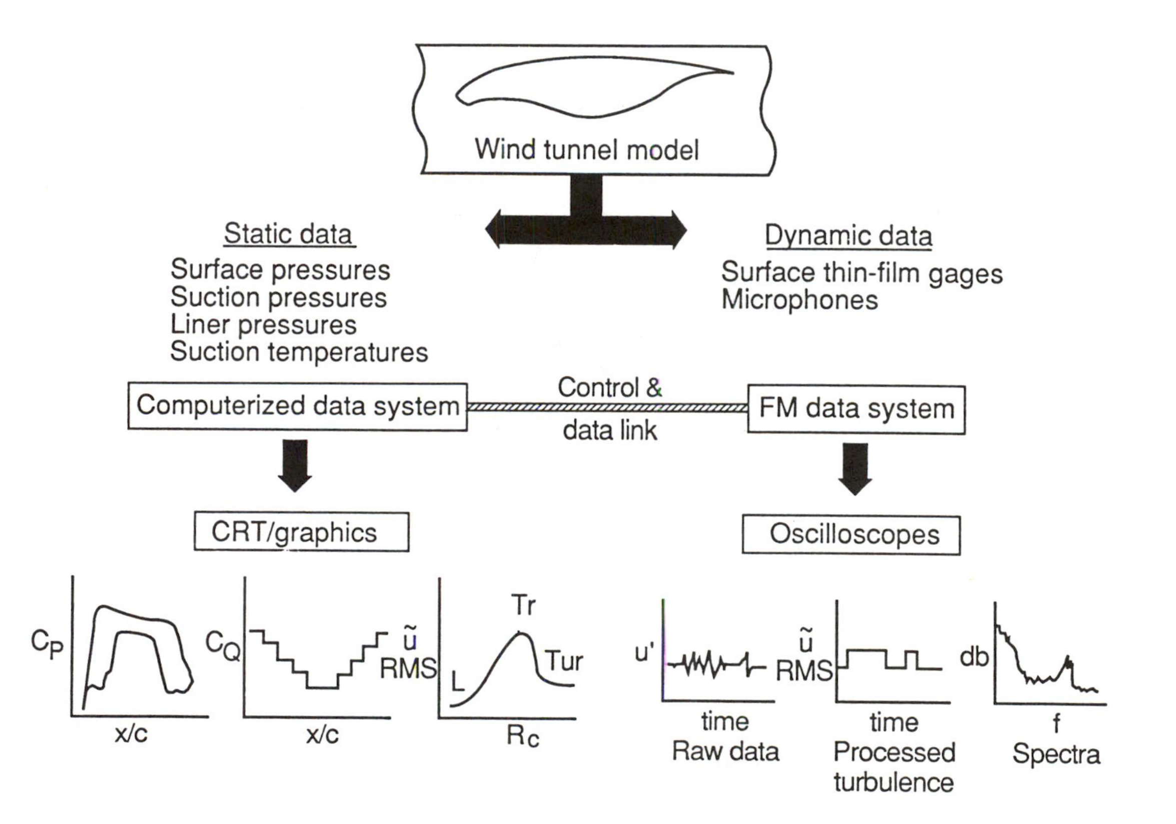


Figure 88. LFC airfoil and liner real-time data system.

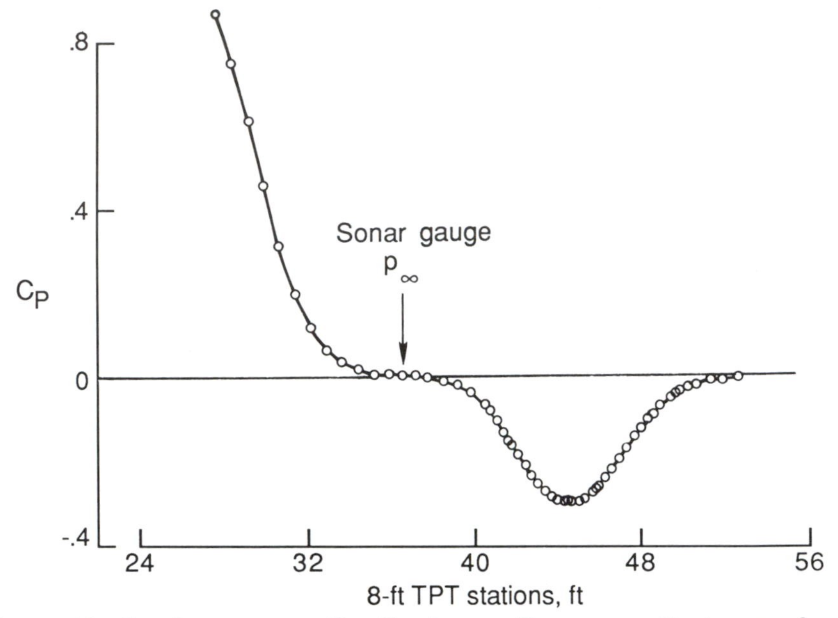


Figure 89. Static pressure distribution on liner opposite top surface of model indicating location of $p_\infty$ measurement.

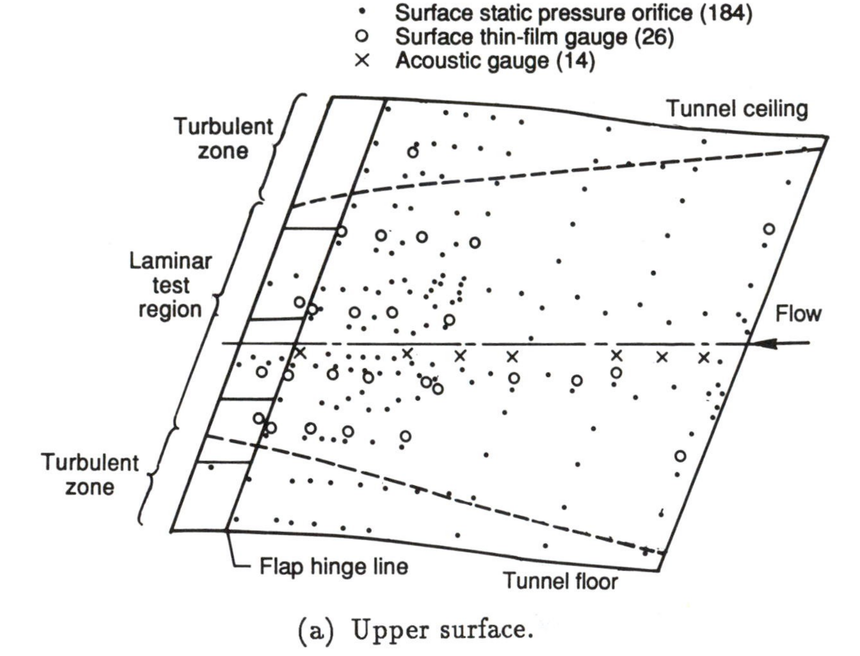


(a) Upper surface.

Figure 90. Upper and lower wing surfaces of model showing location of static pressure orifices and surface thin-film and acoustic gages.

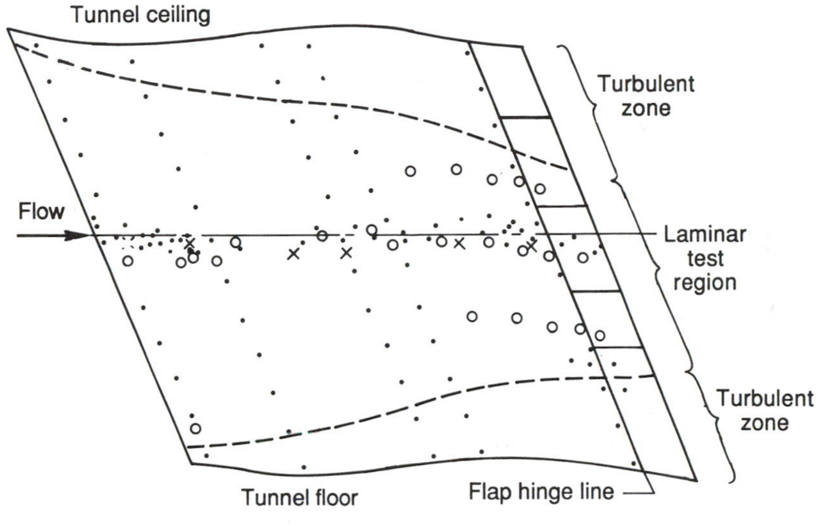


(b) Lower surface.

Figure 90.- Concluded

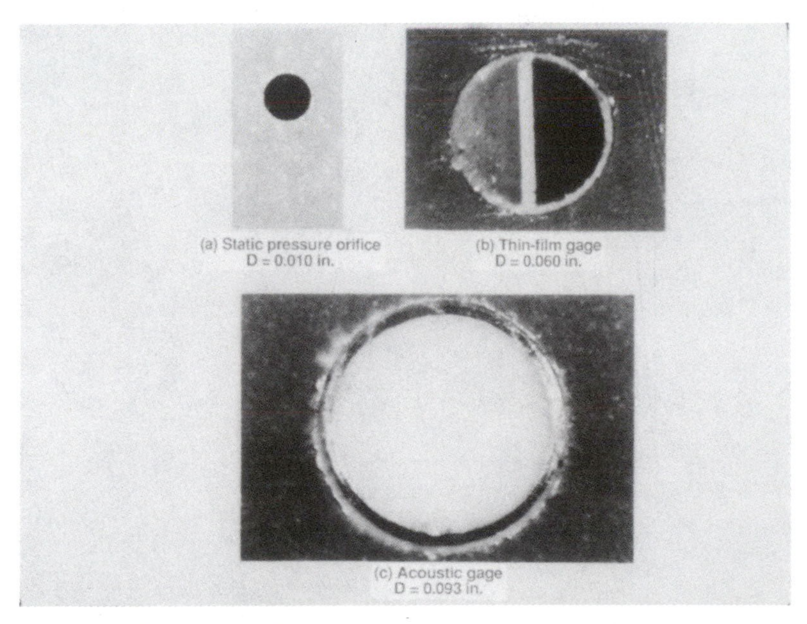


Figure 91. Enlarged photographs of a static pressure orifice and surface thin film and acoustic gages.

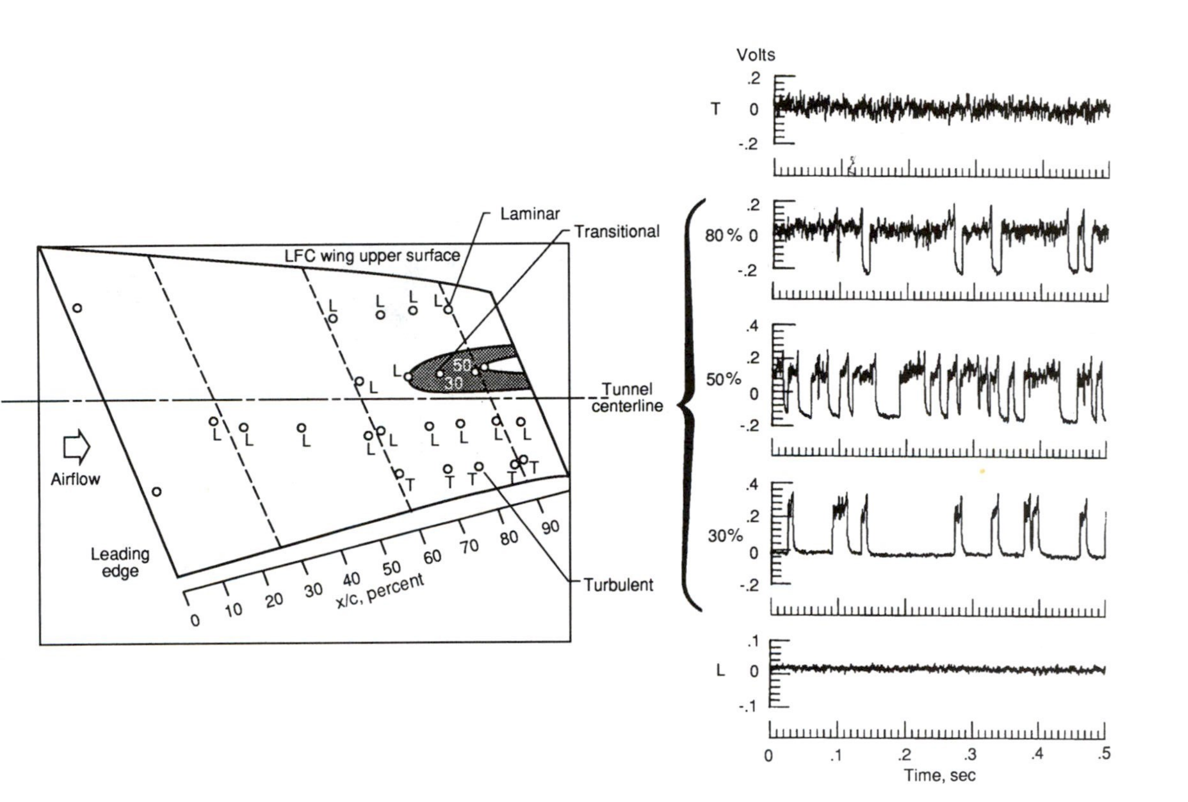


Figure 92. Examples of thin film signals and associated transition pattern.

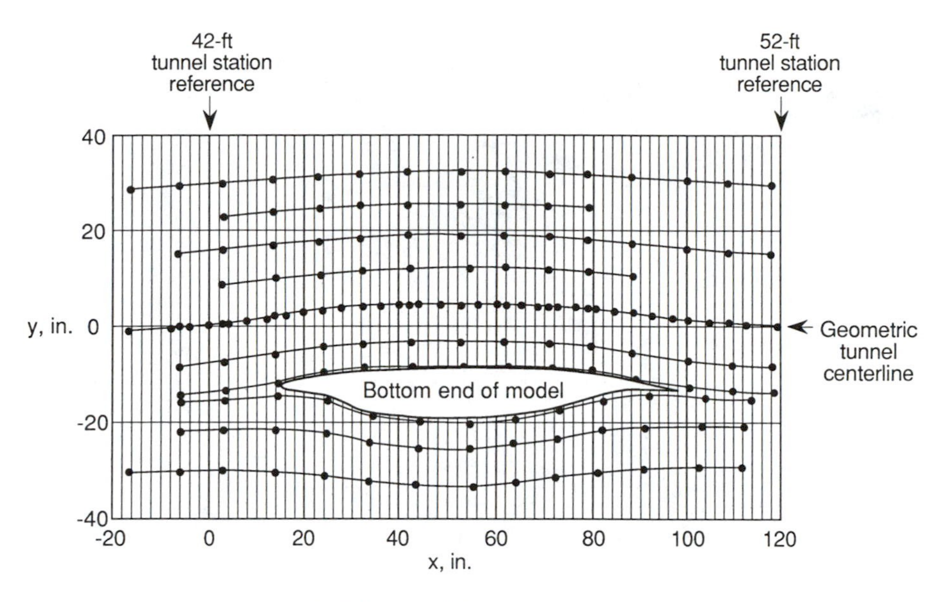


(a) Floor orifices.

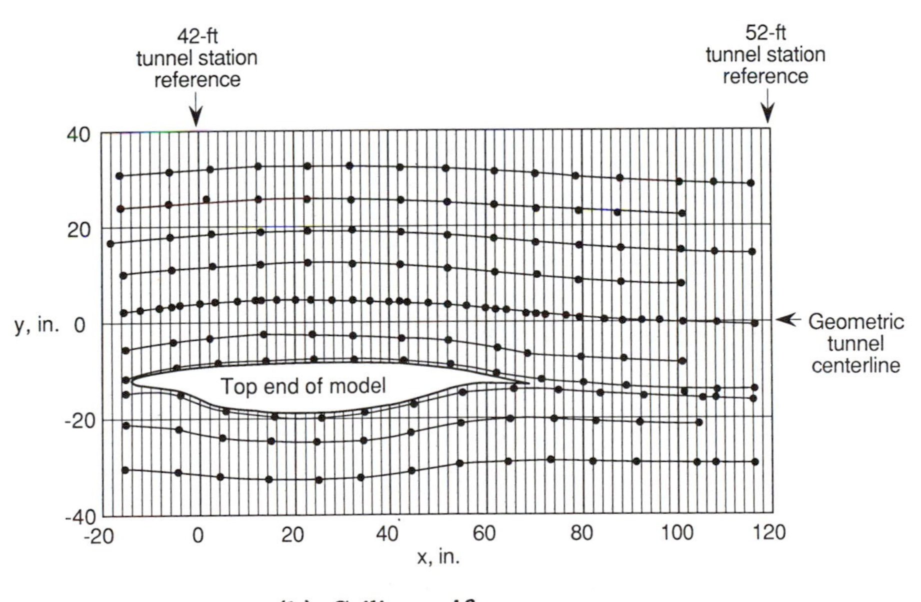


(b) Ceiling orifices.

Figure 93. Plots showing location of static pressure orifices on the floor and ceiling near the ends of the model.

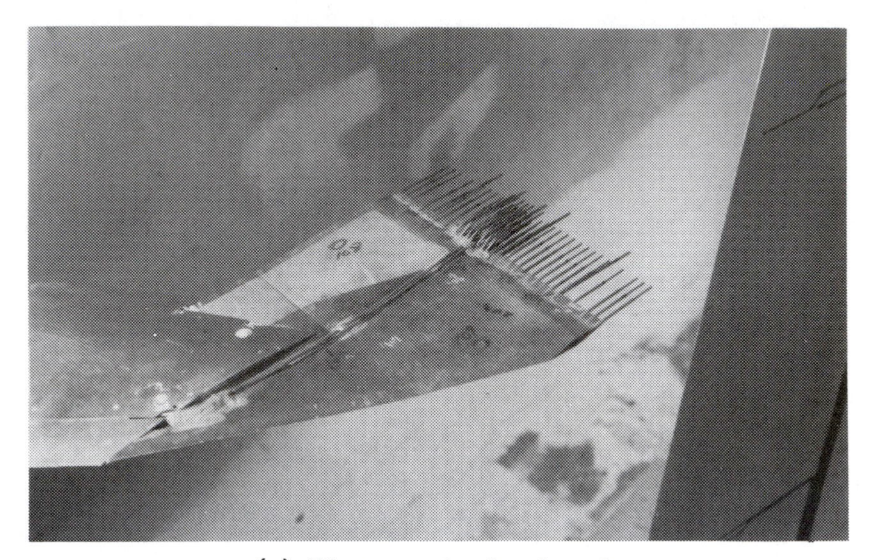

(a) Photograph of wake rake.

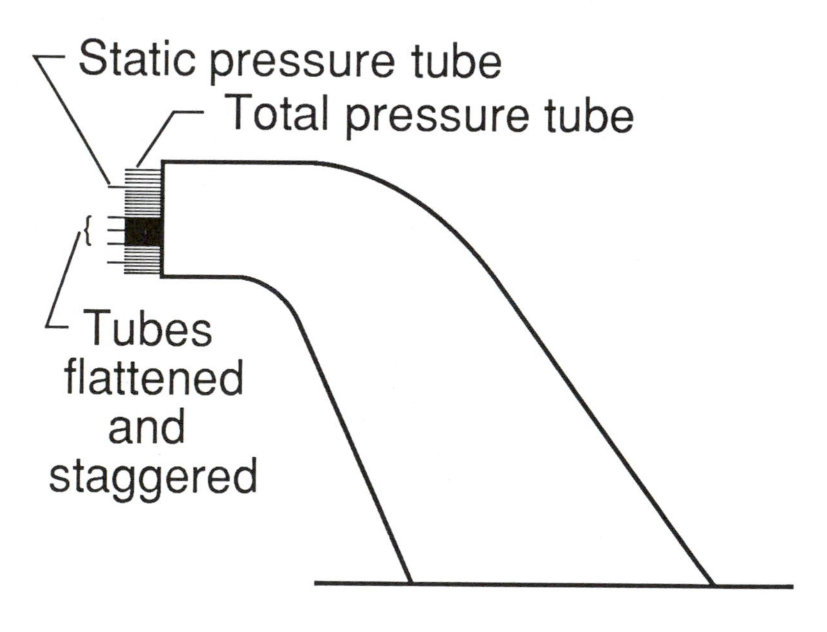


(b) Sketch of wake rake.

Figure 94. Photograph and sketch of the wake rake mounted downstream of the trailing edge.

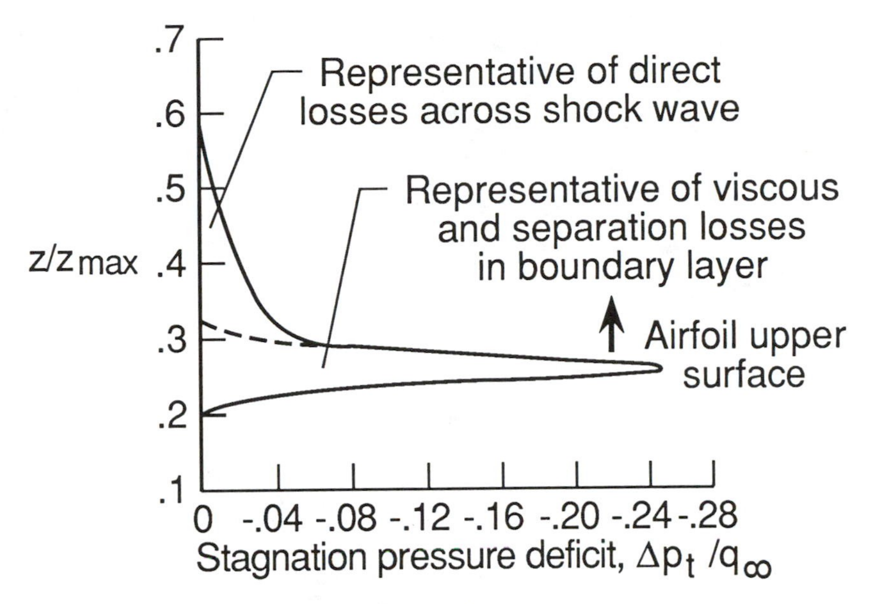


(a) Schematic of wake profile.

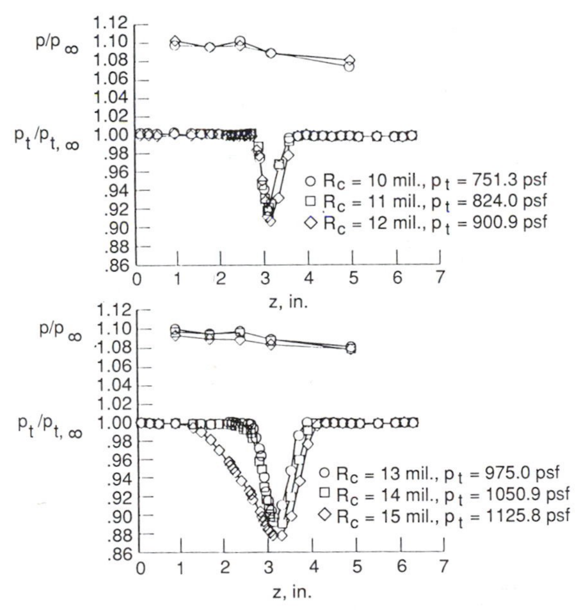


(b) Total pressure and static pressure for six Reynolds numbers.

Figure 95. Schematic of stagnation-pressure-deficit profile and measured pressure and velocity profiles.

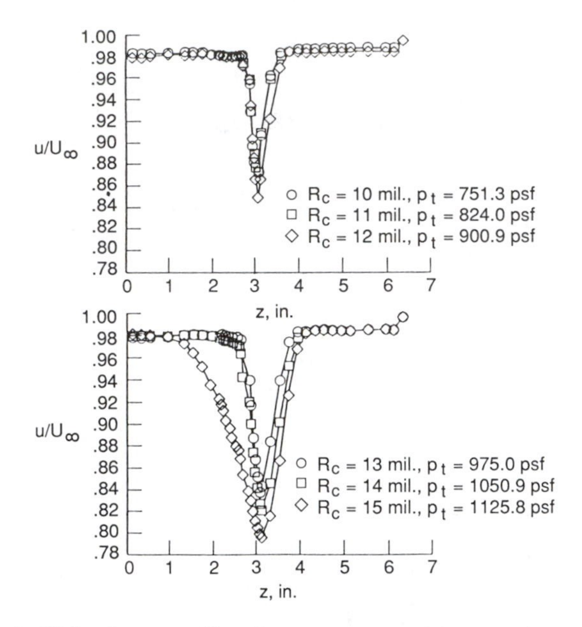


(c) Velocity profiles for six Reynolds numbers.

Figure 95.- Concluded

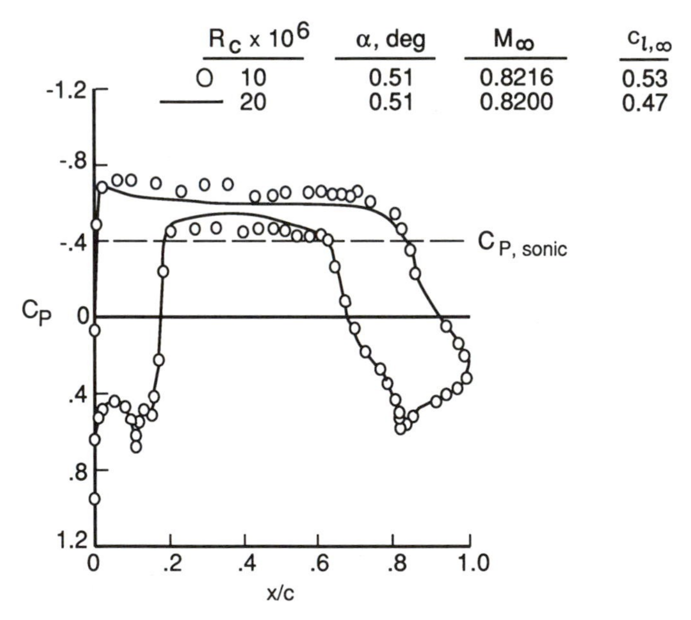


Figure 96. Comparison of measured pressure distribution for slotted LFC model, $M_\infty = 0.82$ and $R_c = 10 \times 10^6$ with that predicted for $M_\infty = 0.82$ and $R_c = 20 \times 10^6$.

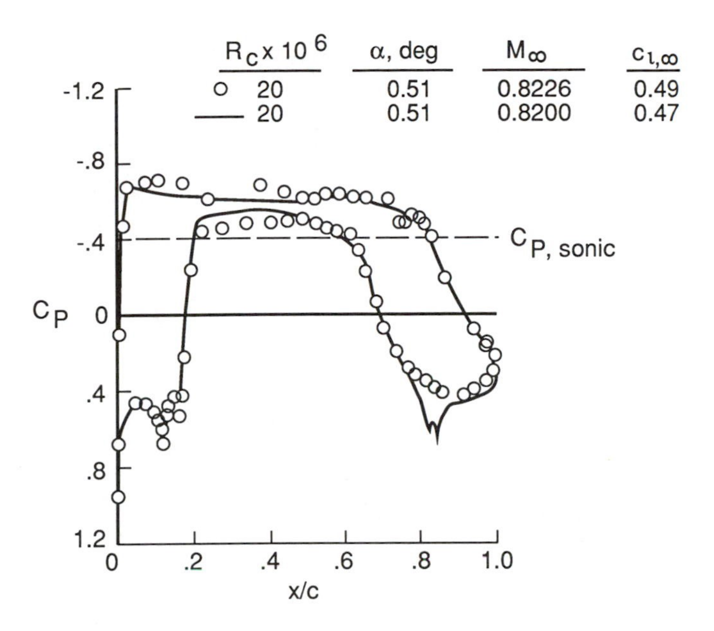


Figure 97. Comparison of measured and predicted pressure distributions for slotted LFC models, $M_\infty = 0.82$ and $R_c = 20 \times 10^6$.

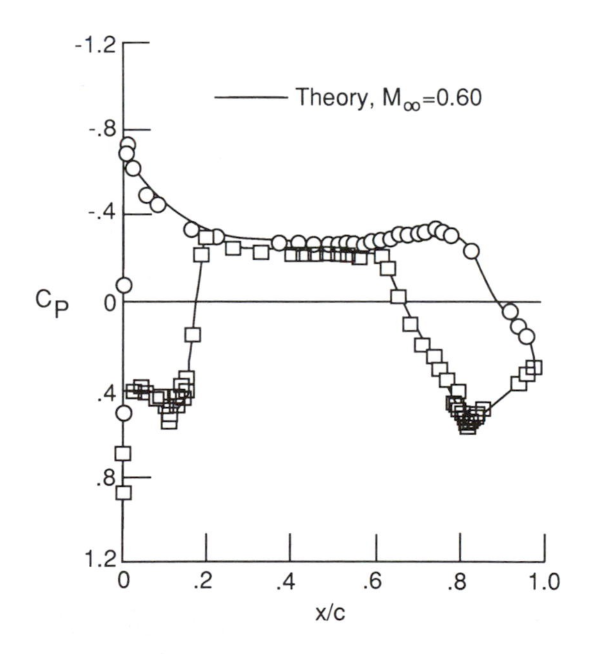


Figure 98. Comparison of measured and predicted pressure distributions for slotted model, $M_\infty = 0.6, R_c = 10 \times 10^6$.

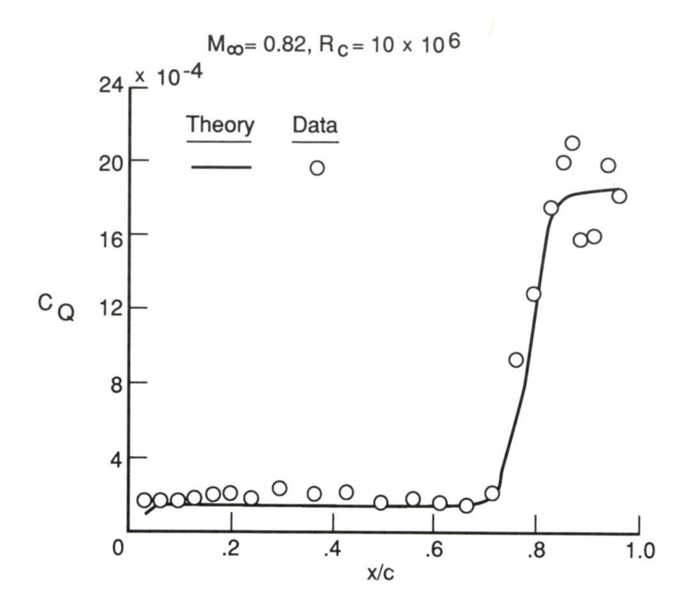


Figure 99. Comparison of measured and predicted upper-surface suction distribution for slotted model, $M_\infty = 0.82, R_c = 10 \times 10^6$.

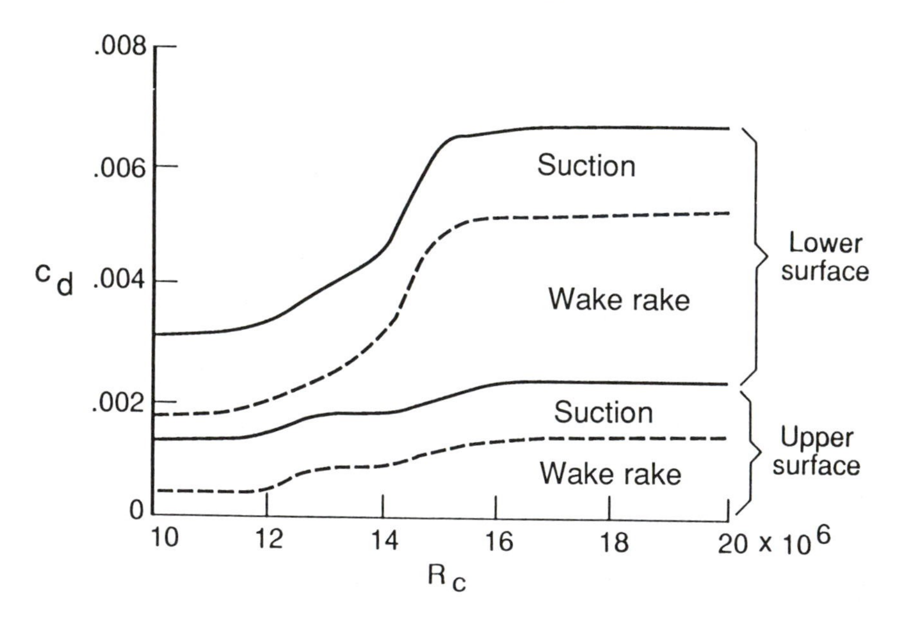


Figure 100. Measured variation of drag components with chord Reynolds number on LFC airfoil with slotted suction surface, $M_\infty = 0.82$.

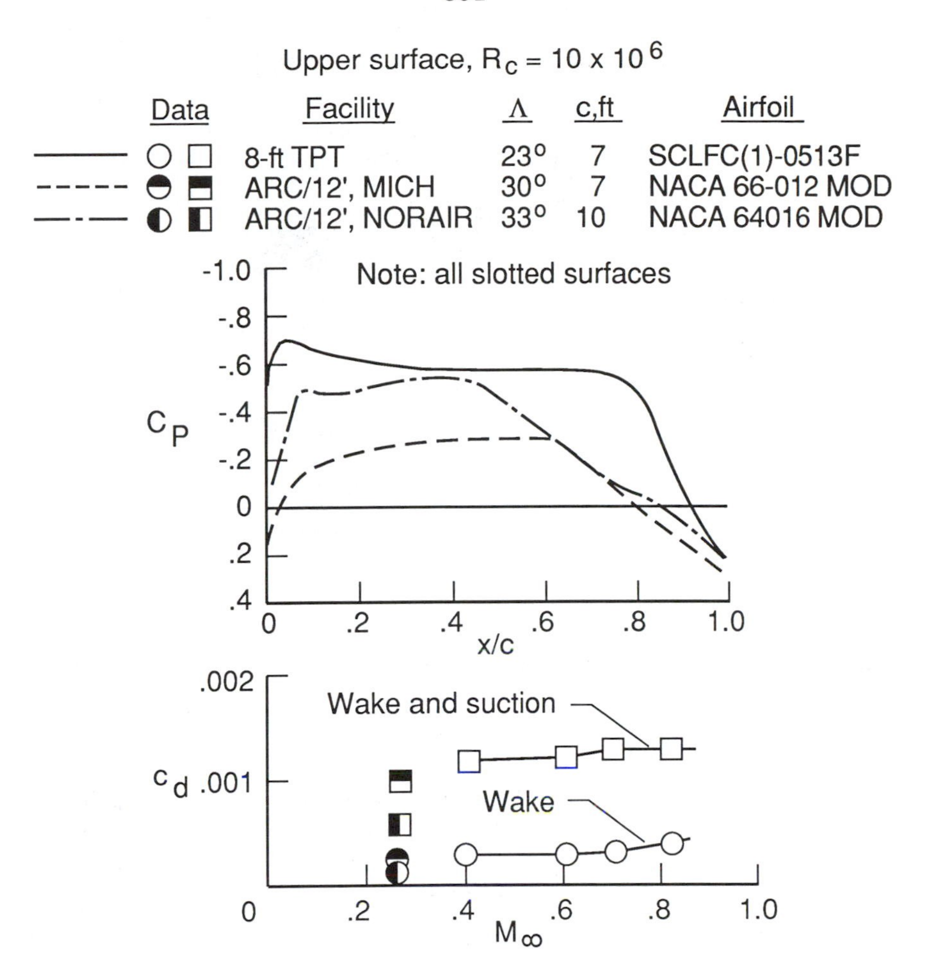


Figure 101. Comparison of measured pressure distributions and total drag for upper surface only on LFC wind tunnel models over speed range.

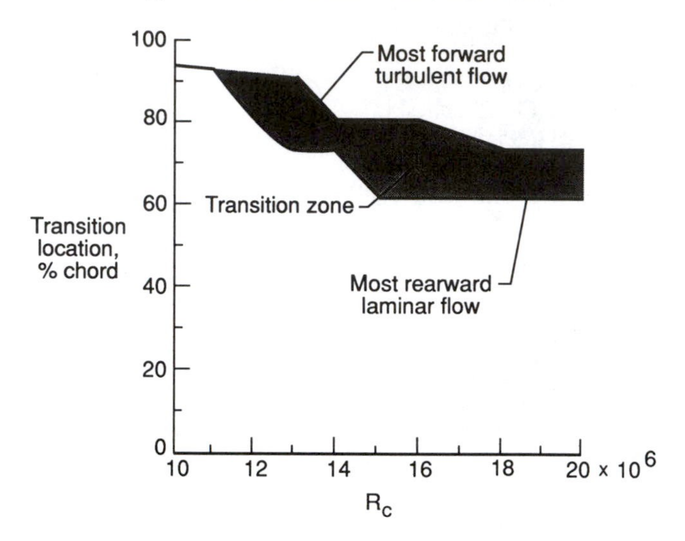


Figure 102. Variation of transition location with $R_c$ for slotted model upper surface, $M_\infty = 0.82$.

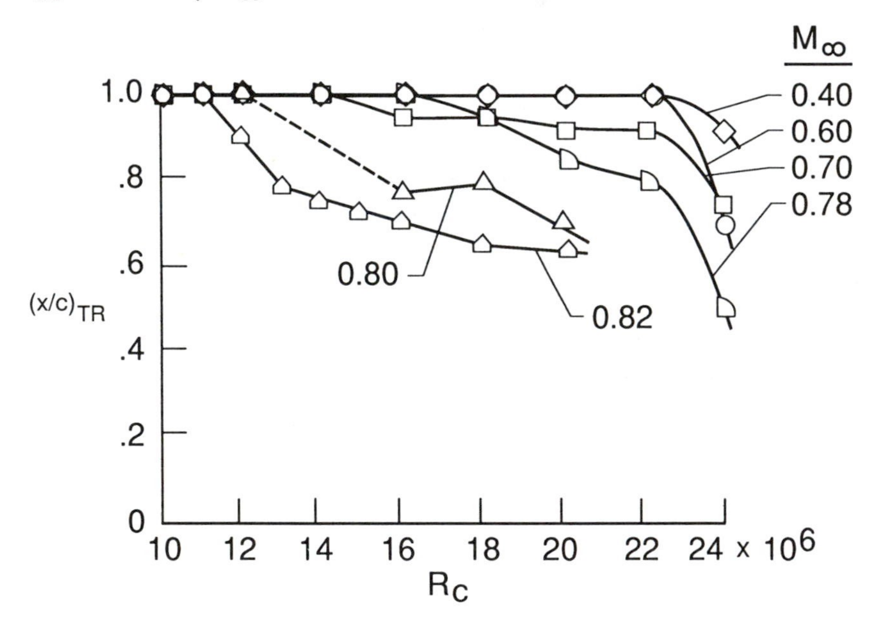


Figure 103. Transition location as a function of $R_c$ for Mach numbers from 0.4 to 0.82.

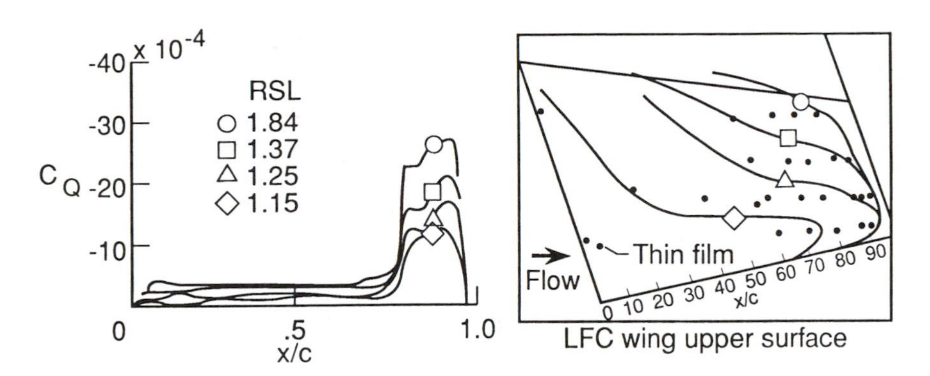


Figure 105. Effect of suction level on transition pattern for slotted model, $M_\infty = 0.6$, $R_c = 10 \times 10^6$.

| Data | $M_\infty$ | $\Lambda$, deg | $c_{l,\infty}$ | LFC Wing/Exp. |
|---|---|---|---|---|
| ○ | .4, .6, .7, .82 | 23 | .55 | LFC/8' TPT |
| ■ | .79 | 30 | ~.3 | X-21/Flight |
| ◆ | .72 | 0 | ~.3 | F-94/Flight |
| ▽△ | .25 | 30,33 | ~.3 | ARC/12'PWT |
| ◺ | .25 | 30 | ~.3 | NORAIR/7'x10' |
| D | .15 | 0 | ~.3 | LRC/LTPT |
| □ | .25 | 30 | ~.3 | Un. of Mich./5'x7' |

$R_{TR}$

$M_\infty$

Figure 104. Variation of maximum transition Reynolds number with Mach number for the LFC airfoil compared to results from other wind tunnel and flight experiments.

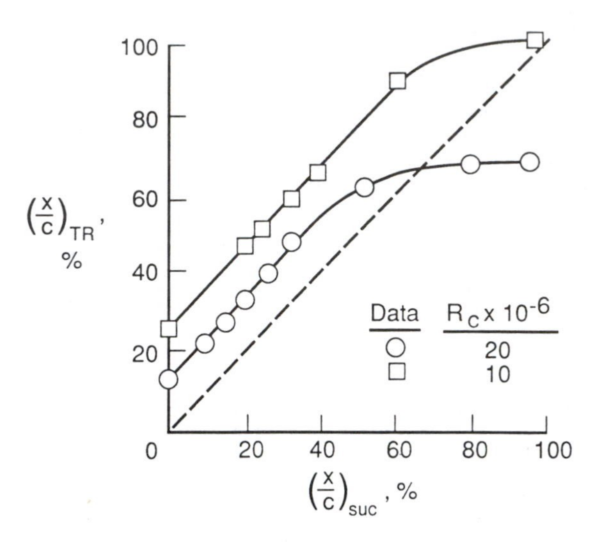


Figure 106. Variation of transition location on the upper surface with chordwise extent of suction.

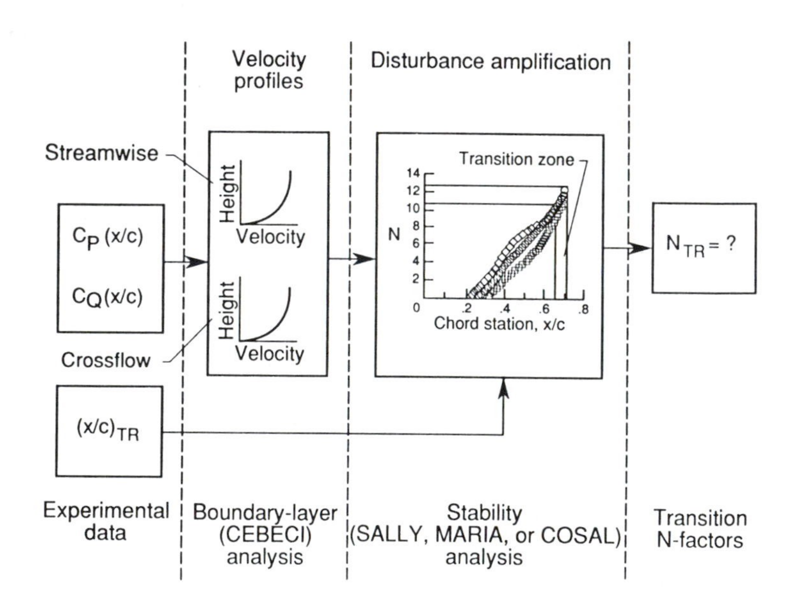


Figure 107. The $N$-factor correlation method.

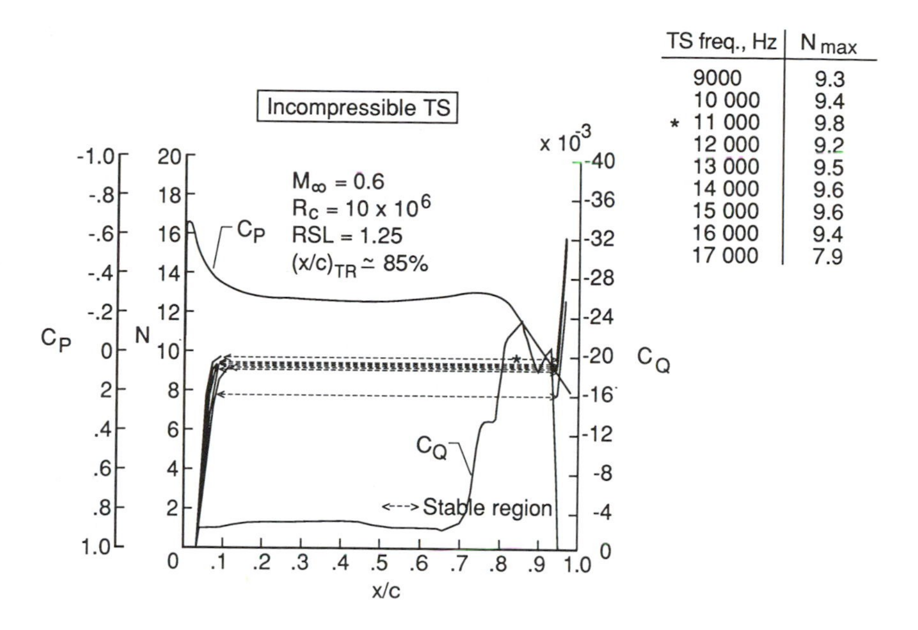


Figure 108. Incompressible TS calculations at $M_\infty = 0.6$, $R_c = 10 \times 10^6$ and RSL = 1.25.

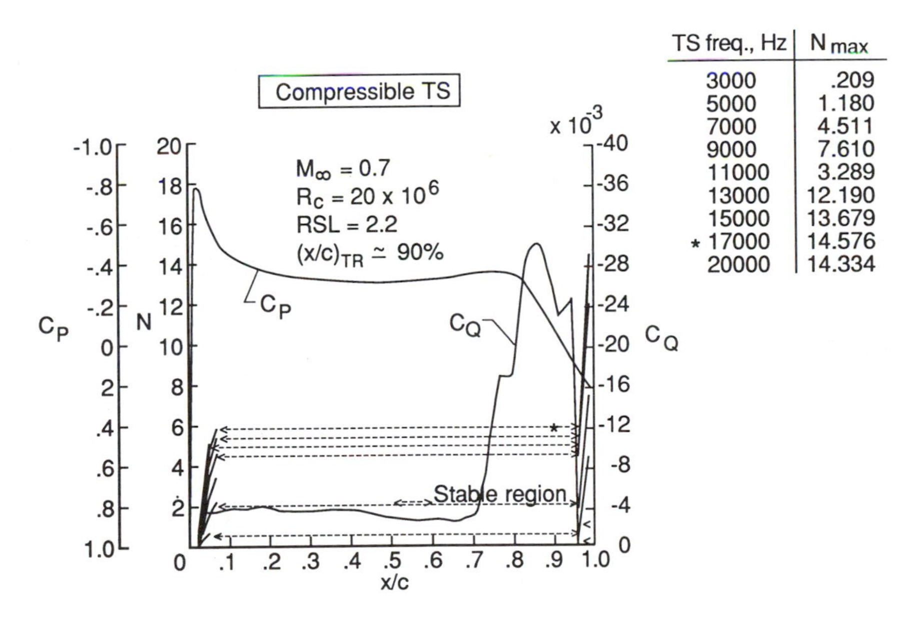


Figure 109. Compressible TS calculations at $M_\infty = 0.7$, $R_c = 20 \times 10^6$ and RSL = 2.2.

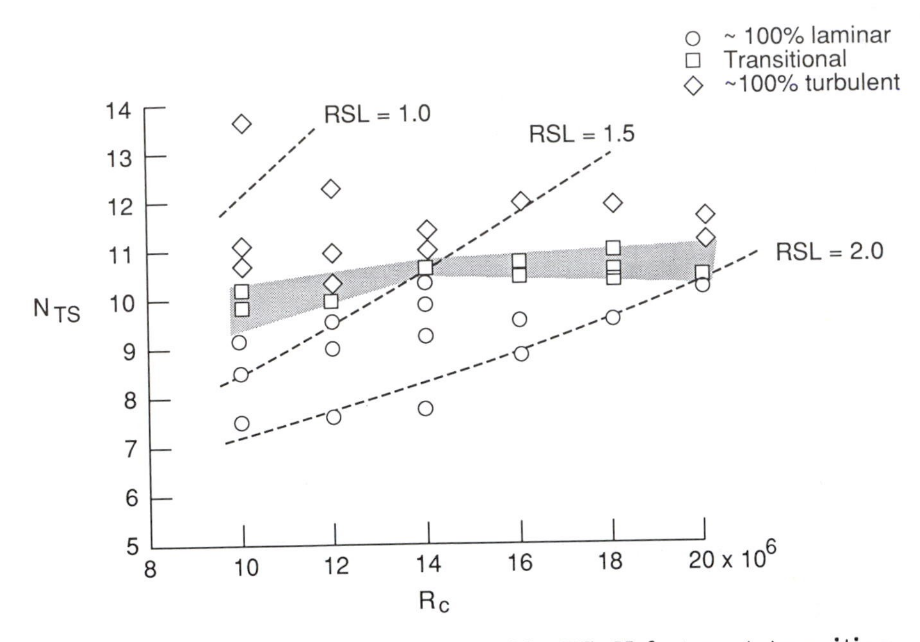


Figure 110. Variation of incompressible TS $N$-factor at transition with chord Reynolds number and relative suction level (RSL) at $M_\infty = 0.6$.

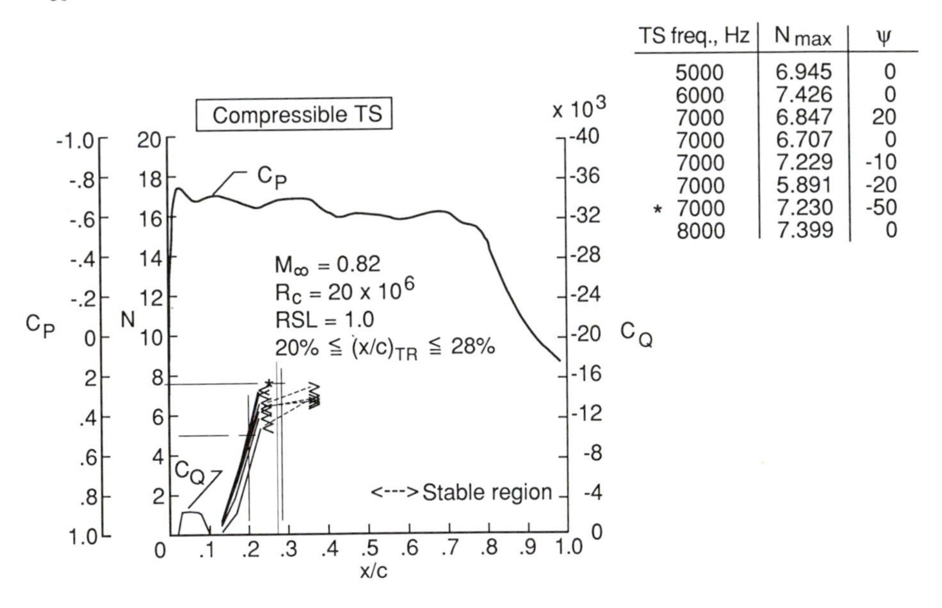

| TS freq., Hz | N max | ψ |
|---|---|---|
| 5000 | 6.945 | 0 |
| 6000 | 7.426 | 0 |
| 7000 | 6.847 | 20 |
| 7000 | 6.707 | 0 |
| 7000 | 7.229 | -10 |
| 7000 | 5.891 | -20 |
| * 7000 | 7.230 | -50 |
| 8000 | 7.399 | 0 |



Figure 111. Compressible TS calculations at $M_\infty = 0.82$, $R_c = 20 \times 10^6$ with suction to 0.08c.

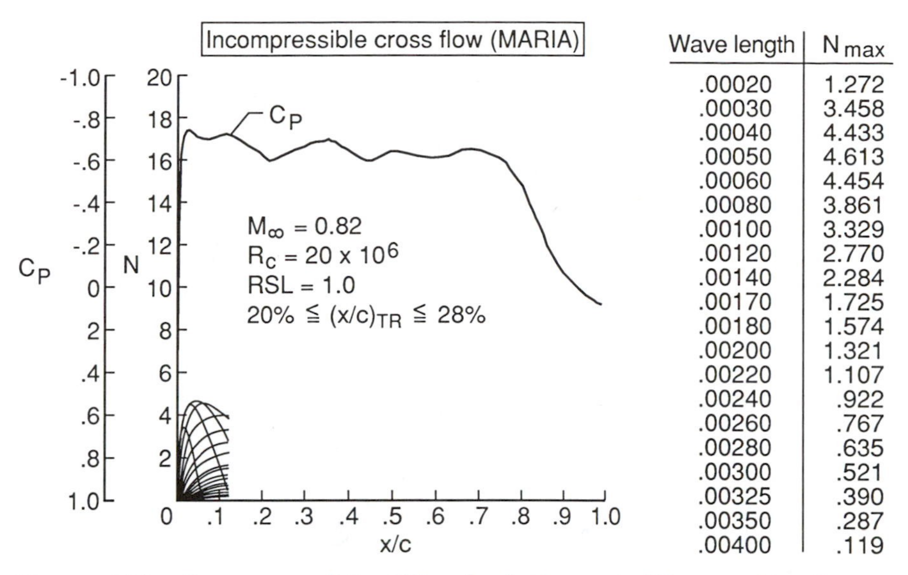


| Wave length | N max |
|---|---|
| .00020 | 1.272 |
| .00030 | 3.458 |
| .00040 | 4.433 |
| .00050 | 4.613 |
| .00060 | 4.454 |
| .00080 | 3.861 |
| .00100 | 3.329 |
| .00120 | 2.770 |
| .00140 | 2.284 |
| .00170 | 1.725 |
| .00180 | 1.574 |
| .00200 | 1.321 |
| .00220 | 1.107 |
| .00240 | .922 |
| .00260 | .767 |
| .00280 | .635 |
| .00300 | .521 |
| .00325 | .390 |
| .00350 | .287 |
| .00400 | .119 |

Figure 112. Incompressible CF calculations at $M_\infty = .82$, $R_c = 20 \times 10^6$ with suction to 0.08c.

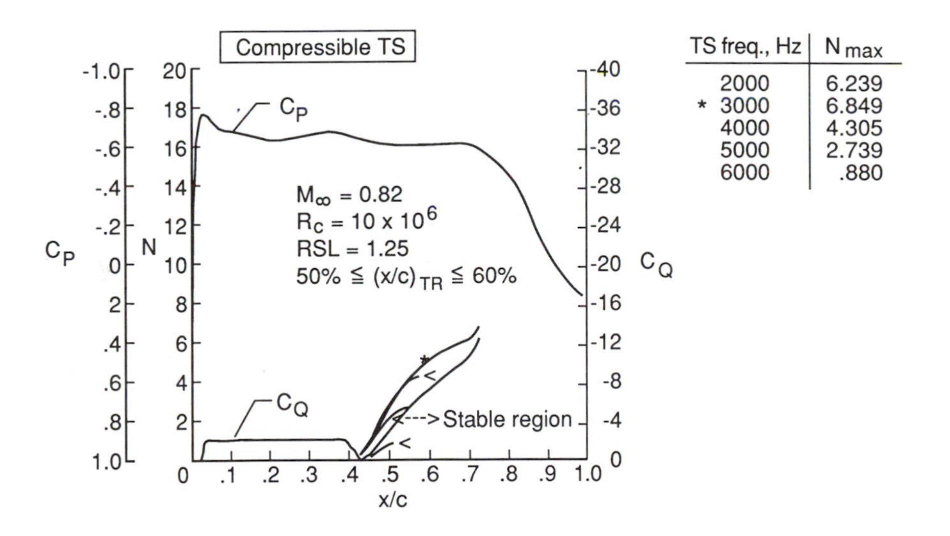


| TS freq., Hz | N max |
|---|---|
| 2000 | 6.239 |
| * 3000 | 6.849 |
| 4000 | 4.305 |
| 5000 | 2.739 |
| 6000 | .880 |

Figure 113. Compressible TS calculations at $M_\infty = .82$, $R_c = 20 \times 10^6$ with suction to 0.40c.

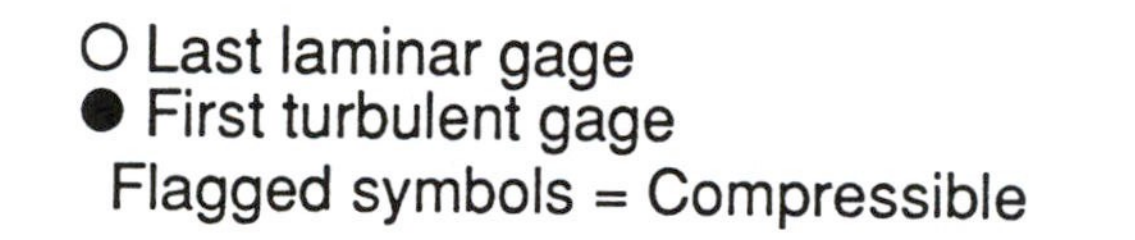


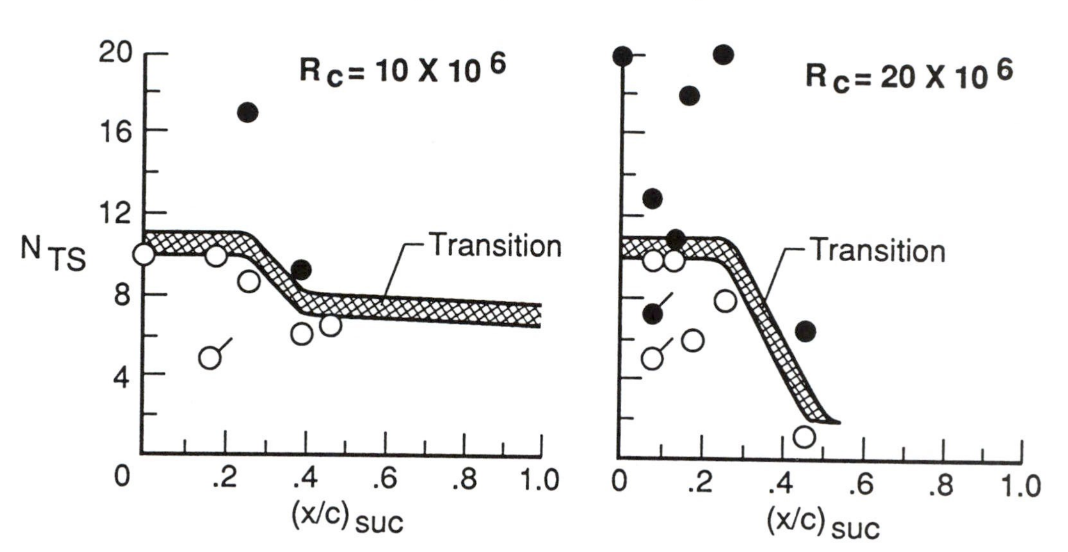


Figure 114. Effect of chordwise suction extent on calculated incompressible TS $N$-factor for slotted LFC airfoil, $M_\infty = 0.82$.

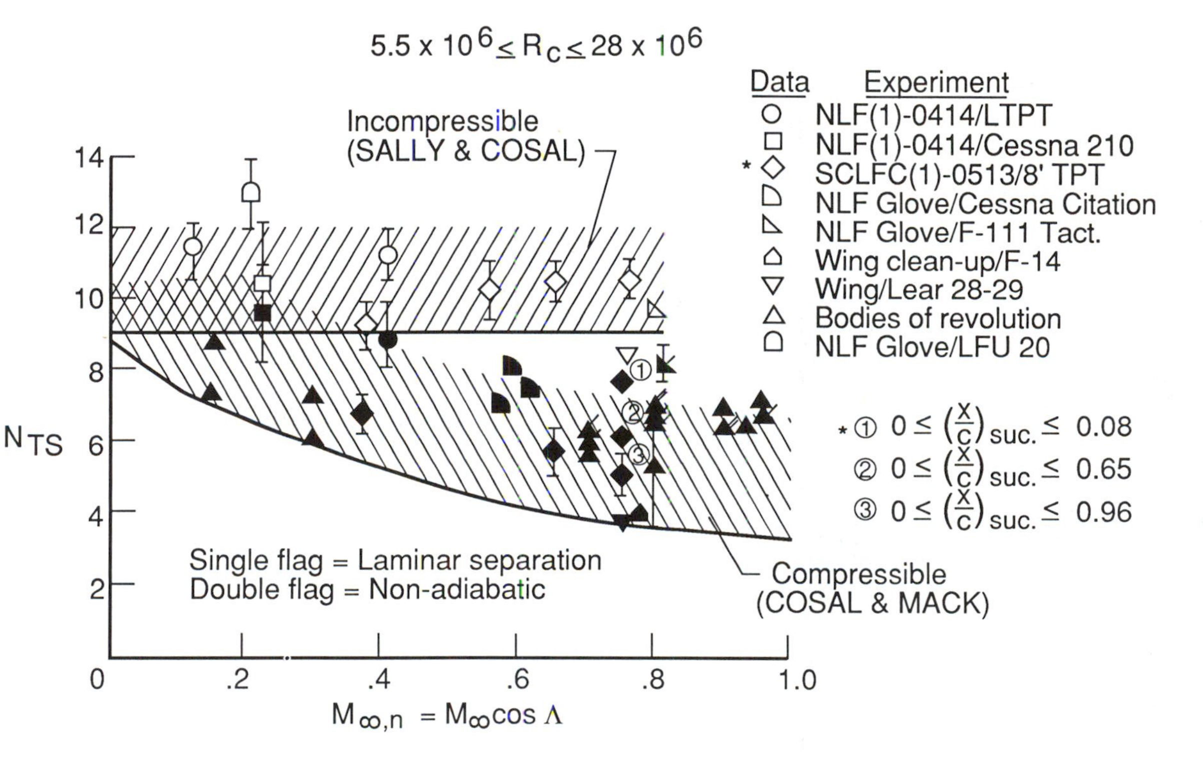


Figure 115. Lower-bound Tollmien-Schlichting $N$-factor at the measured transition location versus $M_\infty \cos \Lambda$. Solid symbols for compressible calculations.

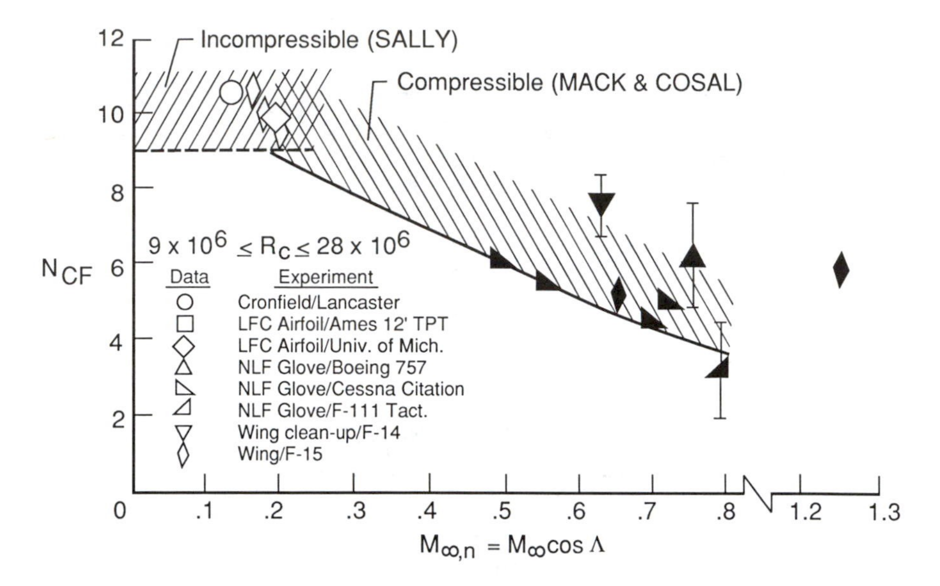


Figure 116. Lower-bound crossflow $N$-factors at the measured transition location versus $M_\infty \cos \Lambda$. Solid symbols for compressible cal-

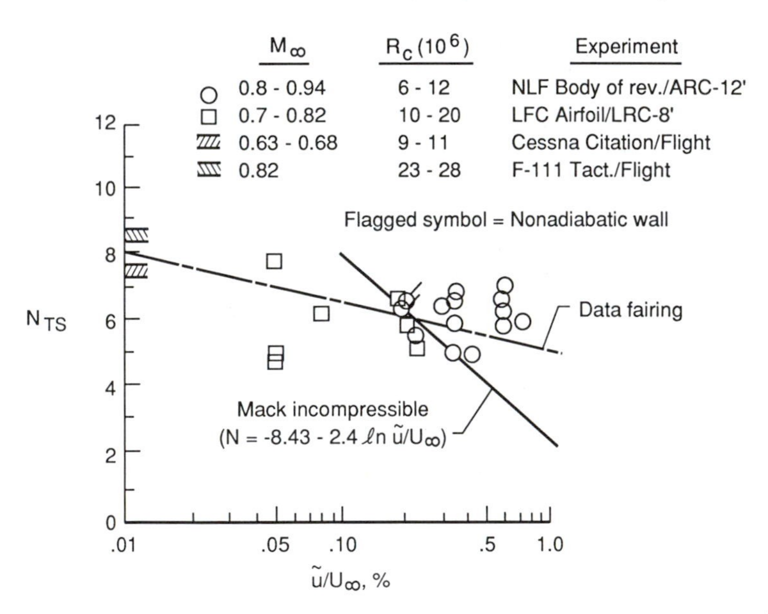


Figure 117. Theoretical and experimental results for $N_{TS}$ as a function of free-stream turbulence level.

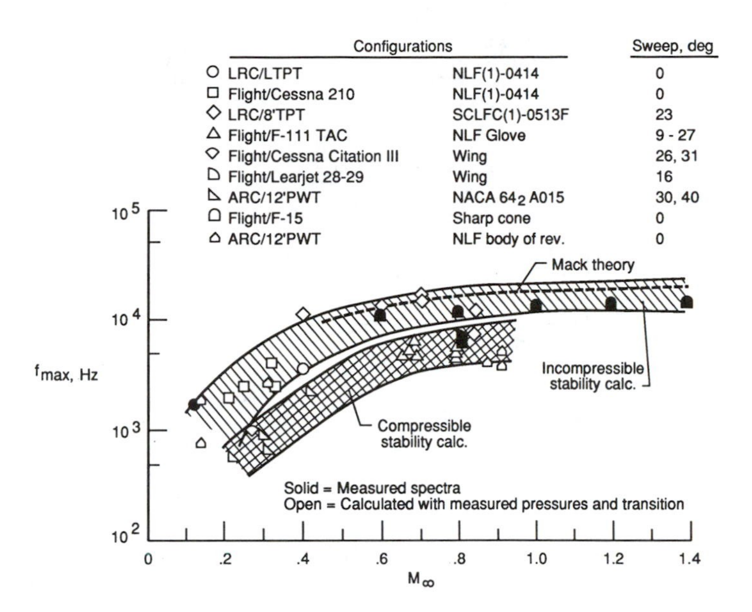


Figure 118. The variation of measured and calculated maximum TS-frequencies at transition with $M_\infty$ for wind tunnel and flight data.

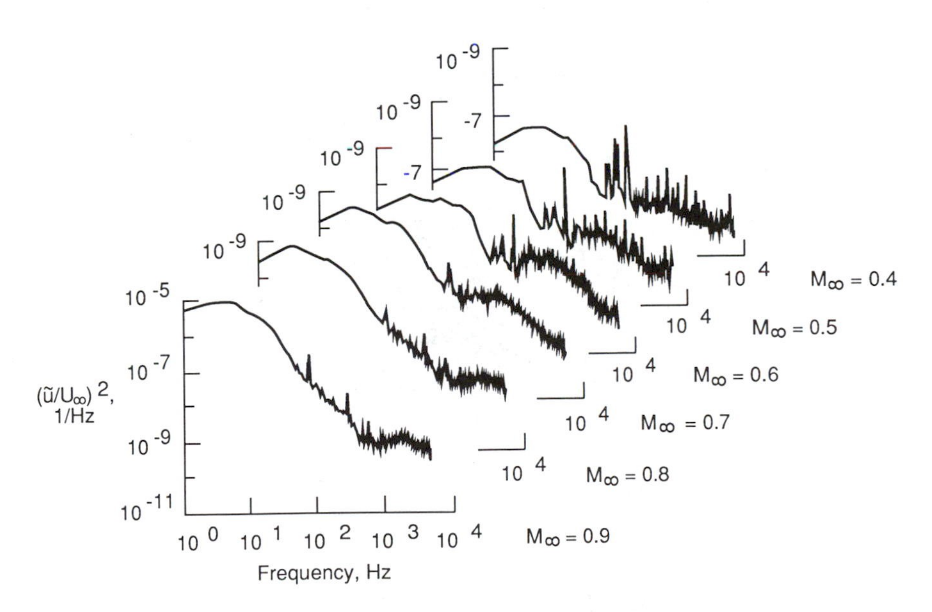


Figure 119. Velocity spectra for a range of Mach numbers in the 8-ft TPT.

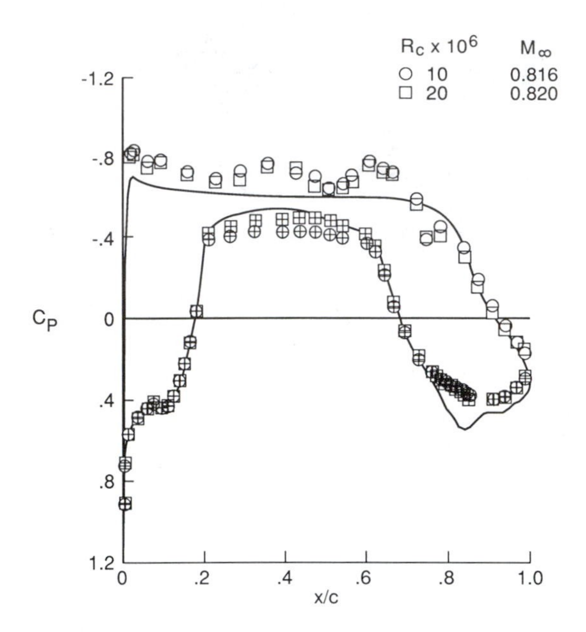


Figure 120. Effect of Reynolds number on experimental chordwise pressure distribution. Open symbols denote upper surface.

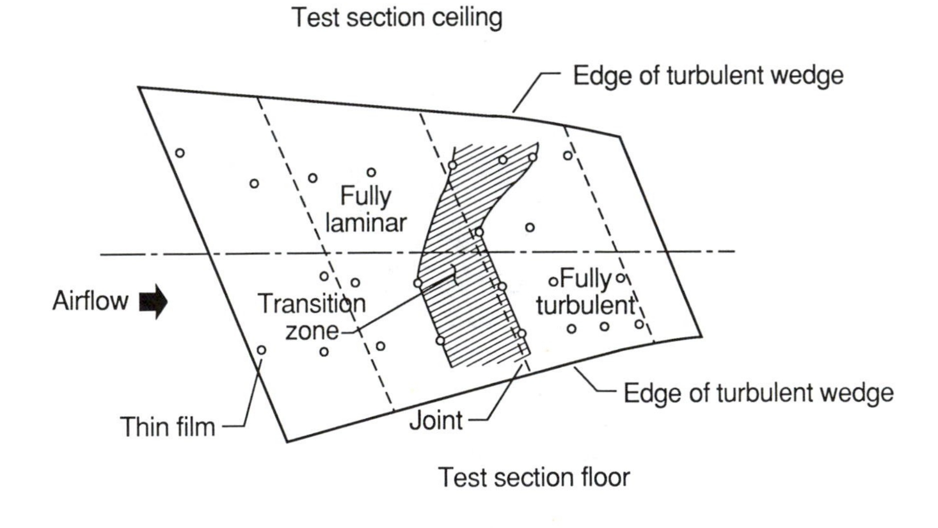


(a) $R_c = 10 \times 10^6$, upper surface.

Figure 121. Transition patterns on original perforated LFC model with nonsuction forward-lower-surface panel, $M_\infty = 0.82$.

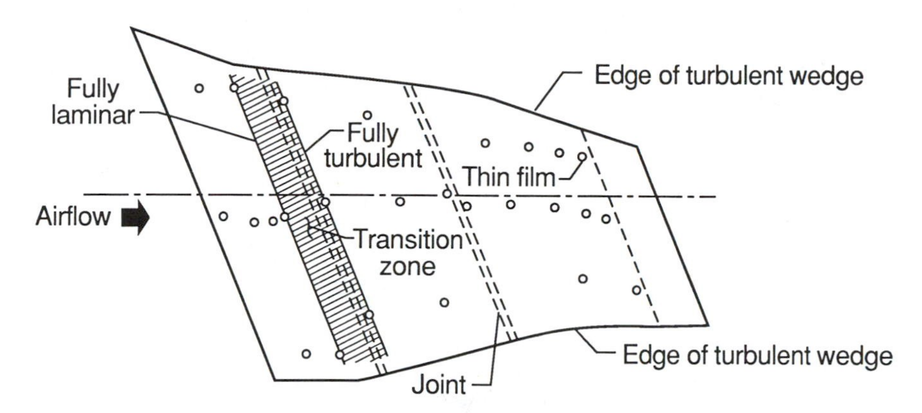


(b) $R_c = 10 \times 10^6$, lower surface.

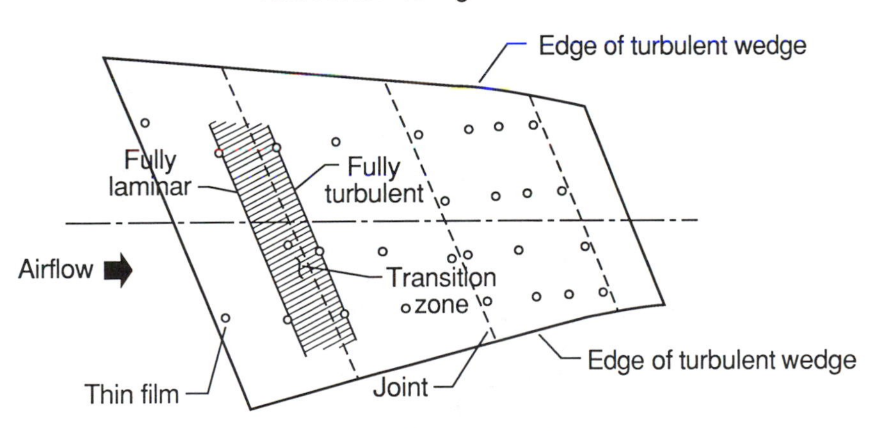


(c) $Rc = 20 \times 10^6$, upper surface.

Figure 121.- Continued

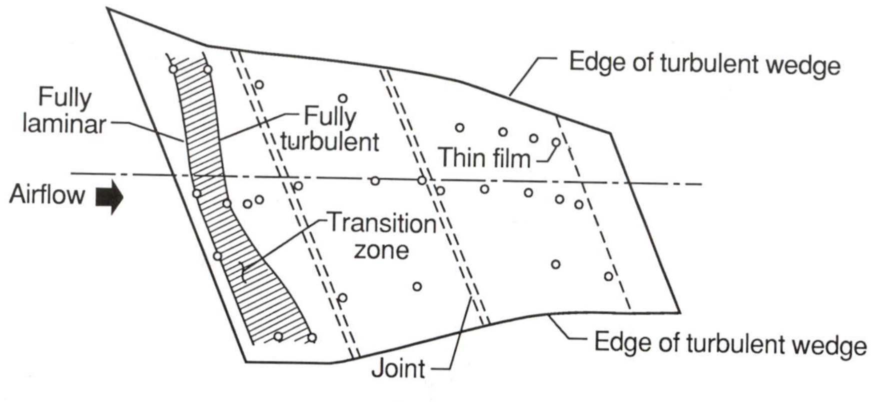


(d) $R_c = 20 \times 10^6$, lower surface.

Figure 121.- Concluded

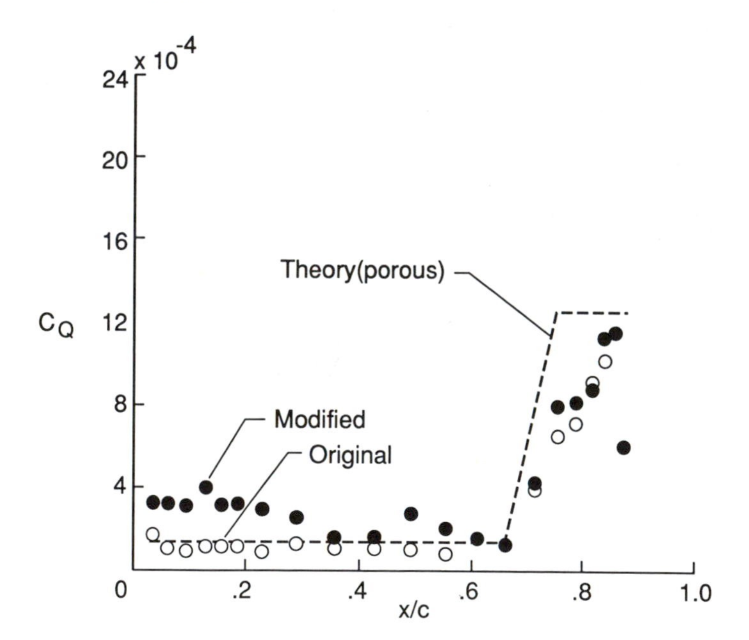


Figure 122. Effects of modifications to the suction system of perforated model, $R_c = 10 \times 10^6$.

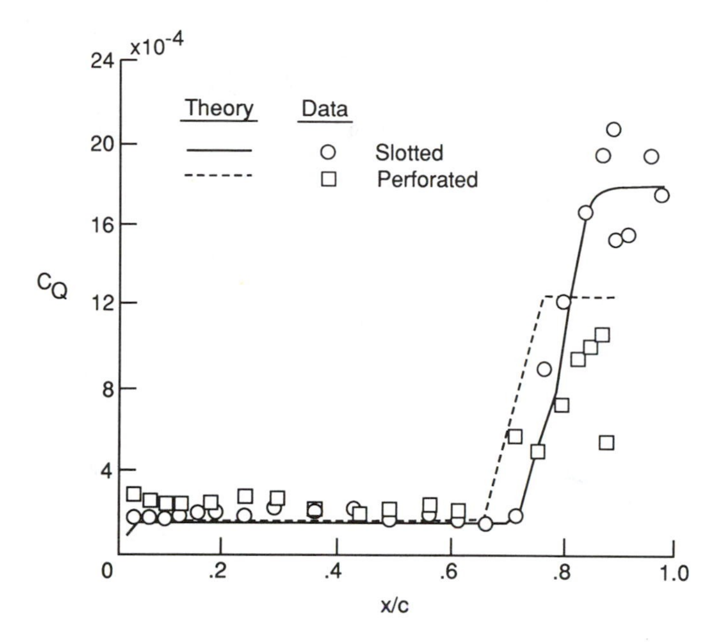


Figure 123. Comparison of suction distributions for final configurations of slotted and modified perforated models, $R_c = 10 \times 10^6, M_\infty = 0.82$.

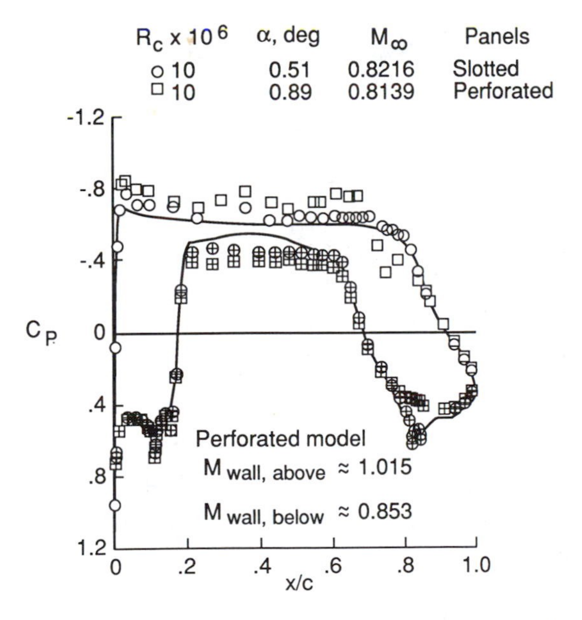


Figure 124. Comparison of pressure distributions for slotted and modified perforated models at $R_c = 10 \times 10^6, M_\infty \approx 0.82$.

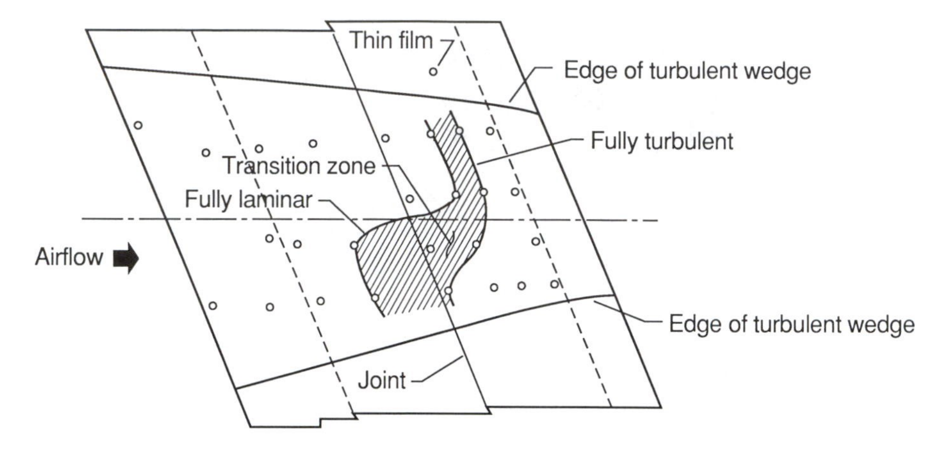


Figure 125. Transition pattern on upper surface - perforated model, $R_c = 9.5 \times 10^6$.

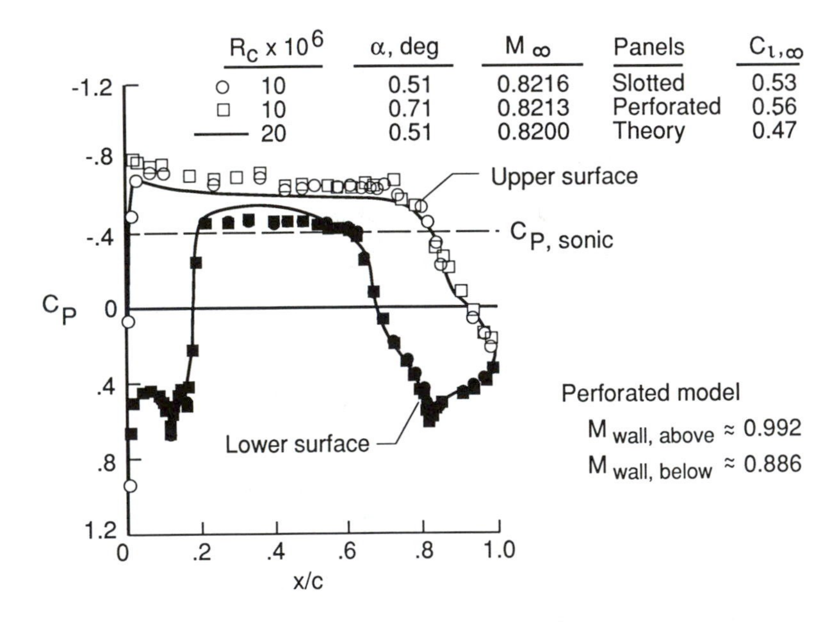


Figure 126. Comparison of the predicted and best measured pressure distributions for slotted and perforated LFC models at $R_c = 10 \times 10^6, M_\infty \approx 0.82$.

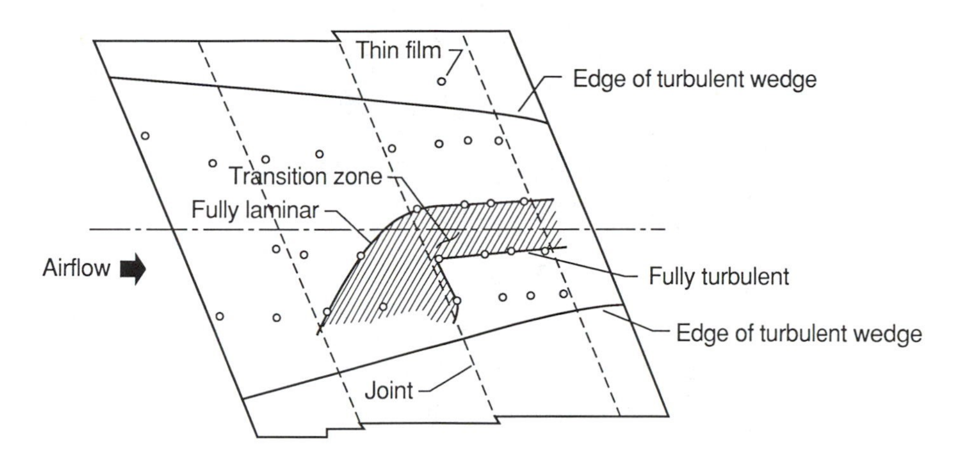


Figure 127. Transition pattern on upper surface of perforated model at $R_c = 10 \times 10^6, M_\infty \approx 0.82$.

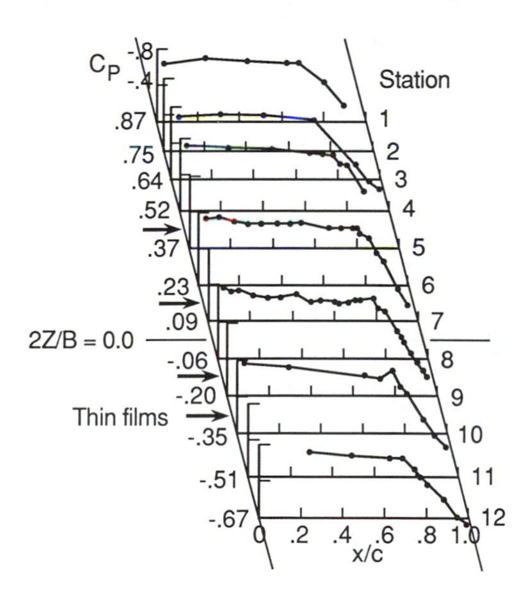


Figure 128. Chordwise pressure distributions at a number of spanwise locations on the upper surface of the perforated model, $M = 0.82$, $R_c = 10 \times 10^6, M_\infty \approx 0.82$ .

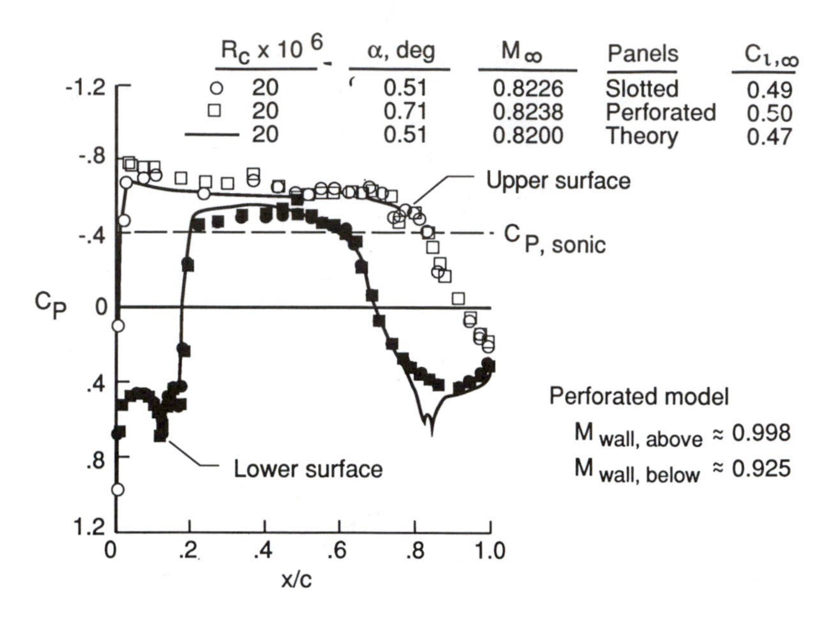


Figure 129. Comparison of the predicted and best measured pressure distributions for slotted and perforated LFC models at $R_c = 20 \times 10^6, M_\infty \approx 0.82$.

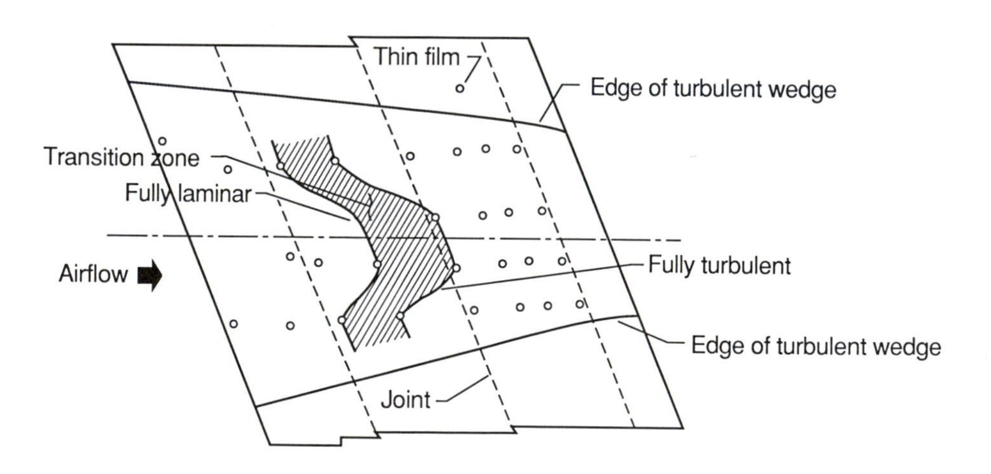


Figure 130. Transition pattern on upper surface of perforated model at $R_c = 20 \times 10^6, M_\infty \approx 0.82$ .

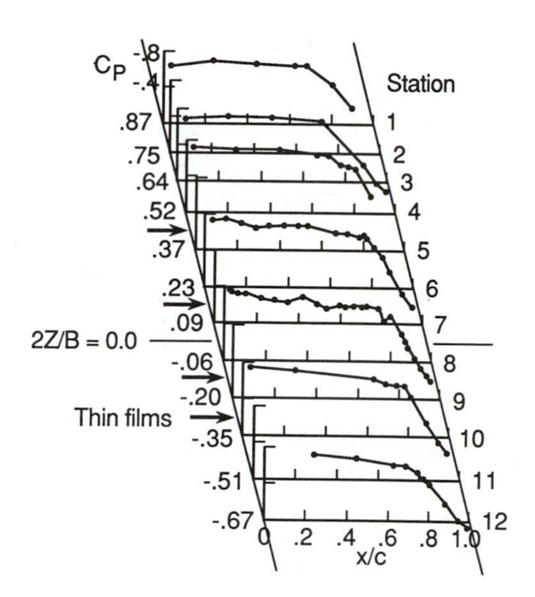


Figure 131. Chordwise pressure distributions at a number of spanwise locations on the upper surface of the perforated model, $M_\infty \approx 0.82$ and $R_c = 20 \times 10^6$.

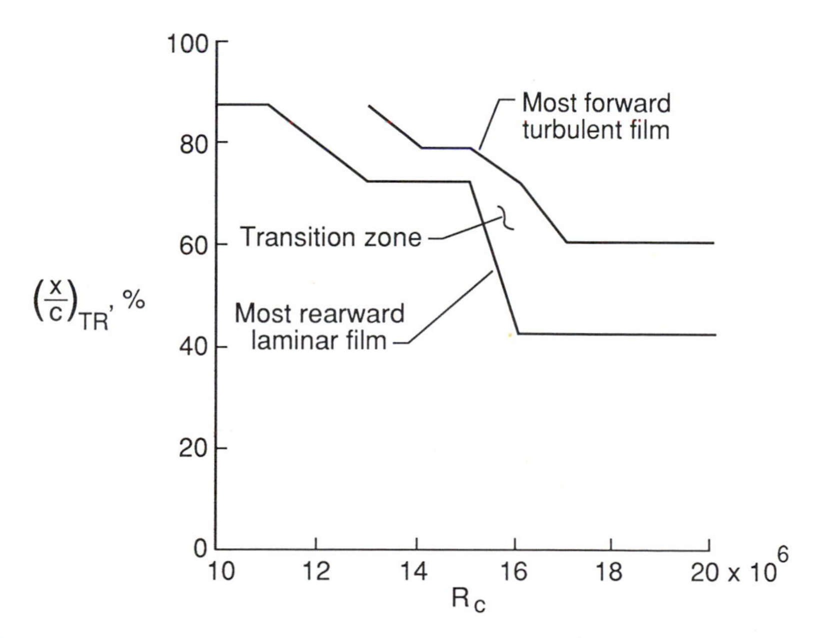


Figure 132. Transition location on the upper surface of perforated model as a function of chord Reynolds number.

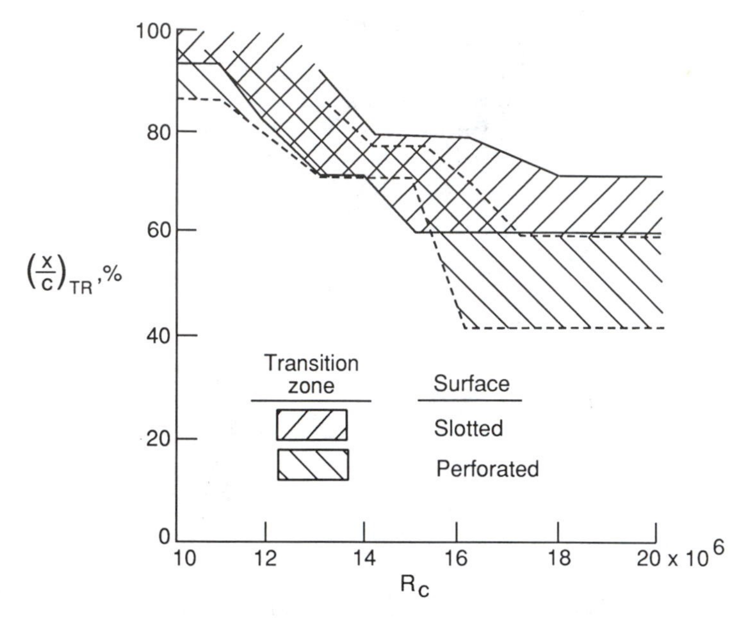


Figure 133. Transition location versus $R_c$ for upper surfaces of slotted and perforated models at $M_\infty \approx 0.82$.

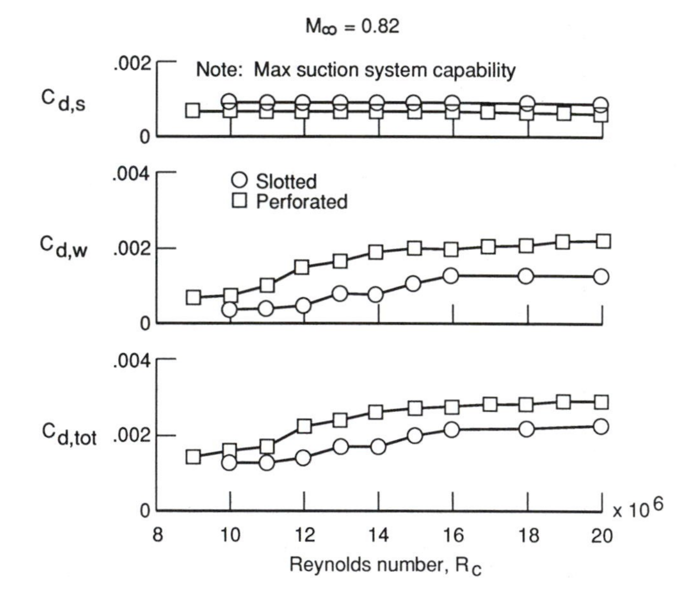


Figure 134. Comparison of upper surface drag coefficients for slotted and perforated surfaces at $M_\infty \approx 0.82$.

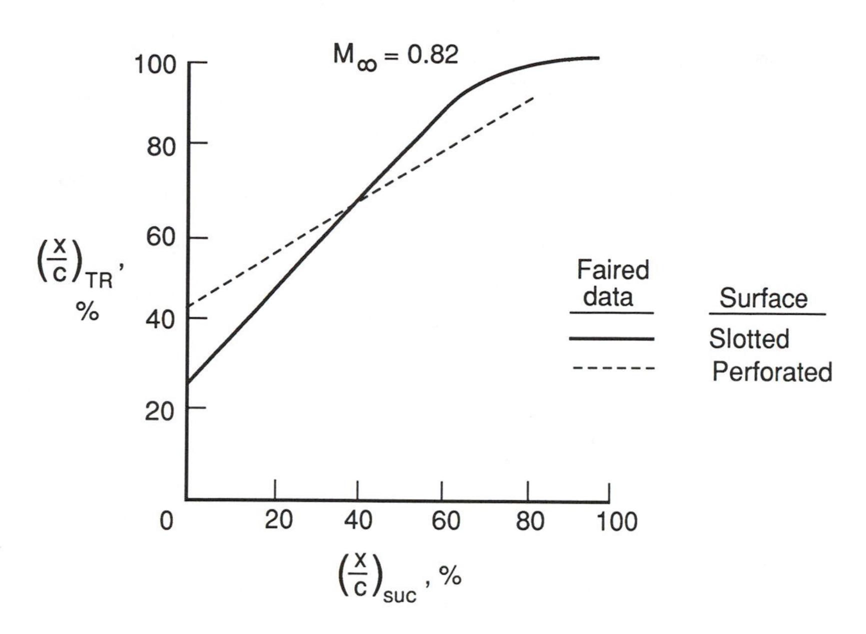


Figure 135. Variation of transition location with extent of suction for the upper surface of both the slotted and perforated panels. Lower surface suction applied from 0 to 25% of chord.